Anonymous

Masterpieces of Prose

Selected from the works of the greatest English and American writers: from Chaucer to Ruskin and Longfellow

Anonymous

Masterpieces of Prose
Selected from the works of the greatest English and American writers: from Chaucer to Ruskin and Longfellow

ISBN/EAN: 9783337075637

Printed in Europe, USA, Canada, Australia, Japan

Cover: Foto ©berggeist007 / pixelio.de

More available books at **www.hansebooks.com**

A TRUE CALEDONIAN.

MASTERPIECES OF PROSE

SELECTED FROM THE WORKS OF
THE GREATEST ENGLISH AND AMERICAN WRITERS

FROM CHAUCER TO RUSKIN AND LONGFELLOW

PROFUSELY ILLUSTRATED

BOSTON
D. LOTHROP COMPANY
1893

COPYRIGHT, 1893,
BY
D. LOTHROP COMPANY.

PREFACE.

THERE is no royal road to literary success. He who writes entertainingly writes successfully; he who writes nobly writes for immortality. The world's masterpieces of literature are not confined to any one age; neither do they always need the halo of time or the flavor of antiquity to gain acceptance. Genius works as well in one generation as in another, and the writer of to-day is capable of as good and as lasting work as was he who created in the time of Addison, or he who labored in the days of Bacon.

So these "Masterpieces of Prose," gathered in this volume, are selected, rather as specimens than as a completed galaxy, from the writers of to-day and from those to whom age has brought the verdict "classic." In the main, they are but brief extracts, chosen rather to afford a taste than a full banquet with each author. If they shall serve as an encouragement to more extended reading, or a closer acquaintance with any or all of the writers represented in this volume, the labor of choice and selection will not have been in vain, and the idea of grouping into a single volume representative extracts from a whole library of English authors will have proved itself a wise and happy suggestion.

Both England and America have been drawn upon for sources of selection, and in giving this volume to the public due credit and hearty thanks are herewith accorded to the following American publishers by whose permission, or under arrangement with whom, copyrighted matter has been allowed to appear in this volume: Houghton, Mifflin & Co., Roberts Brothers, Estes & Lauriat, and Lee & Shepard, of Boston; Harper & Brothers, Charles Scribner's Sons, G. P. Putnam's Sons, D. Appleton & Co., and Fords, Howard & Hulbert, of New York; J. B. Lippincott Company and T. B. Peterson & Brothers, of Philadelphia.

CONTENTS

Contentment	Isaak Walton	9
The All-conquering Power of Truth	John Milton	10
At Rugby	Thomas Arnold	11
My Garden	Mary Abigail Dodge	12
(From "Country Living and Thinking.")		
On Affairs in America	Lord Chatham	13
A Country Parish	Charles Kingsley	14
Limitations of Free Speech	Lord Erskine	15
Deacon Marble's Trout	Henry Ward Beecher	16
(From "Norwood.")		
Christian in Doubting Castle	John Bunyan	17
(From "Pilgrim's Progress.")		
The Pleasures of Private Life	George Washington	21
On the Death of an Old Friend	Charles Lamb	22
Upon Riches	Geoffrey Chaucer	23
(From "Tales of Melibeus.")		
Sam Mends the Clock	Harriet Beecher Stowe	24
(From "Oldtown Folks.")		
The Irish Bard	Edmund Spenser	28
London	Thomas Carlyle	29
Compensation	Ralph Waldo Emerson	30
(From "Essays.")		
The Good Wife	Thomas Fuller	32
(From "The Holy and Profane State.")		
Solitude	Henry David Thoreau	33
(From "Walden.")		
Spring Prospects	Henry David Thoreau	34
(From "Early Spring in Massachusetts.")		
On Mr. Foot's Resolution	Robert Y. Hayne	35
Reply to Hayne	Daniel Webster	36
The Constitution and the Union	Daniel Webster	38
On Pride	Sir Thomas Browne	41
Ben Jonson	John Dryden	42

CONTENTS.

On Studies	Francis Bacon	43
On Bacon	Ben Jonson	44
Sir Roger De Coverley	Joseph Addison	44
(From " Sir Roger De Coverley.")		
The Strength of True Love	Sir Richard Steele	47
(From " The Tatler.")		
Talk	Oliver Wendell Holmes	48
(From " Autocrat of the Breakfast Table.")		
Wouter Van Twiller	Washington Irving	49
(From " Knickerbocker's History of New York.")		
Letter to Mrs. Thrale	Samuel Johnson	51
The Dominie and Meg Merrilies	Sir Walter Scott	51
(From " Guy Mannering.")		
The Story of " Waverley "	Sir Walter Scott	56
The Temperance Preacher	Pansy (Mrs. G. R. Alden)	57
(From " John Remington, Martyr.")		
Garden Ethics	Charles Dudley Warner	58
(From " My Summer in a Garden.")		
Little Pearl in the Forest	Nathaniel Hawthorne	59
(From " The Scarlet Letter.")		
Footprints of Angels	Henry W. Longfellow	61
(From " Hyperion.")		
The Good Man } The Good Woman }	Samuel Richardson	63
John and Lorna	R. D. Blackmore	64
(From " Lorna Doone.")		
Carlyle to his Mother	Thomas Carlyle	65
On the Middle Station of Life	David Hume	66
Melons	Bret Harte	68
(From " Mrs. Skaggs's Husbands, and other Sketches.")		
On Refusal to Negotiate with Napoleon	William Pitt	70
Parody on the Speeches of Charles II.	Andrew Marvell	72
The Destiny of the Republic	Alexander Hamilton Stephens	73
News from the Front	M. E. M. Davis	73
(From " In War-Times at La Rose Blanche.")		
The Gettysburg Address	Abraham Lincoln	75
The Midnight Sun	Bayard Taylor	76
(From " Northern Travel.")		
The Whistle	Benjamin Franklin	78
Inaugural Address	George Washington	79

CONTENTS.

The Essential Principles of Government .	Thomas Jefferson	79
Conciliation . . .	Edmund Burke .	81
On the Kansas-Nebraska Bill . .	Charles Sumner .	83
Speech before the Virginia Convention	Patrick Henry . .	87
The Last Train North . . .	George Washington Cable .	89
(From " Dr. Sevier.")		
The Grasshopper and the Ant . .	George T. Lanigan	92
(From " Æsop's Fables.")		
The Little Women's Romance . .	Louisa M. Alcott	92
(From " Little Women.")		
The Wonderful Tar-baby Story . .	Joel Chandler Harris .	97
(From " Uncle Remus.")		
How Mr. Rabbit was too sharp for Mr. Fox	Joel Chandler Harris .	98
(From " Uncle Remus.")		
Torture . . .	Edgar Allan Poe	100
(From " The Pit and the Pendulum.")		
To Thomas Murray	Thomas Carlyle . .	103
To his Mother . . .	Arthur Penrhyn Stanley .	104
Nil Nisi Bonum . . .	William Makepeace Thackeray	105
Lorna Doone	R. D. Blackmore . .	106
(From " Lorna Doone.")		
The Tyranny of Andros . . .	John Fiske . .	110
(From " Beginnings of New England.")		
The Long Path	Oliver Wendell Holmes	111
(From " The Autocrat of the Breakfast Table.")		
Lucy and the " Rajah " . , .	Charles Reade .	111
(From " Love me Little, Love me Long.")		
Twenty-three . . .	Charles Dickens .	114
(From " A Tale of Two Cities.")		
" De Baptizin' in Elkhorn Creek " . .	James Lane Allen	116
(From " Two Kentucky Gentlemen of the Old School.")		
Scotchmen	Oliver Goldsmith .	117
On England's Foreign Policy .	John Bright .	118
Virtue Alone Beautiful . .	John G. Whittier	120
Cuvier . . .	Andrew P. Peabody	121
(From " Phi Beta Kappa Oration.")		
Is Gardening a Pleasure ? . . .	Oliver Bell Bunce	122
(From " Bachelor Bluff.")		
The Rose of Glengary . . .	John Esten Cooke	124
(From " The Last of the Foresters.")		

CONTENTS.

The Fishwife	Norman Macleod	125
An English Sunset	Mrs. Sara Coleridge	126
Secession	Samuel Sullivan Cox	126
Covetousness	Robert South	131
Bergerson and Moe	Hjalmar Hjorth Boyesen	131
(From "Vagabond Tales.")		
Palm Sunday	Sir Samuel Romilly	133
Mr. Barkis	Charles Dickens	134
(From "David Copperfield.")		
Spring in New England	Thomas Wentworth Higginson	137
(From "April Days.")		
Continental Congress	Richard Hildreth	138
The Siege of Leyden	John Lothrop Motley	141
Popular Culture	John Morley	143
A Question of Supremacy	Frederick William Shelton	144
(From "Up the River.")		
On the Art of Living with others	Arthur Helps	145
The Invention of Gunpowder	Edward Gibbon	146
(From "Memoirs of my Life and Writings.")		
A Lesson in Patriotism	Edward Everett Hale	148
(From "A Man without a Country.")		
To his Daughter	Norman Macleod	150
On History	Thomas Babbington Macaulay	151
(From "Essay on History.")		
Democracy	James Russell Lowell	152
(From Address given at Birmingham, England, October, 1884.)		
Selfishness versus Nobility	James Anthony Froude	153
(From "The Science of History.")		
Hieronymus and Tiddlekins	Katherine Sherwood Bonner McDowell	155
(From "Harper's Magazine.")		
Obedience to Law	Ralph Waldo Emerson	161
(From "Spiritual Laws.")		
War the Destroyer	Charles James Fox	161
The Gift of Gold	George Eliot	162
(From "Silas Marner.")		
Of Kings' Treasuries	John Ruskin	168
(From "Sesame and Lilies.")		
Old Concord	Margaret Sidney	170
(From "Old Concord; Her Highways and Byways.")		
Protection	Richard Cobden	175

CONTENTS.

To his Wife	Sir Richard Steele	176
On the War of 1812	Henry Clay	177
A Talent for Music	Henry Harland (Sidney Luska)	178
(*From "My Uncle Florimond."*)		
The Militia Bill	John Randolph	181
On Conversation	Thomas De Quincey	182
Tourists on the Continent	William Makepeace Thackeray	184
The Miracle of Nature	Charles Kingsley	184
(*From "My Winter Garden."*)		
Mrs. Potiphar's "Cabinet Shop"	George William Curtis	186
(*From "Potiphar Papers."*)		
An Appeal for Union	Henry Clay	189
Justice for the Slave	Wendell Phillips	191
Raleigh's Last Words to his Wife	Walter Raleigh	192
Death, the Conqueror	Jeremy Taylor	193
Greatness and Ability	Theodore Parker	195
Annie and Lawrence	Frank R. Stockton	197
(*From "The Late Mrs. Null."*)		
The Ethics of Laughter	Henry W. Shaw (Josh Billings)	197
(*From "Josh Billings: his works."*)		
Roger Williams	George Bancroft	199
The Death of Col. Newcome	William Makepeace Thackeray	200
(*From "The Newcomes."*)		
To Grosvenor C. Bedford	Robert Southey	202
Making a Friend	George MacDonald	203
(*From "Annals of a Quiet Neighborhood."*)		
To Lady Holland	Sydney Smith	206
To Bernard Barton	Charles Lamb	206
Every Man Great	William Ellery Channing	207
(*From "Address on Self-Culture."*)		
The Alhambra by Moonlight	Washington Irving	209
(*From "The Alhambra."*)		
Nipped in the Bud	Richard Malcolm Johnston	210
(*From "Dukesborough Tales."*)		
To Robert Ainslie	Robert Burns	215
The Advent of Peace	Thomas Paine	215
(*From "The Crisis."*)		
Homer's Inventive Power	Alexander Pope	217
(*From "Preface to the Iliad."*)		
To H. S. Williams	Charlotte Brontë	218

CONTENTS.

Mr. Casaubon's Romance . . . George Eliot		218
(*From "Middlemarch."*)		
The White Rose Road . . . Sarah Orne Jewett		220
(*From "Strangers and Wayfarers."*)		
Miss Maloney on the Chinese Question . Mary Mapes Dodge .		223
(*From "Theophilus and Others."*)		
To Mrs. Jane Lawder . . . . Oliver Goldsmith		225
The Death of Little Nell . . . Charles Dickens		226
(*From "Old Curiosity Shop."*)		
On American Institutions . . . . James Abram Garfield		227
On the War . . . . . Stephen Arnold Douglas		228
For Freedom of Trade . . Frank R. Hurd .		229
John Keats to William Reynolds . John Keats .		229
Mrs. Carlyle to her Husband . Mrs. Thomas Carlyle		230
Sweetness and Light . . . Matthew Arnold .		231
Fashionable Life at Kinkaird House . Thomas Carlyle .		232
Petition of Thugs . . Walter Savage Landor		233
The Battle of Tlascala . . . William Hickling Prescott .		234
(*From "Conquest of Mexico."*)		
Aunt Maria and the Autophone . . Thomas Frederick Crane		239
(*From "Harper's Magazine."*)		
Joel at Work . . . Margaret Sidney .		242
(*From "Five Little Peppers Midway."*)		
Gradle . . . Thomas Hood .		243
Kin Beyond Sea . William Ewart Gladstone .		244
A True Caledonian . Charles Lamb .		250
Sight and Insight . Thomas Starr King .		252
An Apology for English . . Roger Ascham		253
The Justice of Rienzi the Tribune . Lord Lytton		254
(*From "Rienzi."*)		
An Encounter with the Iroquois . James Fenimore Cooper		257
(*From "The Last of the Mohicans."*)		
Unselfishness . . Lydia Maria Child .		266
Philip and Leigh . Blanche Willis Howard		266
(*From "One Summer."*)		
To Miss Mitford . Benjamin Robert Haydon .		269
In Praise of Poetry . Sir Philip Sidney		271
The Footprint on the Shore . . Daniel De Foe .		273
(*From "Robinson Crusoe."*)		
The Rights of Man . . . Thomas Jefferson		276
(*From "Preamble to Declaration of Independence."*)		

CONTENTS.

To William Robertson	*David Hume*	276
"Stay"	*Miss Mulock*	277
(*From "John Halifax, Gentleman."*)		
The True Track	*Josiah Gilbert Holland*	278
(*From "Timothy Titcomb's Letters to Young People."*)		
Spiritual Emancipation	*Henry James*	279
(*From "Democracy and its Issue."*)		
A Sudden Hurricane	*William Gilmore Simms*	280
(*From "The Partisan."*)		
Italian Life	*Lady Mary Wortley Montagu*	286
Progress	*William E. Gladstone*	287
(*From "The Might of Right."*)		
Personal Influence	*Thomas Hughes*	289
(*From "True Manliness."*)		
Mr. Tarbox and Zoséphine	*George W. Cable*	291
(*From "Au Large."*)		
An Encounter with a Panther	*Charles Brockden Brown*	294
(*From "Edgar Huntley."*)		
To a Child	*Thomas Hood*	296
The Premier Gladstone	*Theodore L. Cuyler*	298
(*From "Right to the Point."*)		
The Fourth of July	*John Adams*	299
The Private Character of Webster	*Rufus Choate*	300
The Sabbath	*Frederick W. Robertson*	302
(*From "Well-Springs of Wisdom."*)		
Joan	*Frances Hodgson Burnett*	303
(*From "That Lass o' Lowrie's."*)		
A Question of Loving	*Thomas Hardy*	304
(*From "Far from the Madding Crowd."*)		
Reform	*John Milton*	309
Country Hospitality	*Jonathan Swift*	310
The American Indian	*Elbridge S. Brooks*	311
(*From "The Story of the American Indian."*)		
Burr and Blennerhasset	*William Wirt*	313
Isabella of Spain	*William Hickling Prescott*	315
(*From "History of Ferdinand and Isabella."*)		
The Legend of the Date-tree	*Charles Étienne Arthur Gayarré*	317
(*From "History of Louisiana."*)		
A Scene in the Forecastle	*Herman Melville*	318
(*From "Omoo."*)		

CONTENTS.

Captain Cuttle's Island	Charles Dickens	321
(From " Dombey and Son.")		
Barberry Island	Sophie Swett	326
(From "Good Company.")		
The Town Pump	Nathaniel Hawthorne	327
(From " Twice-Told Tales.")		
Adam and Dinah	George Eliot	329
(From " Adam Bede.")		
Indolence	Samuel Smiles	331
(From " The Art of Living.")		
The Tournament	Sir Walter Scott	332
(From " Ivanhoe.")		
The Culture of the Puritans	John Gorham Palfrey	338
(From " A History of New England.")		
Bess and the Snake	William Gilmore Simms	340
(From " Yemassee.")		
Home	Samuel Smiles	343
(From " The Art of Living.")		
The Tower of London	Charles F. Browne (Artemus Ward)	344
(From " Punch, 1866.")		
Tilly Bones	Elizabeth Whitfield Bellamy	347
(From " The Manhattan Magazine.")		
In Venice	Samuel Rogers	350
The Standard of Speech	Noah Webster	353
(From " Dissertations on the English Language.")		
Dramatic Realism	Charles Dickens	356
Whittier with the Children	Margaret Sidney	357
(From " Whittier with the Children.")		

LIST OF ILLUSTRATIONS.

A True Caledonian	*Frontis.*	A promise of Spring		139
"If he would find content"	9	Experts in the art of war		147
John Milton	10	The quick-witted youth		156
Thomas Arnold	11	The Pop family		157
Into my garden	12	A neighbor's boy		159
Charles Kingsley	14	Bonaparte		163
The rectory at Eversley	15	Mr. Ruskin's house, Brantwood		169
The haunt of the trout	16	The Wayside		171
John Bunyan	17	Visitors' memorial on the site of Thoreau's hut		173
Bedford Jail, where the Pilgrim's Progress was written	19	On the road to "Nine Acre Corner"		174
The Stuart Portrait of Washington	21	Gregory surprises Mr. Finkelstein at the hand organ		179
Charles Lamb	22			
Geoffrey Chaucer	23	Following the leader		183
Harriet Beecher Stowe	24	A sight to make one's pulses throb		185
Mending the clock	25	Guests of the Potiphars		187
Edmund Spenser	28	Sir Walter Raleigh		192
Thomas Carlyle	29	The old mill		199
Emerson's home in Concord, Mass.	30	George MacDonald		203
Ralph Waldo Emerson	31	Miss Julia Louisa Wilkins		211
Henry David Thoreau	33	Homer		217
Daniel Webster	37	The broad open country		220
Francis Bacon	43	The waiter-man		224
Joseph Addison	45	James Abram Garfield		227
Sir Walter Scott	52	William E. Gladstone		245
Hawthorne's study at "Wayside"	60	A typical Caledonian		251
Henry W. Longfellow	61	Arrayed for defense		254
Melons	69	Ready for war		258
Alexander Hamilton Stephens	73	The honeysuckle grew all about		267
Abraham Lincoln	75	At milking-time		270
Bayard Taylor	76	Weathering a gale		273
Benjamin Franklin	77	A sudden storm		281
Statue of Franklin, Independence Hall, Pa.	78	One type of Italian beauty		286
Thomas Jefferson	80	Rugby School		290
Sumner's Study	83	Bathsheba		305
Already in the field	85	Milton at the organ		309
The Alcott home	93	Under sail		319
Dean Stanley	104	"The stormy winds do blow"		324
Gadshill. — The home of Charles Dickens	114	Nathaniel Hawthorne		328
Considering the next text	116	The Tower of London		346
John G. Whittier	120	Tilly on the plantation		348
A fisher lad	127	A leap from the Rialto		351
Mr. Barkis	134	"Sweet Kenoza from the shore, and Watching Hills beyond"		359
In a sea of glory	137			

MASTERPIECES OF PROSE

CONTENTMENT.

I KNEW a man that had health and riches, and several houses, all beautiful and ready furnished, and would often trouble himself and family to be removing from one house to another; and being asked by a friend why he removed so often from one house to another, replied, "It was to find content in some of them." But his friend, knowing his temper, told him, "If he would find content in any of his houses, he must leave himself behind him; for content would never dwell but in a meek and quiet soul." And this may appear, if we read and consider what our Saviour says in St. Matthew's Gospel, for He there says, "Blessed be the merciful, for they shall obtain mercy. Blessed be the pure in heart, for they shall see God. Blessed be the poor in spirit, for theirs is the kingdom of heaven. And blessed be the meek, for they shall possess the earth." Not that the meek shall not also obtain mercy, and see God, and be comforted, and at last come to the kingdom of heaven; but, in the meantime, he, and he

"IF HE WOULD FIND CONTENT."

only, possesses the earth, as he goes toward that kingdom of heaven, by being humble and cheerful, and content with what his good God has allotted him. He has no turbulent, repining, vexatious thoughts that he deserves better; nor is vexed when he sees others possessed of more honor or more riches than his wise God has allotted for his share; but he possesses what he has with a meek and contented quietness, such a quietness as makes his very dreams pleasing, both to God and himself.

ISAAK WALTON.

THE ALL-CONQUERING POWER OF TRUTH.

JOHN MILTON.

Though all the winds of doctrine were let loose to play upon the earth, so Truth be in the field, we do injuriously, by licensing and prohibiting, to misdoubt her strength. Let her and falsehood grapple; who ever knew Truth put to the worst in a free and open encounter? Her confuting is the best and surest suppressing. He who hears what praying there is for light and clear knowledge to be sent down among us, would think of other matters to be constituted beyond the discipline of Geneva, famed and fabricked already to our hands. Yet when the new life which we beg for shines in upon us, there be who envy and oppose, if it come not first in at their casements. What a collusion is this, when as we are exhorted by the wise man to use diligence, "to seek for wisdom as for hidden treasures," early and late, that another order shall enjoin us to know nothing but by statute! When a man hath been laboring the hardest labor in the deep mines of knowledge, hath furnished out his findings in all their equipage, drawn forth his reasons, as it were a battle ranged, scattered and defeated all objections in his way, calls out his adversary into the plain, offers him the advantage of wind and sun, if he please, only that he may try the matter by dint of argument; for his opponents then to skulk, to lay ambushments, to keep a narrow bridge of licensing where the challenger should pass, though it be valor enough in soldiership, is but weakness and cowardice in the wars of Truth. For who knows not that Truth is strong, next to the Almighty? She needs no policies, nor stratagems, nor licensings, to make her victorious; those are the shifts and the defenses that error uses against her power; give her but room, and do not bind her when she sleeps.

JOHN MILTON.

AT RUGBY.

RUGBY, October 12, 1835.

. . . I cannot tell you how I enjoyed our fortnight at Rugby before the school opened. It quite reminded me of Oxford, when Mary and I used to sit out in the garden under the enormous elms of the school-field, which almost overhang the house, and saw the line of our battlemented roofs and the pinnacles and cross of our chapel cutting the unclouded sky. And I had divers happy little matches at cricket with my own boys in the school-field, on the very cricket-ground of the "eleven," that is, of the best players in the school, on which, when the school is assembled, no profane person may encroach. . . . It would overpay me for far greater uneasiness and labor than I have ever had at Rugby, to see the feeling both towards the school and towards myself personally with which some of our boys have been lately leaving us. One stayed with us in the house for his last week at Rugby, dreading the approach of the day which should take him to Oxford, although he was going up to a most delightful society of old friends ; and, when he actually came to take his leave, I really think that the parting was like that of a father and his son. And it is delightful to me to find how glad all the better boys are to come back here after they have left it, and how much they seem to enjoy staying with me ; while a sure instinct keeps at a distance all whose recollections of the place are connected with uncomfortable reflections. Meantime I write nothing, and read barely enough to keep my mind in the state of a running stream, which I think it ought to be if it would form and feed other minds ; for it is ill drinking out of a pond, whose stock of water is

THOMAS ARNOLD.

merely the remains of the long past rains of the winter and spring, evaporating and diminishing with every successive day of drought. . . .

THOMAS ARNOLD.

MY GARDEN.

INTO MY GARDEN.

SQUASHES. — They appeared above ground large-lobed and vigorous. Large and vigorous appeared the bugs all gleaming in green and gold, like the wolf on the fold, and stopped up all the stomata and ate up all the parenchyma, till my squash-leaves looked as if they had grown for the sole purpose of illustrating netveined organizations. In consternation I sought again my neighbor the Englishman. He answered me he had 'em on his, too, lots of 'em. This reconciled me to mine. Bugs are not inherently desirable, but a universal bug does not indicate special want of skill in any one. So I was comforted. But the Englishman said they must be killed. He had killed his. Then I said I would kill mine, too. How should it be done? Oh, put a shingle near the vine at night, and they would come upon it to keep dry, and go out early in the morning and kill 'em. But how to kill them? Why, take 'em right between your thumb and finger and crush 'em!

As soon as I could recover breath I informed him confidentially that, if the world were one great squash, I wouldn't undertake to save it in that way. He smiled a little, but I think he was not overmuch pleased. I asked him why I couldn't take a bucket of water and dip shingle in it and drown them. He said, well, I could try it. I did try it, first wrapping my hand in a cloth to prevent contact with any stray bug. To my amazement, the moment they touched the water they all spread unseen wings and flew away, safe and sound. I should not have been much more surprised to see Halicarnassus soaring over the ridge-pole. I had not the slightest idea that they could fly. Of course I gave up the design of drowning them. I called a council of war. One said I must put a newspaper over them and fasten it down at the edges; then they couldn't get in. I timidly suggested that the squashes couldn't get out. Yes, they could, he said — they'd grow right through the paper. Another said I must surround them with round boxes with the bottoms broken out ; for, though they could fly, they couldn't steer, and when they flew up they just dropped down anywhere, and as there was on the whole a good deal more land on the outside of the boxes than on the inside, the chances were in favor of their dropping on the outside. Another said that ashes must be sprinkled on them. A fourth said lime was an infallible remedy. I began with the paper, which I secured with no little difficulty ; for the wind — the same wind, strange to say — kept blowing the dirt at me, and the paper away from me ; but I consoled myself by remembering the numberless rows of squash pies that should crown my labors, and May took heart from Thanksgiving. The next day I peeped under the paper, and the bugs were a solid phalanx. I reported at headquarters, and they asked me if I killed the bugs before I put the paper down. I said no, I supposed it would stifle them — in fact, I did not think anything about it, but if I had thought anything, that was what I thought. I was not pleased to find I had been cultivating the bugs and furnishing them with free lodgings. I went home, and tried all the remedies in succession. I could hardly decide which agreed best with the structure and habits of the bugs, but they throve on all. Then I tried them all at once and all o'er with a mighty uproar. Presently the bugs went away. I am not sure that they would not have gone just as soon if I had let them alone. After they were gone, the vines scrambled out and put forth some beautiful deep-golden blossoms. When they fell off, that was the end of them. Not a squash — not one — not a single squash, not even a pumpkin. They were all false blossoms. . . .

MARY ABIGAIL DODGE (*Gail Hamilton*).

ON AFFAIRS IN AMERICA.

MY LORDS, this ruinous and ignominious situation, where we cannot act with success, nor suffer with honor, calls upon us to remonstrate in the strongest and loudest language of truth; to rescue the ear of majesty from the delusions which surround it. The desperate state of our arms abroad is in part known. No man thinks more highly of them than I do. I love and honor the English troops. I know their virtues and their valor. I know they can achieve anything except impossibilities ; and I know that the conquest of English America is an impossibility.

You cannot, I venture to say, you cannot conquer America. Your armies in the last war effected everything that could be effected ; and what was it ? It cost a numerous army, under the command of a most able general (Lord Amherst), now a noble Lord in this House, a long and laborious campaign, to expel five thousand Frenchmen from French America. My Lords, you cannot conquer America. What is your present situation there ? We do not know the worst ; but we know that in three campaigns we have done nothing and suffered much. Besides the sufferings, perhaps total loss of the Northern force, the best appointed army that ever took the field, commanded by Sir William Howe, has retired from the American lines. He was obliged to relinquish his attempt, and with great delay and danger, to adopt a new and distant plan of operations. We shall soon know, and in any event have reason to lament, what may have happened since. As to conquest, therefore, my Lords, I repeat, it is impossible. You may swell every expense and every effort still more extravagantly; pile and accumulate every assistance you can buy or borrow; traffic and barter with every little pitiful German prince that sells and sends his subjects to the shambles of a foreign prince; your efforts are forever vain and impotent, doubly so from this mercenary aid on which you rely; for it irritates, to an incurable resentment, the minds of your enemies, to overrun them with the mercenary sons of rapine and plunder, devoting them and their possessions to the rapacity of hireling cruelty! If I were an American, as I am an Englishman, while a foreign troop was landed in my country, I never would lay down my arms — never — never — never.

LORD CHATHAM.

A COUNTRY PARISH.

CHARLES KINGSLEY.

EVERSLEY, 1842.

PETER! — Whether in the glaring saloons of Almack's, or making love in the equestrian stateliness of the park, or the luxurious recumbency of the ottoman, whether breakfasting at one, or going to bed at three, thou art still Peter, the beloved of my youth, the staff of my academic days, the regret of my parochial retirement! — Peter! I am alone! Around me are the everlasting hills, and the everlasting bores of the country! My parish is peculiar for nothing but want of houses and abundance of peat bogs; my parishioners remarkable only for aversion to education, and a predilection for fat bacon. I am wasting my sweetness on the desert air — I say my sweetness, for I have given up smoking, and smell no more. O, Peter, Peter, come down and see me! O, that I could behold your head towering above the fir-trees that surround my lonely dwelling. Take pity on me! I am like a kitten in the washhouse copper with the lid on! And, Peter, prevail on some of your friends here to give me a day's trout-fishing, for my hand is getting out of practice. But, Peter, I am, considering the oscillations and perplex circumgurgitations of this piecemeal world, an improved man. I am much more happy, much more comfortable, reading, thinking, and doing my duty — much more than ever I did before in my life. Therefore I am not discontented with my situation, or regretful that I buried my first-class in a country curacy, like the girl who shut herself up in a band-box on her wedding night (*vide* Rogers' "Italy"). And my lamentations are not general (for I do not want an inundation of the froth and tide-wash of Babylon the Great), but particular, being solely excited by want of thee, O Peter, who art very pleasant to me, and wouldst be more so if thou wouldst come and eat my mutton, and drink my wine, and admire my sermons, some Sunday at Eversley.

THE RECTORY AT EVERSLEY.

CHARLES KINGSLEY.

LIMITATIONS OF FREE SPEECH.

GENTLEMEN, I cannot conclude without expressing the deepest regret at all attacks upon the Christian religion by authors who profess to promote the civil liberties of the world. For under what other auspices than Christianity have the lost and subverted liberties of mankind in former ages been reasserted? By what zeal, but the warm zeal of devout Christians, have English liberties been redeemed and consecrated? Under what other sanctions, even in our own days, have liberty and happiness been spreading to the uttermost corners of the earth? What work of civilization, what Commonwealth of greatness, has this bald religion of nature ever established? We see, on the contrary, the nations that have no other light than that of nature to direct them, sunk in barbarism, or slaves to arbitrary government; whilst under the Christian dispensation, the great career of the world has been slowly but clearly advancing, lighter at every step from the encouraging prophecies of the gospel, and leading, I trust, in the end, to universal and eternal happiness. Each generation of mankind can see but a few revolving links of this mighty and mysterious chain; but by doing our several duties in our allotted stations, we are sure that we are fulfilling the purposes of our existence. You, I trust, will fulfill yours this day.

LORD ERSKINE.

DEACON MARBLE'S TROUT.

HE was a curious trout. I believe he knew Sunday just as well as Deacon Marble did. At any rate, the deacon thought the trout meant to aggravate him. The deacon, you know, is a little waggish. He often tells about that trout. Says he. "One Sunday morning, just as I got along by the willows, I heard an awful splash, and not ten feet from shore I saw the trout, as long as my arm, just curving over like a bow, and going down with something for breakfast. 'Gracious'! says I, and I almost jumped out of the wagon. But my wife Polly, says she, 'What on airth are you thinkin' of, Deacon? It's Sabbath day, and you're goin' to meetin'! It's a pretty business for a deacon!' That sort o' cooled me off. But I do say, that, for about a minute, I wished I wasn't a deacon. But 'twouldn't made any difference, for I came down next day to mill on purpose, and I came down once or twice more, and nothin' was to be seen, tho' I tried him with the most temptin' things. Wal, next Sunday I came along ag'in, and, to save my life, I couldn't keep off worldly and wanderin' thoughts. I tried to be sayin' my catechism,

THE HAUNT OF THE TROUT.

but I couldn't keep my eyes off the pond as we came up to the willows. I'd got along in the catechism, as smooth as the road, to the Fourth Commandment, and was sayin' it out loud for Polly, and jist as I was sayin': 'What is required in the Fourth Commandment?' I heard a splash, and there was the trout, and, afore I could think, I said: 'Gracious, Polly, I must have that trout.' She almost riz right up, 'I knew you wa'n't sayin' your catechism hearty. Is this the way you answer the question about keepin' the Lord's day? I'm ashamed, Deacon Marble,' says she. 'You'd better change your road, and go to meetin' on the road over the hill. If I was a deacon, I wouldn't let a fish's tail whisk the whole catechism out of my head;' and I had to go to meetin' on the hill road all the rest of the summer."

HENRY WARD BEECHER.

CHRISTIAN IN DOUBTING CASTLE.

JOHN BUNYAN.

Now there was, not far from the place where they lay, a castle, called Doubting Castle, the owner whereof was Giant Despair; and it was in his grounds they now were sleeping. Wherefore he, getting up in the morning early, and walking up and down in his fields, caught Christian and Hopeful asleep in his grounds. Then, with a grim and surly voice, he bid them awake; and asked them whence they were and what they did on his grounds. They told him they were pilgrims, and that they had lost their way. Then said the giant, "You have this night trespassed on me, by trampling in and lying on my grounds, and therefore you must go along with me." So they were forced to go, because he was stronger than they. They also had but little to say, for they knew themselves in a fault. The giant, therefore, drove them before him, and put them into his castle, into a very dark dungeon, nasty and stinking to the spirits of these two men. Here, then, they lay from Wednesday morning till Saturday night, without one bit of bread, or drop of drink, or light, or any to ask how they did. They were therefore here in evil case, and were far from friends and acquaintance. Now in this place Christian had double sorrow, because it was through his unadvised counsel that they were brought into this distress.

Now Giant Despair had a wife, and her name was Diffidence. So when he was gone to bed, he told his wife what he had done; to wit, that he had taken a couple of prisoners and cast them into his dungeon, for trespassing on his grounds. Then he asked her also what he had best to do further to them. So she asked him what they were, whence they came, and whither they were bound; and he told her. Then she counselled him that when he arose in the morning he should beat them without any mercy. So, when he arose, he getteth him a grievous crab-tree cudgel, and goes down into the dungeon to them, and there first falls to rating of them as if they were dogs, although they gave him never a word of distaste; then he falls upon them, and beats them fearfully, in such sort they were not able to help themselves, or to turn them upon the floor. This done, he withdraws and leaves them, there to condole their misery and to mourn under their distress. So all that day they spent the time in nothing but sighs and bitter lamentations. The next night she, talking with her husband about them further, and understanding they were yet alive, did advise him to counsel them to make away themselves. So, when morning was come, he goes to them in a surly manner, as before, and perceiving them to be very sore with the stripes that he had given them the day before, he told them, that since they were never like to come out of that place, their only way would be forthwith to make an end of themselves, either with knife, halter or poison. "For why," said he, "should you choose life, seeing it is attended with so much bitterness?" But they desired him to let them go. With that he looked ugly upon them, and rushing to them, had doubtless made an end of them himself, but that he fell into one of his fits (for he sometimes, in sunshiny weather, fell into fits), and lost for a time the use of his hand; wherefore he withdrew, and left them as before, to consider what to do. Then did the prisoners consult between themselves, whether it was best to take his counsel or no; and thus they began to discourse:

"Brother," said Christian, "what shall we do? The life that we now live is miserable. For my part, I know not whether is best, to live thus, or to die out of hand. 'My soul chooseth strangling rather than life,' and the grave is more easy for me than this dungeon. Shall we be ruled by the giant?"

HOPE. Indeed, our present condition is dreadful, and death would be far more welcome to me than thus for ever to abide; but yet, let us consider, the Lord of the country to which we are going hath said, "Thou shalt do no murder:" no not to another man's person; much more then, are we forbidden to take his counsel to kill ourselves. Besides, he that kills another can but commit murder upon his body; but for one to kill himself is to kill body and soul at once. And moreover, my brother, thou talkest of ease in the grave; but hast thou forgotten the hell whither for certain the murderers go? For "no murderer hath eternal life." And let us consider again, that all the law is not in the hand of Giant Despair. Others, so far as I can understand, have been taken by him, as well as we; and yet have escaped out of his hand. Who knows but that God that made the world may cause that Giant Despair may die? or that at some time or other he may forget to lock us in? or that he may in a short time have another of his fits before us, and

CHRISTIAN IN DOUBTING CASTLE.

may lose the use of his limbs? — and if ever that should come to pass again, for my part I am resolved to pluck up the heart of a man, and to try my utmost to get from under his hand. I was a fool that I did not try to do it before; but, however, my brother, let's be patient, and endure a while; the time may come that may give us a happy release; but let us not be our own murderers.

With these words, Hopeful at present did moderate the mind of his brother; so they continued together (in the dark) that day, in their sad and doleful condition.

Well, towards evening, the giant goes down into the dungeon again, to see if his prisoners had taken his counsel; but when he came there he found them alive; and, truly, alive was all, for now, what for want of bread and water, and by reason of the wounds they received when he beat them, they could do little but breathe. But, I say, he found them alive; at which he fell into a grievous rage, and told them that, seeing they had disobeyed his counsel, it should be worse with them than if they had never been born.

At this they trembled greatly, and I think that Christian fell into a swoon; but, coming a little to himself again, they renewed their discourse about the giant's counsel; and whether yet they had best to take it or no. Now Christian again seemed to be for doing it, but Hopeful made his second reply as followeth:

BEDFORD JAIL, WHERE THE PILGRIM'S PROGRESS WAS WRITTEN.

"My brother," said he, "rememberest thou not how valiant thou hast been heretofore? Apollyon could not crush thee, nor could all that thou didst hear, or see, or feel, in the Valley of the Shadow of Death. What hardship, terror, and amazement hast thou already gone through! and art thou now nothing but fears? Thou seest that I am in the dungeon with thee, a far weaker man by nature than thou art: also, this giant has wounded me as well as thee, and hath also cut off the bread and water from my mouth; and with thee I mourn without the light. But let's exercise a little more patience: remember how thou playedest the man at Vanity Fair, and wast neither afraid of the chain, nor cage, nor yet of bloody death. Wherefore, let us (at least to avoid the shame, that becomes not a Christian to be found in) bear up with patience as well as we can."

Now night being come again, and the giant and his wife being in bed, she asked him concerning the prisoners, and if they had taken his counsel. To which he replied, "They are sturdy rogues; they choose rather to bear all hardship, than to make away themselves." Then said she, "Take them into the castle-yard to-morrow, and show them the bones and skulls of those that thou hast already dispatched, and make them believe, ere a week comes to an end, thou also wilt tear them in pieces as thou hast done their fellows before them."

So when the morning was come, the giant goes to them again, and takes them into the castle-yard, and shows them as his wife had bidden him. "These," said he, "were pilgrims as you are, once, and they trespassed in my grounds, as you have done; and when I thought fit, I tore them in pieces; and so within ten days I will do you. Go, get you down to your den again;" and with that he beat them all the way thither. They lay, therefore, all day on Saturday in a lamentable case, as before. Now when night was come, and when Mrs. Diffidence and her husband, the giant, were got to bed, they began to renew their discourse of their prisoners; and, withal, the old giant wondered that he could neither by his blows nor his counsel, bring them to an end. And with that his wife replied, "I fear," said she, "that they live in hope that some will come to relieve them, or that they have picklocks about them, by the means of which they hope to escape." "And sayest thou so, my dear?" said the giant; "I will therefore search them in the morning."

Well, on Saturday, about midnight, they began to pray, and continued in prayer till almost break of day.

Now, a little before it was day, good Christian, as one half amazed, brake out in this passionate speech: "What a fool," quoth he, "am I, thus to lie in a stinking dungeon, when I may as well walk at liberty! I have a key in my bosom, called Promise, that will, I am persuaded, open any lock in Doubting Castle." Then said Hopeful, "That is good news; good brother, pluck it out of thy bosom, and try."

Then Christian pulled it out of his bosom, and began to try at the dungeon door, whose bolt (as he turned the key) gave back, and the door flew open with ease, and Christian and Hopeful both came out. Then he went to the outward door that leads into the castle-yard, and, with his key, opened that door also. After, he went to the iron gate, for that must be opened too; but that lock went damnable hard, yet the key did open it. Then they thrust open the gate to make their escape with speed, but that gate, as it opened, made such a creaking that it waked Giant Despair, who, hastily rising to pursue his prisoners, felt his limbs to fail, for his fits took him again, so that he could by no means go after them. Then they went on, and came to the King's highway, and so were safe, because they were out of his jurisdiction.

Now, when they were gone over the stile, they began to contrive with themselves what they should do at that stile, to prevent those that should come after from falling into the hands of Giant Despair. So they consented to erect there a pillar, and to engrave upon the side thereof this sentence: "Over this stile is the way to Doubting Castle, which is kept by Giant Despair, who despiseth the King of the Celestial Country, and seeks to destroy his holy pilgrims." Many therefore that followed after read what was written, and escaped the danger.

JOHN BUNYAN.

THE STUART PORTRAIT OF WASHINGTON.

THE PLEASURES OF PRIVATE LIFE.

Under the shadow of my own vine and my own fig-tree, free from the bustle of a camp, and the busy scenes of public life, I am solacing myself with those tranquil enjoyments, of which the Soldier, who is ever in pursuit of fame, the Statesman, whose watchful days and sleepless nights are spent in devising schemes to promote the welfare of his own, perhaps the ruin of other countries, as if the globe was insufficient for us all — and the Courtier, who is always watching the countenance of his Prince, in hopes of catching a gracious smile — can have very little conception. I have not only retired from all public employments, but I am retiring within myself, and shall be able to view the solitary walk, and tread the paths of private life, with a heartfelt satisfaction. Envious of none, I am determined to be pleased with all; and, this being the order of my march, I will move gently down the stream of life until I sleep with my fathers.

GEORGE WASHINGTON.

ON THE DEATH OF AN OLD FRIEND.

CHARLES LAMB.

Colebrook Row, Islington,
January 20, 1826.

I called upon you this morning, and found that you were gone to visit a dying friend. I had been upon a like errand. Poor Norris . . . in him I have a loss the world cannot make up. He was my friend and my father's friend all the life I can remember. I seem to have made foolish friends ever since. Those are friendships which outlive a second generation. Old as I am waxing, in his eyes I was still the child he first knew me. To the last he called me Charley. I have none to call me Charley now. He was the last link that bound me to the Temple. You are but of yesterday. In him seem to have died the old plainness of manners and singleness of heart. Letters he knew nothing of, nor did his reading extend beyond the pages of the *Gentleman's Magazine*. Yet there was a pride of literature about him from being amongst books (he was librarian), and from some scraps of doubtful Latin which he had picked up in his office of entering students, that gave him very diverting airs of pedantry. Can I forget the erudite look with which when he had been in vain trying to make out a black-letter text of Chaucer in the Temple Library, he laid it down and told me that — "in these old books, Charley, there is sometimes a deal of very indifferent spelling," — and seemed to console himself in the reflection! His jokes, for he had his jokes, are now ended ; but they were old trusty perennials, staples that pleased after *decies repetita*, and were always as good as new. One song he had, which was reserved for the night of Christmas-day, which we always spent in the Temple. It was an old thing, and spoke of the flat bottoms of our foes and the possibility of their coming over in darkness, and alluded to threats of an invasion many years blown over; and when he came to the part

> "We'll still make 'em run, and we'll still make 'em sweat,
> In spite of the devil and *Brussels Gazette*,"

his eyes would sparkle as with the freshness of an impending event. And what is the *Brussels Gazette* now? I cry while I enumerate these trifles. "How shall we tell them in a stranger's ear?" . . .

CHARLES LAMB.

UPON RICHES.

IN getting of your riches, and in using of 'em, ye shulden alway have three things in your heart, that is to say, our Lord God, conscience, and good name. First, ye shulden have God in your heart, and for no riches ye shulden do nothing which may in any manner displease God that is your creator and maker; for, after the word of Solomon, it is better to have a little good with the love of God, than to have muckle good and lese the love of his Lord God; and the prophet saith, that better it is to ben a good man and have little good and treasure, than to be holden a shrew and have great riches. And yet I say furthermore, that ye shulden always do your business to get your riches, so that ye get 'em with a good conscience. And the apostle saith, that there nis thing in this world, of which we shulden have so great joy, as when our conscience beareth us good witness; and the wise man saith, The substance of a man is full good when sin is not in a man's conscience.

GEOFFREY CHAUCER.

Afterward, in getting of your riches and in using of 'em, ye must have great business and great diligence that your good name be alway kept and conserved; for Solomon saith, that better it is and more it availeth a man to have a good name than for to have great riches; and therefore he saith in another place, Do great diligence (saith he) in keeping of thy friends and of thy good name, for it shall longer abide with thee than any treasure, be it never so precious; and certainly he should not be called a gentleman that, after God and good conscience all things left, ne doth his diligence and business to keepen his good name; and Cassiodore saith, that it is a sign of a gentle heart, when a man loveth and desireth to have a good name.

GEOFFREY CHAUCER.

SAM MENDS THE CLOCK.

HARRIET BEECHER STOWE.

"WHY, ye see, Miss Lois," he would say, "clocks can't be druv; that's jest what they can't. Some things can be druv, and then ag'in some things can't, and clocks is that kind. They's jest got to be humored. Now this 'ere's a 'mazin' good clock; give me my time on it, and I'll have it so 'twill keep straight on to the Millennium."

"Millennium!" says Aunt Lois, with a snort of infinite contempt. "Yes, the Millennium," says Sam, letting fall his work in a contemplative manner. "That 'ere's an interestin' topic. Now Parson Lothrop, he don't think the Millennium will last a thousand years. What's your 'pinion on that p'int, Miss Lois?" "My opinion is," said Aunt Lois, in her most nipping tones, "that if folks don't mind their own business, and do with their might what their hand finds to do, the Millennium won't come at all."

"Wal, you see, Miss Lois, it's just here — one day is with the Lord as a thousand years, and a thousand years as one day."

"I should think you thought a day was a thousand years, the way you work,' said Aunt Lois.

"Wal," says Sam, sitting down with his back to his desperate litter of wheels, weights and pendulums, and meditatively caressing his knee as he watched the sailing clouds in abstract meditation, "ye see, ef a thing's ordained, why, it's got to be, ef you don't lift a finger. That 'ere's so now, ain't it?"

"Sam Lawson, you are about the most aggravating creature I ever had to do with. Here you've got our clock all to pieces, and have been keeping up a perfect hurrah's nest in our kitchen for three days, and there you sit maundering and

MENDING THE CLOCK.

talking with your back to your work, fussin' about the Millennium, which is none of your business, or mine, as I know of! Do either put that clock together or let it alone!"

"Don't you be a grain uneasy, Miss Lois. Why, I'll have your clock all right in the end, but I can't be druv. Wal, I guess I'll take another spell on't to-morrow or Friday."

Poor Aunt Lois, horror-stricken, but seeing herself actually in the hands of the imperturbable enemy, now essayed the task of conciliation. "Now do, Lawson, just finish up this job, and I'll pay you down, right on the spot; and you need the money."

"I'd like to 'blige ye, Miss Lois; but ye see money ain't everything in this world. Ef I work tew long on one thing, my mind kind o' gives out, ye see; and besides, I've got some 'sponsibilities to 'tend to. There's Mrs. Captain Brown, she made me promise to come to-day and look at the nose o' that 'ere silver teapot o' hern; it's kind o' sprung a leak. And then I 'greed to split a little oven-wood for the Widdah Pedee, that lives up on the Shelburn road. Must visit the widdahs in their affliction, Scriptur' says. And then there's Hepsy: she's allers a castin' it up at me that I don't do nothin' for her and the chil'en; but then, lordy massy, Hepsy hain't no sort o' patience. Why, jest this mornin' I was tellin' her to count up her marcies, and I 'clare for't if I didn't think she'd a throwed the tongs at me. That 'ere woman's temper railly makes me consarned. Wal, good day, Miss Lois. I'll be along again to-morrow or Friday or the first o' next week." And away he went with long loose strides down the village street, while the leisurely wail of an old fuguing tune floated back after him,—

> "Thy years are an
> Eternal day,
> Thy years are an
> Eternal day."

"An eternal torment," said Aunt Lois, with a snap. "I'm sure, if there's a mortal creature on this earth that I pity, it's Hepsy Lawson. Folks talk about her scolding — that Sam Lawson is enough to make the saints in Heaven fall from grace. And you can't do anything with him: it's like charging bayonet into a wool-sack."

HARRIET BEECHER STOWE.

EDMUND SPENSER.

THE IRISH BARD.

THERE is amongst the Irish a certain kind of people called *Bards*, which are to them instead of poets, whose profession is to set forth the praises or dispraises of men, in their poems or rithmes; the which are had in so high regard and estimation amongst them, that none dare displease them for fear to run into reproach through their offense, and to be made infamous in the mouths of all men. For their verses are taken up with a general applause, and usually sung at all feasts and meetings by certain other persons, whose proper function that is, who also receive for the same great rewards and reputation amongst them. . . .

Such poets as in their writings do labor to better the manners of men, and through the sweet bait of their numbers to steal into the young spirits a desire of honor and virtue, are worthy to be had in great respect. But these Irish bards

are for the most part of another mind, and so far from instructing young men in moral discipline, that they themselves do more deserve to be sharply disciplined: for they seldom use to choose unto themselves the doings of good men for the arguments of their poems, but whomsoever they find to be most licentious of life, most bold and lawless in his doings, most dangerous and desperate in all parts of disobedience and rebellious disposition; him they set up and glorify in their rithmes, him they praise to the people, and to young men make an example to follow.
EDMUND SPENSER.

LONDON.

THOMAS CARLYLE.

. . . I CANNOT say that this huge blind monster of a city is without some sort of charm for me. It leaves one alone to go his own road unmolested. Deep in your soul you take up your protest against it, defy it, and even despise it, but need not divide yourself from it for that. Worthy individuals are glad to hear your thoughts if it have any sincerity; they do not exasperate themselves or you about it; they have not even time for such a thing. Nay, in stupidity itself, on a scale of this magnitude, there is an impressiveness, almost a sublimity; one thinks how, in the words of Schiller, "the very gods fight against it in vain"; how it lies on its unfathomable foundation there, inert, yet peptic, nay, eupeptic, and is a Fact in the world, let theory object as it will. Brown-stout, in quantities that would float a seventy-four goes down the throats of men; and the roaring flood of life pours on;—over which Philosophy and Theory are but a poor shriek of remonstrance, which oftenest times were wiser, perhaps, to hold its peace. . . .
THOMAS CARLYLE.

COMPENSATION.

Human labor, through all its forms, from the sharpening of a stake to the construction of a city or an epic, is one immense illustration of the perfect compensation of the universe. The absolute balance of Give and Take, the doctrine that everything has its price — and if that price is not paid, not that thing but something else is obtained, and that it is impossible to get anything without its price — is not less sublime in the columns of a ledger than in the budgets of States, in the laws of light and darkness, in all the action and reaction of nature. I cannot doubt that the high laws which each man sees implicated in those processes with which he is conversant — the stern ethics which sparkle on his chisel-edge, which are measured out by his plumb and foot-rule, which stand as manifest in the footing of the shop-bill as in the history of a State — do recommend to him his trade, and though seldom named, exalt his business to his imagination.

The league between virtue and nature engages all things to assume a hostile front to vice. The beautiful laws and substances of the world persecute and whip the traitor. He finds that things are arranged for truth and benefit, but there is no den in the wide world to hide a rogue. Commit a crime, and the earth is made of glass. Commit a crime, and it seems as if a coat of snow fell on the ground, such as reveals in the woods the track of every partridge and fox and squirrel and mole. You cannot recall the spoken word, you cannot wipe out the foot-track, you cannot draw up the ladder, so as to leave no inlet or clue. Some damning circumstance always transpires. The laws and substances of nature — water, snow, wind, gravitation — become penalties to the thief. On the other hand, the law holds with equal sureness for all right action.

EMERSON'S HOME IN CONCORD, MASS.

RALPH WALDO EMERSON.

Love, and you shall be loved. All love is mathematically just, as much as the two sides of an algebraic equation. The good man has absolute good, which like fire turns everything to its own nature, so that you cannot do him any harm; but as the royal armies sent against Napoleon, when he approached, cast down their colors and from enemies became friends, so disasters of all kinds, as sickness, offense, poverty, prove benefactors:

> "Winds blow and waters roll
> Strength to the brave, and power and deity,
> Yet in themselves are nothing." . . . .

Our strength grows out of our weakness. The indignation which arms itself with secret forces does not awaken until we are pricked and stung and sorely assailed. A great man is always willing to be little. Whilst he sits on the cushion of advantages, he goes to sleep. When he is pushed, tormented, defeated, he has a chance to learn something; he has been put on his wits, on his manhood; he has gained facts; learns his ignorance; is cured of the insanity of conceit; has got moderation and real skill. The wise man throws himself on the side of his assailants. It is more his interest than it is theirs to find his weak point. The wound cicatrizes and falls off from him like a dead skin, and when they would triumph, lo! he has passed on invulnerable. Blame is safer than praise. I hate to be defended in a newspaper. As long as all that is said is said against me, I feel a certain assurance of success. But as soon as honeyed words of praise are spoken for me, I feel as one that lies unprotected before his enemies. In general, every evil to which we do not succumb is a benefactor. As the Sandwich Islander believes that the strength and valor of the enemy he kills passes into himself, so we gain the strength of the temptation we resist.

The same guards which protect us from disaster, defect, and enmity, defend us, if we will, from selfishness and fraud. Bolts and bars are not the best of our institutions, nor is shrewdness in trade a mark of wisdom. Men suffer all their life long, under the foolish superstition that they can be cheated. But it is as impossible for a man to be cheated by any one but himself, as for a thing to be and not to be at the same time. There is a third silent party to all our bargains. The nature and soul of things takes on itself the guarantee of the fulfillment of every contract, so that honest service cannot come to loss. If you serve an ungrateful master, serve him the more. Put God in your debt. Every stroke shall be repaid. The longer the payment is withholden, the better for you; for compound interest on compound interest is the rate and usage of this exchequer.

RALPH WALDO EMERSON.

THE GOOD WIFE.

She commandeth her husband in any equal matter, by constant obeying him.

She never crosseth her husband in the spring-tide of his anger, but stays till it be ebbing-water. Surely men, contrary to iron, are worst to be wrought upon when they are hot.

Her clothes are rather comely than costly, and she makes plain cloth to be velvet by her handsome wearing it.

Her husband's secrets she will not divulge: especially she is careful to conceal his infirmities.

In her husband's absence she is wife and deputy husband, which makes her double the files of her diligence. At his return he finds all things so well, that he wonders to see himself at home when he was abroad.

Her children, though many in number, are none in noise, steering them with a look whither she listeth.

The heaviest work of her servants she maketh light, by orderly and seasonably enjoining it.

In her husband's sickness she feels more grief than she shows.

THOMAS FULLER.

SOLITUDE.

THERE can be no very black melancholy to him who lives in the midst of Nature, and has his senses still. There was never yet such a storm but it was Æolian music to a healthy and innocent ear. Nothing can rightly compel a simple and brave man to a vulgar sadness. While I enjoy the friendship of the seasons I trust that nothing can make life a burden to me. The gentle rain which waters my beans and keeps me in the house to-day is not drear and melancholy, but good for me too. Though it prevents my hoeing them, it is of far more worth than my hoeing. If it should continue so long as to cause the seeds to rot in the ground and destroy the potatoes in the lowlands, it would still be good for the grass on the uplands, and being good for the grass, it would be good for me. Sometimes, when I compare myself with other men, it seems as if I were more favored by the gods than they, beyond any deserts that I am conscious of; as if I had a warrant and surety at their hands which my fellows have not, and were especially guided and guarded. I do not flatter myself, but if it be possible they flatter me. I have never felt lonesome, or in the least oppressed by a sense of solitude, but once, and that was a few weeks

HENRY DAVID THOREAU.

after I came to the woods, when, for an hour, I doubted if the near neighborhood of man was not essential to a serene and healthy life. To be alone was something unpleasant. But I was at the same time conscious of a slight insanity in my mood, and seemed to foresee my recovery. In the midst of a gentle rain, while these thoughts prevailed, I was suddenly sensible of such sweet and beneficent society in Nature, in the very pattering of the drops, and in every sound and sight around my house, an infinite and unaccountable friendliness all at once like an atmosphere sustaining me, as made the fancied advantages of human neighborhood insignificant, and I have never thought of them since. Every little pine-needle expanded and swelled with sympathy and befriended me. I was so distinctly made aware of the presence of something kindred to me, even in scenes which we are accustomed to call wild and dreary, and also that the nearest of blood to me and humanest was not a person nor a villager, that I thought no place could ever be strange to me again.

HENRY DAVID THOREAU.

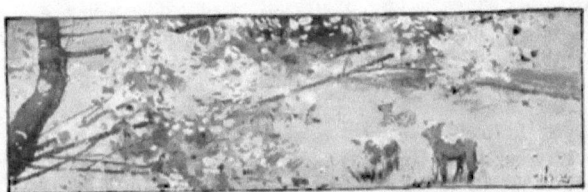

SPRING PROSPECTS.

WE talk about spring as at hand before the end of February, and yet it will be two good months, one sixth part of the whole year, before we can go a-Maying. There may be a whole month of solid and uninterrupted winter yet, plenty of ice, and good sleighing. We may not even see the bare ground, and hardly the water; and yet we sit down and warm our spirits annually with the distant prospect of spring. As if a man were to warm his hands by stretching them towards the rising sun, and rubbing them. We listen to the February cock-crowing and turkey-gobbling as to a first course or prelude. The bluebird, which some wood-chopper or inspired walker is said to have seen in that sunny interval between the snowstorms, is like a speck of clear blue sky seen near the end of a storm, reminding us of an ethereal region, and a heaven which we had forgotten. Princes and magistrates are often styled serene; but what is their turbid serenity to that ethereal serenity which the bluebird embodies. His most serene Birdship! His soft warble melts in the ear as the snow is melting in the valleys around. The bluebird comes, and with his warble drills the ice, and sets free the rivers and ponds and frozen ground. As the sand flows down the slopes a little way, assuming the forms of foliage when the frost comes out of the ground, so this little rill of melody flows a short way down the concave of the sky.

HENRY DAVID THOREAU.

ON MR. FOOT'S RESOLUTION.

SIR, at that day the whole country was divided on this very question. It formed the line of demarcation between the federal and republican parties; and the great political revolution which then took place turned upon the very questions involved in these resolutions. That question was decided by the people, and by that decision the Constitution was, in the emphatic language of Mr. Jefferson, "saved at its last gasp." I should suppose, sir, it would require more self-respect than any gentleman here would be willing to assume, to treat lightly doctrines derived from such high sources. Resting on authority like this, I will ask, gentlemen, whether South Carolina has not manifested a high regard for the Union, when, under a tyranny ten times more grievous than the alien and sedition laws, she has hitherto gone no further than to petition, remonstrate and to solemnly protest against a series of measures which she believes to be wholly unconstitutional and utterly destructive of her interests. Sir, South Carolina has not gone one step further than Mr. Jefferson himself was disposed to go, in relation to the present subject of our present complaints — not a step further than the statesmen from New England were disposed to go under similar circumstances; no further than the Senator from Massachusetts himself once considered as within "the limits of a constitutional opposition." The doctrine that it is the right of a State to judge of the violations of the Constitution on the part of the Federal Government, and to protect her citizens from the operations of unconstitutional laws, was held by the enlightened citizens of Boston, who assembled in Faneuil Hall, on the 25th of January, 1809. They state, in that celebrated memorial, that "they looked only to the State Legislature, which was competent to devise relief againt the unconstitutional acts of the General Government. That your power (say they) is adequate to that object, is evident from the organization of the confederacy." . . .

Thus it will be seen, Mr. President, that the South Carolina doctrine is the Republican doctrine of '98 — that it was promulgated by the fathers of the faith — that it was maintained by Virginia and Kentucky in the worst of times — that it constituted the very pivot on which the political revolution of that day turned — that it embraces the very principles, the triumph of which, at that time, saved the Constitution at its last gasp, and which New England statesmen were not unwilling to adopt when they believed themselves to be the victims of unconstitutional legislation. Sir, as to the doctrine that the Federal Government is the exclusive judge of the extent as well as the limitations of its power, it seems to me to be utterly subversive of the sovereignty and independence of the States. It makes but little difference, in my estimation, whether Congress or the Supreme Court are invested with this power. If the Federal Government, in all, or any, of its departments, is to prescribe the limits of its own authority, and the States are bound to submit to the decision, and are not to be allowed to examine and decide for themselves when the barriers of the Constitution shall be overleaped, his is practically "a government without limitation of powers." The States are at once reduced to mere petty corporations, and the people are entirely at your mercy. I have but one word more to add. In all the efforts that have been made by South Carolina to resist the unconstitutional laws which Congress has extended over them, she has kept steadily in view the preservation of the Union, by the only means by which she believes it can be long preserved — a firm, manly, and steady resistance against usurpation. The measures of the Federal Government have, it is true, prostrated her interests, and will soon involve the whole South in irretrievable ruin. But even this evil, great as it is, is not the chief ground of our complaints. It is the principle involved in the contest — a principle which, substituting the discretion of Congress for the limitations of the Constitution, brings the States and the people to the feet of the Federal Government, and leaves them nothing they can call their own. Sir, if the measures of the Federal Government were less oppressive, we should still strive against this usurpation. The South is acting on a principle she has always held sacred — resistance to unauthorized taxation. These, sir, are the principles which induced the immortal Hampden to resist the payment of a tax of twenty shillings. Would twenty shillings have ruined his fortune? No! but the payment of half of twenty shillings, on the principle on which it was demanded, would have made him a slave. Sir, if acting on these high motives — if animated by that ardent love of liberty which has always been the most prominent trait in the Southern character, we would be hurried beyond the bounds of a cold and calculating prudence — who is there, with one noble and generous sentiment in his bosom, who would not be disposed, in the language of Burke, to exclaim, "You must pardon something to the spirit of liberty"?

ROBERT Y. HAYNE.

REPLY TO HAYNE.

MR. PRESIDENT, I have thus stated the reasons of my dissent to the doctrines which have been advanced and maintained. I am conscious of having detained you and the Senate much too long. I was drawn into the debate with no previous deliberation, such as is suited to the discussion of so grave and important a subject. But it is a subject of which my heart is full, and I have not been willing to suppress the utterance of its spontaneous sentiments. I cannot, even now, persuade myself to relinquish it, without expressing once more my deep conviction, that, since it respects nothing less than the union of States, it is of most vital and essential importance to the public happiness. I profess, sir, in my career hitherto, to have kept steadily in view the prosperity and honor of the whole country, and the preservation of our Federal Union. It is to that Union we owe our safety at home, and our consideration and dignity abroad. It is to that Union that we are chiefly indebted for whatever makes us most proud of our country. That Union we reached

only by the discipline of our virtues in the severe school of adversity. It had its origin in the necessities of disordered finance, prostrate commerce and ruined credit. Under its benign influences, these great interests immediately awoke, as from the dead, and sprang forth with newness of life. Every year of its duration has teemed with fresh proofs of its utility and its blessings; and although our territory has stretched out wider and wider, and our population spread farther and farther, they have not outrun its protection or its benefits. It has been to us all a copious fountain of national, social, and personal happiness.

I have not allowed myself, sir, to look beyond the Union, to see what might lie hidden in the dark recess behind. I have not coolly weighed the chances of preserving liberty when the bonds that unite us together shall be broken asunder. I have not accustomed myself to hang over the precipice of disunion, to see whether, with my short sight, I can fathom the depth of the abyss below; nor could I regard him as a safe counsellor in the affairs of this Government, whose thoughts should be mainly bent on considering, not how the Union may be best preserved, but how tolerable might be the condition of the people when it should be broken up and destroyed. While the Union lasts we have high, exciting, gratifying prospects spread out before us, for us and our children. Beyond that I seek not to penetrate the veil. God grant that in my day at least that curtain may not rise! God grant that on my vision never may be opened what lies behind! When my eyes shall be turned to behold for the last time the sun in heaven, may I not see him shining on the broken and dishonored fragments of a once glorious Union; on States dissevered, discordant, belligerent; on a land rent with civil feuds, or drenched, it may be, in fraternal blood! Let their last feeble and lingering glance rather behold the gorgeous ensign of the Republic, now known and honored throughout the earth, still full high advanced, its arms and trophies streaming in their original luster, not a stripe erased or polluted, not a single star obscured, bearing for its motto, no such miserable interrogatory as "What is all this worth?" nor those other words of delusion and folly, "Liberty first and Union afterward;" but everywhere, spread all over in characters of living light, blazing on all its ample folds, as they float over the sea and over the land, and in every wind under the whole heavens, that other sentiment, dear to every true American heart — Liberty and Union, now and forever, one and inseparable!

DANIEL WEBSTER.

THE CONSTITUTION AND THE UNION.

MR. PRESIDENT, I should much prefer to have heard from every member on this floor declarations of opinion that this Union could never be dissolved, than the declaration of opinion by anybody, that in any case, under the pressure of any circumstances, such a dissolution was possible. I hear with distress and anguish the word "secession," especially when it falls from the lips of those who are patriotic, and known to the country, and known all over the world for their political services. Secession! Peaceable secession! Sir, your eyes and mine are never destined to see that miracle. The dismemberment of this vast country without convulsion! The breaking up of the fountains of the great deep without ruffling the surface! Who is so foolish — I beg everybody's pardon — as to expect to see any such thing? Sir, he who sees these States, now revolving in harmony around a common center, and expects to see them quit their places and fly off without convulsion, may look the next hour to see the heavenly bodies rush from their spheres, and jostle against each other in the realms of space, without causing the wreck of the universe. There can be no such thing as a peaceable secession. Peaceable secession is an utter impossibility. Is the great Constitution under which we live, covering this whole country, is it to be thawed and melted away by secession, as the snows on the mountain melt under the influence of a vernal sun, disappear almost unobserved, and run off? No, sir! No, sir! I will not state what might produce the disruption of the Union; but, sir, I see as plainly as I can see the sun in heaven what that disruption itself must produce; I see that it must produce war, and such a war as I will not describe, in its twofold character.

Peaceable secession! Peaceable secession! The concurrent agreement of all the members of this great Republic to separate! A voluntary separation, with alimony on one side and on the other. Why, what would be the result? Where is the line to be drawn? What States are to secede? What is to remain American? What am I to be? An American no longer? Am I to become a sectional man, a local man, a separatist, with no country in common with the gentlemen who sit around me here, or who fill the other house of Congress? Heaven forbid! Where is the flag of the Republic to remain? Where is the eagle still to tower? or is he to cower, and shrink, and fall to the ground? Why, sir, our ancestors, our fathers and our grandfathers, those of them that are yet living amongst us with prolonged lives, would rebuke and reproach us; and our children and our grandchildren would cry out shame upon us, if we of this generation should dishonor these ensigns of the power of the Government and the harmony of that Union which is every day felt among us with so much joy and gratitude. What is to become of the army? What is to become of the navy? What is to become of the public lands? How is each of the thirty States to defend itself? I know, although the idea has not been stated distinctly, there is to be, or it is supposed possible that there will be, a Southern Confederacy. I do not mean, when I allude to this statement, that any one seriously contemplates such a state of things. I do not mean to say that it is true, but I have heard it suggested elsewhere, that the idea has been entertained, that, after the dissolution of this Union, a Southern Confederacy might be formed. I am sorry, sir, that it has ever been thought of, talked of, in the wildest flights of human imagination. But the idea, so far as it exists, must be of a separation, assigning the slave States to one side, and the free States to the other. Sir, I may express myself too strongly, perhaps, but there are impossibilities in the natural as well as in the physical world, and I hold the idea of the separation of these States, those that are free to form one government, and those that are slave-holding to form another, as such an impossibility. We could not separate the States by any such line, if we were to draw it. We could not sit down here to-day and draw a line of separation that would satisfy any five men in the country. There are natural causes that would keep and tie us together, and there are social and domestic relations which we could not break if we would, and which we should not if we could.

Sir, nobody can look over the face of this country at the present moment, nobody can see where its population is the most dense and growing, without being ready to admit, and compelled to admit, that ere long the strength of America will be in the Valley of the Mississippi. Well, now, sir, I beg to inquire what the wildest enthusiast has to say on the possibility of cutting that river in two, and leaving free States at its source and on its branches, and slave States down near its mouth, each forming a separate government? Pray, sir, let me say to the people of this country, that these things are worthy of their pondering and of their consideration. Here, sir, are five millions of freemen in the free States north of the river Ohio. Can anybody suppose that this population can be severed, by a line that divides them from the territory of a foreign and alien government, down somewhere, the Lord knows where, upon the lower banks of the Mississippi? What would become of Missouri? Will she join the *arrondissement* of the slave States? Shall the man from the Yellowstone and the Platte be connected, in the new republic, with the man who lives on the southern extremity of the Cape of Florida? Sir, I am ashamed to pursue this line of remark. I dislike it; I have an utter disgust for it. I would rather hear of natural blasts and mildews, war, pestilence, and famine, than to hear gentlemen talk of secession. To break up this great Government! to dismember this glorious country! to astonish Europe with an act of folly such as Europe for two centuries has never beheld in any government or any people! No, sir! no, sir! There will be no secession! Gentlemen are not serious when they talk of secession.

Sir, I hear there is to be a convention held at Nashville. I am bound to believe that if worthy gentlemen meet at Nashville in convention, their object will be to adopt conciliatory counsels; to advise the South to forbearance and moderation, and to advise the North to forbearance and moderation; and to inculcate principles of brotherly love and affection, and attachment to the Constitution of the country as it now is. I believe, if the convention meet at all, it will be for this purpose; for certainly, if they meet for any purpose hostile to the Union, they have been singularly inappropriate in their selection of a place. I remember, sir, that, when the treaty of Amiens was concluded between France and England, a sturdy Englishman and a distinguished orator, who regarded the conditions of the peace as ignominious to England, said in the House of Commons, that if King William could know the terms of that treaty, he would turn in his coffin! Let me commend this saying to Mr. Windham, in all its emphasis and in all its force, to any persons who shall meet at Nashville for the purpose of concerting measures for the overthrow of this Union over the bones of Andrew Jackson. . . .

And now, Mr. President, instead of speaking of the possibility or utility of secession, instead of dwelling in those caverns of darkness, instead of groping with those ideas so full of all that is horrid and horrible, let us come out into the light of the day; let us enjoy the fresh air of Liberty and Union; let us cherish those hopes which belong to us; let us devote ourselves to those great objects that are fit for our consideration and our action; let us raise our conceptions to the magnitude and the importance of the duties that devolve upon us; let our comprehension be as broad as the country for which we act, our aspirations as high as its certain destiny; let us not be pigmies in a case that calls for men. Never did there devolve on any generation of men higher trusts than now devolve upon us, for the preservation of this Constitution and the harmony and peace of all who are destined to live under it. Let us make our generation one of the strongest and brightest links in that golden chain which is destined, I fondly believe, to grapple the people of all the States to this Constitution for ages to come. We have a great, popular, Constitutional Government, guarded by law and by judicature, and defended by the affections of the whole people. No monarchical throne presses these States together, no iron chain of military power encircles them; they live and stand under a Government popular in its form, representative in its character, founded upon principles of equality, and so constructed, we hope, as to last forever. In all its history it has been beneficent; it has trodden down no man's liberty; it has crushed no State. Its daily respiration is liberty and patriotism; its yet youthful veins are full of enterprise, courage, and honorable love of glory and renown. Large before, the country has now, by recent events, become vastly larger. This republic now extends, with a vast breadth across the whole continent. The two great seas of the world wash the one and the other shore. We realize, on a mighty scale, the beautiful description of the ornamental border of the buckler of Achilles:

> " Now, the broad shield complete, the artist crowned
> With his last hand, and poured the ocean round;
> In living silver seemed the waves to roll,
> And beat the buckler's verge, and bound the whole."

DANIEL WEBSTER.

ON PRIDE.

I THANK God amongst those millions of vices I do inherit and hold from Adam, I have escaped one, and that a mortal enemy to charity, the first and father sin, not only of man, but of the devil — pride; a vice whose name is comprehended in a monosyllable, but in its nature not circumscribed with a world; I have escaped it in a condition that can hardly avoid it; those petty acquisitions and reputed perfections that advance and elevate the conceits of other men, add no feathers into mine. I have seen a grammarian tour and plume himself over a single line in Horace, and show more pride in the construction of one ode, than the author in the composure of the whole book. For my own part, besides the jargon and patois of several provinces, I understand no less than six languages; yet I protest I have no higher conceit of myself, than had our fathers before the confusion of Babel, when there was but one language in the world, and none to boast himself either linguist or critic. I have not only seen several countries, beheld the nature of their climes, the chorography of their provinces, topography of their cities, but understood their several laws, customs and policies; yet cannot all this persuade the dullness of my spirit unto such an opinion of myself, as I behold in nimbler and conceited heads, that never looked a degree beyond their nests. I know the names, and somewhat more, of all the constellations in my horizon; yet I have seen a prating mariner that could only name the pointers and the North star, out-talk me, and conceit himself a whole sphere above me. I know most of the plants of my country, and of those about me; yet methinks I do not know so many as when I did but know a hundred, and had scarcely ever simpled further than Cheapside; for indeed heads of capacity, and such as are not full with a handful, or easy measure of knowledge, think they know nothing till they know all; which being impossible, they fall upon the opinion of Socrates, and only know they know not any thing.

SIR THOMAS BROWNE.

BEN JONSON.

As for Jonson, to whose character I am now arrived, if we look upon him while he was himself (for his last plays were but his dotages), I think him the most learned and judicious writer which any theater ever had. He was a most severe judge of himself as well as others. One cannot say he wanted wit, but rather that he was frugal of it. In his works you find little to retrench or alter. Wit, and language, and humor, also in some measure, we had before him; but something of art was wanting to the drama, till he came. He managed his strength to more advantage than any who preceded him. You seldom find him making love in any of his scenes, or endeavoring to move the passions; his genius was too sullen and saturnine to do it gracefully, especially when he knew he came after those who had performed both to such a height. Humor was his proper sphere; and in that he delighted most to represent mechanical people. He was deeply conversant in the ancients, both Greek and Latin, and he borrowed boldly from them; there is scarce a poet or historian among the Roman authors of those times, whom he has not translated in Sejanus and Catiline. But he has done his robberies so openly, that one may see he fears not to be taxed by any law. He invades authors like a monarch; and what would be theft in other poets, is only victory in him. With the spoils of these writers he so represents old Rome to us, in his rites, ceremonies, and customs, that if one of their poets had written either of his tragedies, we had seen less of it than in him. If there was any fault in his language, 'twas that he weaved it too closely and laboriously, in his comedies especially: perhaps, too, he did a little too much Romanize our tongue, leaving the words which he translated almost as much Latin as he found them; wherein, though he learnedly followed their language, he did not enough comply with the idiom of ours. If I would compare him with Shakespeare, I must acknowledge him the more correct poet, but Shakespeare the greater wit. Shakespeare was the Homer, or father of our dramatic poets: Jonson was the Virgil, the pattern of elaborate writing: I admire him, but I love Shakespeare.

JOHN DRYDEN.

ON STUDIES.

Studies serve for delight, for ornament, and for ability. Their chief use for delight is in privateness and retiring ; for ornament, is in discourse ; and for ability, is in the judgment and disposition of business ; for expert men can execute, and perhaps judge of particulars, one by one ; but the general counsels, and the plots and marshalling of affairs, come best from those that are learned. To spend too much time in studies, is sloth ; to use them too much for ornament, is affectation ; to make judgment wholly by their rules, is the humor of a scholar ; they perfect nature, and are perfected by experience — for natural abilities are like natural plants, that need pruning by study ; and studies themselves do give forth directions too much at large, except they be bound in by experience. Crafty men contemn studies, simple men admire them, and wise men use them ; for they teach not their own use ; but that is a wisdom without them, and above them, won by observation. Read not to contradict and confute, nor to believe and take for granted, nor to find talk and discourse, but to weigh and consider. Some books are to be tasted, others to be swallowed, and some few to be chewed and digested : that is, some books are to be read only in parts ; others to be read, but not curiously ; and some few to be read wholly, and with diligence and attention. Some books also may be read by deputy, and extracts made of them by others ; but that would be only in the less important arguments and the meaner sort of books ; else distilled books are, like common distilled waters, flashy things. Reading maketh a full man, conference a ready man, and writing an exact man ; and, therefore, if a man write little, he had need have a great memory ; if he confer little, he had need have a present wit ; and if he read little, he had need have much cunning, to seem to know that he doth not.

Francis Bacon.

FRANCIS BACON.

ON BACON.

One, though he be excellent, and the chief, is not to be imitated alone; for no imitator ever grew up to his author; likeness is always on this side truth. Yet there happened in my time one noble speaker, who was full of gravity in his speaking. His language (where he could spare or pass by a jest) was nobly censorious. No man ever spake more neatly, more pressly, more weightily, or suffered less emptiness, less idleness, in what he uttered. No member of his speech but consisted of his own graces. His hearers could not cough, or look aside from him, without loss. He commanded where he spoke; and had his judges angry and pleased at his devotion. No man had their affections more in his power. The fear of every man that heard him was, lest he should make an end.

My conceit of his person was never increased toward him by his place or honors, but I have and do reverence him for the greatness that was only proper to himself, in that he seemed to me ever, by his work, one of the greatest men, and most worthy of admiration, that had been in many ages. In his adversity I ever prayed that God would give him strength; for greatness he could not want. Neither could I condole in a word or syllable for him, as knowing no accident could do harm to virtue, but rather help to make it manifest.

Ben Jonson.

SIR ROGER DE COVERLEY.

Having often received an invitation from my friend Sir Roger de Coverley to pass away a month with him in the country, I last week accompanied him thither, and am settled with him for some time at his country-house, where I intend to form several of my ensuing speculations. Sir Roger, who is very well acquainted with my humor, lets me rise and go to bed when I please, dine at his own table or in my chamber as I think fit, sit still and say nothing without bidding me be merry. When the gentlemen of the country come to see him, he only shows me at a distance. As I have been walking in his fields, I have observed them stealing a sight of me over a hedge, and have heard the knight desiring them not to let me see them, for that I hated to be stared at.

I am the more at ease in Sir Roger's family, because it consists of sober and staid persons; for as the knight is the best master in the world, he seldom changes his servants; and as he is beloved by all about him, his servants never care for leaving him: by this means his domestics are all in years, and grown old with their master. You would take his valet-de-chambre for his brother, his butler is gray-headed, his groom is one of the gravest men that I have ever seen, and his coachman has the looks of a privy-counsellor. You see the goodness of the master even in the old house-dog, and in a gray pad that is kept in the stable with great care and tenderness out of regard to his past services, though he has been useless for several years.

I could not but observe with a great deal of pleasure the joy that appeared in the countenances of these ancient domestics upon my friend's arrival at his country seat. Some of them could not refrain from tears at the sight of their old master; every one of them pressed forward to do something for him, and seemed discouraged if they were not employed. At the same time the good old knight, with a mixture of the father and the master of the family, tempered the inquiries after his own affairs with several kind questions relating to themselves. This humanity and good nature engages everybody to him, so that when he is pleasant upon any of them, all his family are in good humor, and none so much as the person whom he diverts himself with: on the contrary, if he coughs, or betrays any infirmity of old age, it is easy for a stander-by to observe a secret concern in the looks of all his servants.

My worthy friend has put me under the particular care of his butler, who is a very prudent man, and, as well as the rest of his fellow-servants, wonderfully desirous of pleasing me, because they have often heard their master talk of me as his particular friend.

My chief companion, when Sir Roger is diverting himself in the woods or the fields, is a very venerable man who is ever with Sir Roger, and has lived at his house in the nature of a chaplain above thirty years.

JOSEPH ADDISON.

This gentleman is a person of good sense and some learning, of a very regular life and obliging conversation: he heartily loves Sir Roger, and knows that he is very much in the old knight's esteem, so that he lives in the family rather as a relation than a dependent.

I have observed in several of my papers, that my friend Sir Roger, amidst all his good qualities, is something of a humorist; and that his virtues, as well as imperfections, are as it were tinged by a certain extravagance, which makes them particularly his, and distinguishes them from those of other men. This cast of mind, as it is generally very innocent in itself, so it renders his conversation highly agreeable, and more delightful than the same degree of sense and virtue would appear in their common and ordinary colors. As I was walking with him last night, he asked me how I liked the good man whom I have just now mentioned, and without staying for my answer told me that he was afraid of being insulted with Latin and Greek at his own table; for which reason he desired a particular friend of his at the university to find him out a clergyman rather of plain sense than much learning, of a good aspect, a clear voice, a sociable temper, and, if possible, a man that understood a little of backgammon. "My friend," says Sir Roger, "found me out this gentleman, who, besides the endowments required of him, is, they tell me, a good scholar, though he does not show it. I have given him the parsonage of the parish; and, because I know his value, have settled upon him a good annuity for life. If he outlives me, he shall find that he was higher in my esteem than perhaps he thinks he is. He has now been with me thirty years; and though he does not know I have taken notice of it, has never in all that time asked anything of me for himself, though he is every day soliciting me for something in behalf of one or other of my tenants, his parishioners. There has not been a lawsuit in the parish since he has lived among them; if any dispute arises, they apply themselves to him for the decision; if they do not acquiesce in his judgment, which I think never happened above once or twice at most, they appeal to me. At his first settling with me, I made him a present of all the good sermons which have been printed in English, and only begged of him that every Sunday he would pronounce one of them in the pulpit. Accordingly he has digested them into such a series, that they follow one another naturally, and make a continued system of practical divinity."

As Sir Roger was going on in his story, the gentleman we were talking of came up to us; and upon the knight's asking him who preached to-morrow (for it was Saturday night), told us, the Bishop of St. Asaph in the morning and Dr. South in the afternoon. He then showed us his list of preachers for the whole year, where I saw with a great deal of pleasure Archbishop Tillotson, Bishop Saunderson, Dr. Barrow, Dr. Calamy, with several living authors who have published discourses of practical divinity. I no sooner saw this venerable man in the pulpit, but I very much approved of my friend's insisting upon the qualifications of a good aspect and a clear voice; for I was so charmed with the gracefulness of his figure and delivery, as well as with the discourses he pronounced, that I think I never passed any time more to my satisfaction. A sermon repeated after this manner is like the composition of a poet in the mouth of a graceful actor.

I could heartily wish that more of our country clergy would follow this example; and instead of wasting their spirits in laborious compositions of their own, would endeavor after a handsome elocution, and all those other talents that are proper to enforce what has been penned by great masters. This would not only be more easy to themselves, but more edifying to the people.

JOSEPH ADDISON.

THE STRENGTH OF TRUE LOVE.

A YOUNG gentleman and lady of ancient and honorable houses in Cornwall had from their childhood entertained for each other a generous and noble passion, which had been long opposed by their friends, by reason of the inequality of their fortunes; but their constancy to each other, and obedience to those on whom they depended, wrought so much upon their relations, that these celebrated lovers were at length joined in marriage. Soon after their nuptials the bridegroom was obliged to go into a foreign country, to take care of a considerable fortune which was left him by a relation, and came very opportunely to improve their moderate circumstances. They received the congratulations of all the country on this occasion; and I remember it was a common sentence in every one's mouth, "You see how faithful love is rewarded."

He took this agreeable voyage, and sent home every post fresh accounts of his success in his affairs abroad; but at last, though he designed to return with the next ship, he lamented in his letters that "business would detain him some time longer from home," because he would give himself the pleasure of an unexpected arrival.

The young lady, after the heat of the day, walked every evening on the sea-shore, near which she lived, with a familiar friend, her husband's kinswoman; and diverted herself with what objects they met there, or upon discourses of the future methods of life, in the happy change of their circumstances. They stood one evening on the shore together in a perfect tranquillity, observing the setting of the sun, the calm face of the deep, and the silent heaving of the waves, which gently rolled towards them, and broke at their feet; when at a distance her kinswoman saw something float on the waters, which she fancied was a chest; and with a smile told her, "She saw it first, and if it came ashore full of jewels, she had a right to it." They both fixed their eyes upon it, and entertained themselves with the subject of the wreck, the cousin still asserting her right; but promising, "if it was a prize, to give her a very rich coral for her youngest child." Their mirth soon abated, when they observed, upon the nearer approach, that it was a human body. The young lady, who had a heart naturally filled with pity and compassion, made many melancholy reflections on the occasion. "Who knows," said she, "but this man may be the only hope and heir of a wealthy house; the darling of indulgent parents, who are now in impertinent mirth, and pleasing themselves with the thoughts of offering him a bride they had got ready for him? Or, may he not be the master of a family that wholly depended upon his life? There may, for aught we know, be half a dozen fatherless children and a tender wife, now exposed to poverty by his death.

What pleasure might he have promised himself in the different welcome he was to have from her and them! But let us go away; it is a dreadful sight! The best office we can do, is to take care that the poor man, whoever he is, may be decently buried." She turned away, when a wave threw the carcass on the shore. The kinswoman immediately shrieked out, "Oh, my cousin!" and fell upon the ground. The unhappy wife went to help her friend, when she saw her own husband at her feet, and dropped in a swoon upon the body. An old woman, who had been the gentleman's nurse, came out about this time to call the ladies to supper, and found her child, as she always called him, dead on the shore, her mistress and kinswoman both lying dead by him. Her loud lamentations, and calling her young master to life, soon awaked the friend from her trance; but the wife was gone for ever.

SIR RICHARD STEELE.

TALK.

I REALLY believe some people save their bright thoughts as being too precious for conversation. What do you think an admiring friend said the other day to one that was talking good things — good enough to print? "Why," said he, "you are wasting merchantable literature, a cash article, at the rate, as nearly as I can tell, of fifty dollars an hour." The talker took him to the window, and asked him to look out and tell what he saw.

"Nothing but a very dusty street," he said, "and a man driving a sprinkling-machine through it."

"Why don't you tell the man he is wasting that water? What would be the state of the highways of life, if we did not drive our *thought-sprinklers* through them with the valves open, sometimes?"

Besides, there is another thing about this talking which you forget. It shapes our thoughts for us; — the waves of conversation roll them as the surf rolls the pebbles on the shore. Let me modify the image a little. I rough out my thoughts in talk as an artist models in clay. Spoken language is so plastic, — you can pat and coax, and spread and shave, and rub out, and fill up, and stick on so easily, when you work that soft material, that there is nothing like it for modelling. Out of it come the shapes which you turn into marble or bronze in your immortal books, if you happen to write such. Or, to use another illustration, writing or printing is like shooting with a rifle; you may hit your reader's mind, or miss it; — but talking is like playing at a mark with the pipe of an engine; if it is within reach, and you have time enough, you can't help hitting it.

OLIVER WENDELL HOLMES.

WOUTER VAN TWILLER.

The renowned Wouter (or Walter) Van Twiller was descended from a long line of Dutch burgomasters, who had successively dozed away their lives, and grown fat upon the bench of magistracy in Rotterdam ; and who had comported themselves with such singular wisdom and propriety, that they were never either heard or talked of — which, next to being universally applauded, should be the object of ambition of all magistrates and rulers. There are two opposite ways by which some men make a figure in the world ; one, by talking faster than they think, and the other, by holding their tongues and not thinking at all. By the first, many a smatterer acquires the reputation of a man of quick parts ; by the other, many a dunderpate, like the owl, the stupidest of birds, comes to be considered the very type of wisdom. This, by the way, is a casual remark, which I would not, for the universe, have it thought I applied to Governor Van Twiller. It is true he was a man shut up within himself, like an oyster, and rarely spoke, except in monosyllables ; but then it was allowed he seldom said a foolish thing. So invincible was his gravity that he was never known to laugh or even smile through the whole course of a long and prosperous life. Nay, if a joke were uttered in his presence, that set light-minded hearers in a roar, it was observed to throw him into a state of perplexity. Sometimes he would deign to inquire into the matter, and when, after much explanation, the joke was made as plain as a pike-staff, he would continue to smoke his pipe in silence, and at length, knocking out the ashes, would exclaim, "Well, I see nothing in all that to laugh about."

With all his reflective habits, he never made up his mind on a subject. His adherents accounted for this by the astonishing magnitude of his ideas. He conceived every subject on so grand a scale that he had not room in his head to turn it over and examine both sides of it. Certain it is, that, if any matter were propounded to him on which ordinary mortals would rashly determine at first glance, he would put on a vague, mysterious look, shake his capacious head, smoke some time in profound silence, and at length observe, that "he had his doubts about the matter ;" which gained him the reputation of a man slow of belief and not easily imposed upon. What is more, it gained him a lasting name ; for to this habit of the mind has been attributed his surname of Twiller ; which is said to be a corruption of the original Twijfler, or, in plain English, Doubter. . . .

The very outset of the career of this excellent magistrate was distinguished by an example of legal acumen, that gave flattering presage of a wise and equitable administration. The morning after he had been installed in office, and at the moment that he was making his breakfast from a prodigious earthen dish, filled with milk and Indian pudding, he was interrupted by the appearance of Wandle Schoonhoven, a very important old burgher of New Amsterdam, who complained bitterly of one Barent Bleecker, inasmuch as he refused to come to a settlement of accounts, seeing that there was a heavy balance in favor of the said Wandle.

Governor Van Twiller, as I have already observed, was a man of few words; he was likewise a mortal enemy to multiplying writings — or being disturbed at his breakfast. Having listened attentively to the statements of Wandle Schoonhoven, giving an occasional grunt, as he shoveled a spoonful of Indian pudding into his mouth — either as a sign that he relished the dish, or comprehended the story — he called unto him his constable, and pulling out of his breeches-pocket a huge jack-knife, dispatched it after the defendant as a summons, accompanied by his tobacco-box as a warrant.

This summary process was as effectual in those simple days as was the sealring of the great Haroun Alraschid among the true believers. The two parties being confronted before him, each produced a book of accounts, written in a language and character that would have puzzled any but a High-Dutch commentator, or a learned decipherer of Egyptian obelisks. The sage Wouter took them one after the other, and having poised them in his hands, and attentively counted over the number of leaves, fell straightway into a very great doubt, and smoked for half an hour without saying a word; at length, laying his finger beside his nose, and shutting his eyes for a moment, with the air of a man who has just caught a subtle idea by the tail, he slowly took his pipe from his mouth, puffed forth a column of tobacco-smoke, and with marvellous gravity and solemnity pronounced, that, having carefully counted over the leaves and weighed the books, it was found that one was just as thick and as heavy as the other: therefore, it was the final opinion of the court that the accounts were equally balanced: therefore, Wandle should give Barent a receipt, and Barent should give Wandle a receipt, and the constable should pay the costs.

This decision, being straightway made known, diffused general joy throughout New Amsterdam, for the people immediately perceived that they had a very wise and equitable magistrate to rule over them. But its happiest effect was, that not another lawsuit took place throughout the whole of his administration; and the office of constable fell into such decay, that there was not one of those losel scouts known in the province for many years. I am the more particular in dwelling on this transaction, not only because I deem it one of the most sage and righteous judgments on record, and well worthy the attention of modern magistrates, but because it was a miraculous event in the history of the renowned Wouter — being the only time he was ever known to come to a decision in the whole course of his life.

WASHINGTON IRVING.

LETTER TO MRS. THRALE.

SINCE you have written to me with the attention and tenderness of olden time, your letters give me a great part of the pleasure which a life of solitude admits. You will never bestow any share of your good-will on one who deserves better. Those that have loved longest love best. A sudden blaze of kindness may by a single blast of coldness be extinguished; but that fondness that length of time has connected with many circumstances and occasions, though it may for a while be depressed by disgust or resentment, with or without a cause, is hourly revived by accidental recollection. To those that have lived long together, every thing heard and every thing seen recalls some pleasure communicated or some benefit conferred, some petty quarrel or some slight endearment Esteem of great powers, or amiable qualities newly discovered, may embroider a day or a week, but a friendship of twenty years is interwoven with the texture of life. A friend may be often found and lost; but an *old friend* never can be found, and nature has provided that he cannot easily be lost. . . .

SAMUEL JOHNSON.

THE DOMINIE AND MEG MERRILIES.

THE result of these cogitations was a resolution to go and visit the scene of the tragedy at Warroch Point, where he had not been for many years — not, indeed, since the fatal accident had happened. The walk was a long one, for the Point of Warroch lay on the farther side of the Ellangowan property, which was interposed between it and Woodbourne. Besides, the Dominie went astray more than once, and met with brooks swollen into torrents by the melting of the snow, where he, honest man, had only the summer recollection of little trickling rills.

At length, however, he reached the woods which he had made the object of his excursion, and traversed them with care, muddling his disturbed brains with vague efforts to recall every circumstance of the catastrophe. It will readily be supposed that the influence of local situation and association was inadequate to produce conclusions different from those which he had formed under the immediate pressure of the occurrences themselves. "With many a weary sigh, therefore, and many a groan," the poor Dominie returned from his hopeless pilgrimage, and weariedly plodded his way towards Woodbourne, debating at times in his altered mind a question which was forced upon him by the cravings of an appetite rather of the keenest, namely, whether he had breakfasted that morning or no. It was in this twilight humor, now thinking of the loss of the child, then involuntarily compelled to meditate upon the somewhat incongruous subject of hung-beef, rolls, and butter, that his route, which was different from that which he had taken in the morning, conducted him past the small ruined tower, or rather vestige of a tower, called by the country people the Kaim of Derncleugh.

The reader may recollect the description of this ruin in the twenty-seventh chapter of this narrative, as the vault in which young Bertram, under the auspices of Meg Merrilies, witnessed the death of Hatteraick's lieutenant. The tradition of the country added ghostly terrors to the natural awe inspired by the situation of this place — which terrors the gypsies, who so long inhabited the vicinity, had probably invented, or at least propagated, for their own advantage. It was said, that during the times of the Galwegian independence, one Hanlon Mac-Dingawaie, brother to the reigning chief, Knarth Mac-Dingawaie, murdered his brother and sovereign, in order to usurp the principality from his infant nephew, and that being pursued for vengeance by the faithful allies and retainers of the house, who espoused the cause of the lawful heir, he was compelled to retreat with a few followers whom he had involved in his crime, to this impregnable tower called the Kaim of Derncleugh, where he defended himself until nearly reduced by famine, when, setting fire to the place, he and the small remaining garrison desperately perished by their own swords, rather than fall into the hands of their exasperated enemies. This tragedy, which, considering the wild times wherein it was placed, might have some foundation in truth, was larded with many legends of superstition and *diablerie*, so that most of the peasants of the neighborhood, if benighted, would rather have chosen to make a considerable circuit than pass these haunted walls. The lights, often seen around the tower when used as the rendezvous of the lawless characters by whom it was occasionally frequented, were accounted for, under authority of these tales of witchery, in a manner at once convenient for the private parties concerned, and satisfactory to the public.

SIR WALTER SCOTT.

Now it must be confessed that our friend Sampson, although a profound scholar and mathematician, had not travelled so far in philosophy as to doubt the reality of witchcraft or apparitions. Born indeed at a time when a doubt in the existence of witches was interpreted as equivalent to a justification of their infernal practices, a belief of such legends had been impressed upon the Dominie as an article indivisible from his religious faith ; and perhaps it would have been equally difficult to have induced him to doubt the one as the other. With these feelings, and in a thick misty day, which was already drawing to its close, Dominie Sampson did not pass the Kaim of Derncleugh without some feelings of tacit horror.

What, then, was his astonishment, when, on passing the door — that door which was supposed to have been placed there by one of the latter Lairds of Ellangowan to prevent presumptuous strangers from incurring the dangers of the haunted vault — that door supposed to be always locked, and the key of which was popularly said to be deposited with the presbytery — that door, that very door, opened suddenly, and the figure of Meg Merrilies, well known, though not seen for many a revolving year, was placed at once before the eyes of the startled Dominie ! She stood immediately before him in the footpath, confronting him so absolutely, that he could not avoid her except by fairly turning back, which his manhood prevented him from thinking of.

" I kenn'd ye wad be here," she said, with her harsh and hollow voice : " I ken wha ye seek ; but ye maun do my bidding."

" Get thee behind me ! " said the alarmed Dominie — " Avoid ye ! — *Conjuro te, scelestissima — nequissima — spurcissima — iniquissima — atque miserrima — conjuro te ! ! !*"

Meg stood her ground against this tremendous volley of superlatives, which Sampson awked up from the pit of his stomach, and hurled at her in thunder. " Is the carl daft," she said, " wi his glamor ? "

" *Conjuro,*" continued the Dominie, " *abjuro, contestor, atque viriliter impero tibi.*"

" What in the name of Sathan, are ye feared for, wi' your French gibberish, that would make a dog sick ? Listen, ye stickit stibbler, to what I tell ye, or ye sall rue it while there's a limb o' ye hings to anither ! Tell Colonel Mannering that I ken he's seeking me. He kens, and I ken, that the blood will be wiped out, and the lost will be found,

And Bertram's right and Bertram's might
Shall meet on Ellangowan height.

Hae, there's a letter to him ; I was gaun to send it in another way. I canna write mysell ; but I hae them that will baith write and read, and ride and rin for me. Tell him the time's coming now, and the weird's dreed, and the wheel's turning. Bid him look at the stars as he has looked at them before. Will ye mind a' this ? "

" Assuredly," said the Dominie, " I am dubious — for, woman, I am perturbed at thy words, and my flesh quakes to hear thee."

" They'll do you nae ill though, and may be muckle gude."

" Avoid ye ! I desire no good that comes by unlawful means."

THE DOMINIE AND MEG MERRILIES.

"Fule-body that thou art!" said Meg, stepping up to him with a frown of indignation that made her dark eyes flash like lamps from under her bent brows — "Fule-body! if I meant ye wrang, couldna I clod ye ower that craig, and wad man ken how ye cam by your end mair than Frank Kennedy? Hear ye that, ye worricow?"

"In the name of all that is good," said the Dominie, recoiling, and pointing his long pewter-headed walking-cane like a javelin at the supposed sorceress, — "in the name of all that is good, bide off hands! I will not be handled — woman, stand off, upon thine own proper peril! — desist, I say — I am strong — lo, I will resist!" Here his speech was cut short; for Meg, armed with supernatural strength (as the Dominie asserted), broke in upon his guard, put by a thrust which he made at her with his cane, and lifted him into the vault, "as easily," said he, "as I could sway a Kitchen's Atlas."

"Sit down there," she said, pushing the half-throttled preacher with some violence against a broken chair — "sit down there, and gather your wind and your senses, ye black barrow-tram o' the kirk that ye are! Are ye fou or fasting?"

"Fasting — from all but sin," answered the Dominie, who, recovering his voice, and finding his exorcisms only served to exasperate the intractable sorceress, thought it best to affect complaisance and submission, inwardly conning over, however, the wholesome conjurations which he durst no longer utter aloud. But as the Dominie's brain was by no means equal to carry on two trains of ideas at the same time, a word or two of his mental exercise sometimes escaped, and mingled with his uttered speech in a manner ludicrous enough, especially as the poor man shrunk himself together after every escape of the kind, from terror of the effect it might produce upon the irritable feelings of the witch.

Meg, in the meanwhile, went to a great black cauldron that was boiling on a fire on the floor, and lifting the lid, an odor was diffused through the vault which, if the vapors of a witch's cauldron could in aught be trusted, promised better things than the hell-broth which such vessels are usually supposed to contain. It was in fact the savor of a goodly stew, composed of fowls, hares, partridges, and moorgame, boiled in a large mess with potatoes, onions, and leeks, and from the size of the cauldron, appeared to be prepared for half a dozen people at least.

"So ye hae eat naething a' day?" said Meg, heaving a large portion of this mess into a brown dish, and strewing it savorily with salt and pepper.*

"Nothing," answered the Dominie — "*scelestissima!* — that is — gudewife."

"Hae, then," said she, placing the dish before him, "there's what will warm your heart."

"I do not hunger — *malefica* — that is to say, Mrs. Merrilies!" for he said unto himself, "the savor is sweet, but it hath been cooked by a Canidia or an Ericthoe."

"If ye dinna eat instantly, and put some saul in ye, by the bread and the salt, I'll put it down your throat wi' the cutty spoon, scalding as it is, and whether ye will or no. Gape, sinner, and swallow!"

Sampson, afraid of eye of newt, and toe of frog, tigers' chaudrons, and so forth,

* Gypsy Cookery.

had determined not to venture; but the smell of the stew was fast melting his obstinacy, which flowed from his chops, as it were, in streams of water, and the witch's threats decided him to feed. Hunger and fear are excellent casuists.

"Saul," said Hunger, "feasted with the witch of Endor." "And," quoth Fear, "the salt which she sprinkled upon the food showeth plainly it is not a necromantic banquet, in which that seasoning never occurs." "And besides," says Hunger, after the first spoonful, "it is savory and refreshing viands."

"So ye like the meat?" said the hostess.

"Yea," answered the Dominie, "and I give thee thanks — *sceleratissima !* — which means Mrs. Margaret."

"Aweel, eat your fill; but an ye kenn'd how it was gotten, ye maybe wadna like it sae weel." Sampson's spoon dropped in the act of conveying its load to his mouth. "There's been mony a moonlight watch to bring a' that trade thegither," continued Meg, "the folk that are to eat that dinner thought little o' your game-laws."

"Is that all?" thought Sampson, resuming his spoon, and shoveling away manfully; "I will not lack my food upon that argument."

"Now, ye maun tak a dram."

"I will," quoth Sampson — "*conjuro te* — that is, I thank you heartily," for he thought to himself, in for a penny in for a pound; and he fairly drank the witch's health in a cupful of brandy. When he had put this cope-stone upon Meg's good cheer, he felt, as he said, mightily elevated, and afraid of no evil which could befall unto him.

"Will ye remember my errand now?" said Meg Merrilies. "I ken by the cast o' your ee that ye're anither man than when you cam in."

"I will, Mrs. Margaret," repeated Sampson stoutly; "I will deliver unto him the sealed yepistle, and will add what you please to send by word of mouth."

"Then I'll make it short," says Meg. "Tell him to look at the stars without fail this night, and to do what I desire him in that letter, as he would wish.

> That Bertram's right and Bertram's might
> Should meet on Ellangowan height.

I have seen him twice when he saw na me; I ken when he was in this country first, and I ken what's brought him back again. Up, an' to the gate! ye're ower lang here — follow me."

Sampson followed the sibyl accordingly, who guided him about a quarter of a mile through the woods, by a shorter cut than he could have found for himself; they then entered upon the common, Meg still marching before him at a great pace, until she gained the top of a small hillock which overhung the road.

"Here," she said, "stand still here. Look how the setting sun breaks through yon cloud that's been darkening the lift a' day. See where the first stream o' light fa's — it's upon Donagild's round tower — the auldest tower in the Castle o' Ellangowan — that's no for naething! — See as it's glooming to seaward abune yon sloop in the bay — that's no for naething neither. Here I stood on this very spot," said she, drawing herself up so as not to lose one hair-breadth of her uncommon height, and stretching out her long sinewy arm and clenched hand — "here I stood when I tauld the last Laird o' Ellangowan what was coming on his house; and did that fa' to the ground? Na, it hit even ower sair! And here, where I brake the wand of peace ower him — here I stand again — to bid God bless and prosper the just heir of Ellangowan that will sune be brought to his ain; and the best laird he shall be that Ellangowan has seen for three hundred years. I'll no live to see it, maybe; but there will be mony a blythe ee see it though mine be closed. And now, Abel Sampson, as ever ye lo'ed the house of Ellangowan, away wi' my message to the English Colonel, as if life and death were upon your haste!"

So, saying, she turned suddenly from the amazed Dominie, and regained with swift and long strides the shelter of the wood from which she had issued, at the point where it most encroached upon the common. Sampson gazed after her for a moment in utter astonishment, and then obeyed her directions, hurrying to Woodbourne at a pace very unusual for him, exclaiming three times, *Prodigious! prodigious! pro-di-gi-ous!*"

Sir Walter Scott.

THE STORY OF "WAVERLEY."

. . . Now, to go from one important subject to another, I must account for my own laziness, which I do by referring you to a small anonymous sort of a novel, in three volumes — Waverley — which you will receive by the mail of this day. It was a very old attempt of mine to embody some traits of those characters and manners peculiar to Scotland, the last remnants of which vanished during my own youth, so that few or no traces now remain. I had written great part of the first volume, and sketched other passages, when I mislaid the MS., and only found it by the merest accident as I was rummaging the drawers of an old cabinet; and I took the fancy of finishing it, which I did so fast that the last two volumes were written in three weeks. I had a great deal of fun in the accomplishment of this task, though I do not expect that it will be popular in the south, as much of the humor, if there be any, is local, and some of it even professional. You, however, who are an adopted Scotchman, will find some amusement in it. It has made a very strong impression here, and the good people of Edinburgh are busied in tracing the author, and in finding out originals for the portraits it contains. In the first case, they will probably find it difficult to convict the guilty author, although he is far from escaping suspicion. Jeffrey has offered to make oath that it is mine, and another great critic has tendered his affidavit, *ex contrario;* so that these authorities have divided the gude town. However, the thing has succeeded very well, and is thought highly of. I don't know if it has got to London yet. I intend to maintain my *incognito*. Let me know your opinion about it. . . .

Sir Walter Scott.

THE TEMPERANCE PREACHER.

The Gospel temperance meetings were still in the full tide of power, when Mr. Remington left them one evening in charge of others, and went to answer a call to another part of the city, where there was a political rally.

"Vote as you pray" might almost have been said to be the text of the sermon he preached. A strong, keen, logical sermon addressed to keen-brained men, voters, every one of them; men who listened intently, and weighed carefully the problems which he presented before them, and the facts and figures with which he clinched his arguments.

"You did some good work for the cause to-night, Mr. Remington," said the man who had called for him, when his twenty minutes' speech was concluded. "That speech will give us a dozen more votes at least, and we are getting where a dozen more votes will tell. We are gaining on the enemy, Mr. Remington, as sure as the world. If we could only have all the church members with us, how quickly we would sweep this curse out of the land!"

"Yes," said Mr. Remington, and he could not forbear a sigh. Even in his own beloved church there were men who loved him, and prayed with and for him, and who worked earnestly in the Gospel temperance meetings, and were honest to the heart's core, he knew, yet who would in a few days go to the polls and array themselves against him, and on the side of the saloons which he and they were fighting.

"It is very strange," he said. "It is the problem of the centuries how to understand the honest Christian people of our country on this question. There is nothing like it; we see eye to eye on every other moral question under the sun. I do not understand it; the utmost that I can do is to work and pray and wait."

"I know precisely how you feel," answered his friend. "I have a brother, as good a man as ever lived, and as square as he can be on all other questions, as you say; but when it comes to this, there isn't a bat in the world as blind as he. It is unaccountable to me, except on the principle that the 'god of this world has blinded his eyes.' When I get to thinking about it, and get all wrought up, as you have wrought me up to-night, Mr. Remington, by your speech, the only language in which I can express myself is the old cry, 'O, Lord! how long!'"

"Pansy" (Mrs G. R. Alden).

GARDEN ETHICS.

I BELIEVE that I have found, if not original sin, at least vegetable total depravity in my garden; and it was there before I went into it. It is the bunch- or joint- or snake-grass — whatever it is called. As I do not know the names of all the weeds and plants, I have to do as Adam did in his garden — name things as I find them.

This grass has a slender, beautiful stalk: and when you cut it down, or pull up a long root of it, you fancy it is got rid of; but in a day or two it will come up in the same spot in half a dozen vigorous blades. Cutting down and pulling up is what it thrives on. Extermination rather helps it. If you follow a slender white root, it will be found to run under the ground until it meets another slender white root; and you will soon unearth a network of them, with a knot somewhere, sending out dozens of sharp-pointed, healthy shoots, every joint prepared to be an independent life and plant. The only way to deal with it is to take one part hoe and two parts fingers, and carefully dig it out, not leaving a joint anywhere. It will take a little time, say all summer, to dig out thoroughly a small patch; but if you once dig it out, and keep it out, you will have no further trouble.

I have said it was total depravity. Here it is. If you attempt to pull up and root out sin in you, which shows on the surface — if it does not show, you do not care for it — you may have noticed how it runs into an interior network of sins, and an ever-sprouting branch of these roots somewhere; and that you cannot pull out one without making a general internal disturbance, and rooting up your whole being. I suppose it is less trouble to quietly cut them off at the top — say once a week, on Sunday, when you put on your religious clothes and face — so that no one will see them, and not try to eradicate the network within.

Remark. — This moral vegetable figure is at the service of any clergyman who will have the manliness to come forward and help me at a day's hoeing on my potatoes. None but the orthodox need apply.

I, however, believe in the intellectual, if not the moral, qualities of vegetables, and especially weeds. There was a worthless vine that (or who) started up about midway between a grape-trellis and a row of bean-poles, some three feet from each, but a little nearer the trellis. When it came out of the ground, it looked around to see what it should do. The trellis was already occupied. The bean-pole was empty. There was evidently a little the best chance of light, air, and sole proprietorship on the pole. And the vine started for the pole, and began to climb it with determination. Here was as distinct an act of choice, of reason, as a boy exercises when he goes into a forest, and, looking about, decides which tree he will climb. And, besides, how did the vine know enough to travel in exactly the right direction, three feet, to find what it wanted? This is intellect. The weeds, on the other hand, have hateful moral qualities. To cut down a weed is, therefore, to do a moral action. I feel as if I were destroying a sin. My hoe becomes an instrument of retributive justice. I am an apostle of nature. This view of the matter lends a dignity to the art of hoeing which nothing else does, and lifts it into the region of ethics. Hoeing becomes, not a pastime, but a duty. And you get to regard it so, as the days and the weeds lengthen.

Observation. — Nevertheless, what a man needs in gardening is a cast-iron back, with a hinge in it. The hoe is an ingenious instrument, calculated to call out a great deal of strength at a great disadvantage.

The striped bug has come, the saddest of the year. He is a moral double-ender, iron-clad at that. He is unpleasant in two ways. He burrows in the ground so that you cannot find him, and he flies away so that you cannot catch him. He is rather handsome, as bugs go, but utterly dastardly, in that he gnaws the stem of the plant close to the ground, and ruins it without any apparent advantage to himself. I find him on the hills of cucumbers (perhaps it will be a cholera-year, and we shall not want any), the squashes (small loss), and the melons (which never ripen). The best way to deal with the striped bug is to sit down by the hills and patiently watch for him. If you are spry, you can annoy him. This, however, takes time. It takes all day and part of the night. For he flieth in the darkness, and wasteth at noon-day. If you get up before the dew is off the plants, — it goes off very early, — you can sprinkle soot on the plant (soot is my panacea : if I can get the disease of a plant reduced to the necessity of soot, I am all right) ; and soot is unpleasant to the bug. But the best thing to do is set a toad to catch the bugs. The toad at once establishes the most intimate relations with the bug. It is a pleasure to see such unity among the lower animals. The difficulty is to make the toad stay and watch the hill. If you know your toad, it is all right. If you do not, you must build a tight fence around the plants, which the toad cannot jump over. This, however, introduces a new element. I find that I have a zoölogical garden on my hands. It is an unexpected result of my little enterprise, which never aspired to the completeness of the Paris "*Jardin des Plantes.*"

CHARLES DUDLEY WARNER.

LITTLE PEARL IN THE FOREST.

PEARL had not found the hour pass wearisomely, while her mother sat talking with the clergyman. The great black forest — stern as it showed itself to those who brought the guilt and troubles of the world into its bosom — became the playmate of the lonely infant, as well as it knew how. Somber as it was, it put on the kindliest of its moods to welcome her. It offered her the partridge-berries, the growth of the preceding autumn, but ripening only in the spring, and now red as drops of blood upon the withered leaves. These Pearl gathered, and was pleased with their wild flavor. The small denizens of the wilderness hardly took pains to move out of her path. A partridge, indeed, with a brood of ten behind her, ran forward threateningly, but soon repented of her fierceness, and clucked to her young ones not to be afraid. A pigeon, alone on an old branch, allowed Pearl to come beneath, and uttered a sound as much of greeting as alarm. A squirrel, from the lofty depths of his domestic tree, chattered either in anger or merriment, — for a squirrel is such a choleric and humorous little personage, that it is hard to distinguish between his moods, — so he chattered at the child, and flung down a nut upon her head. It was a last year's nut, and already gnawed by his sharp tooth. A fox, startled from his sleep by her light footstep on the leaves, looked inquisitively at Pearl, as doubting whether it were better to steal off, or renew his nap on the same spot. A wolf, it is said, — but here the tale has surely lapsed into the improbable, — came up, and smelt of Pearl's robe, and offered his savage head to be patted by her hand. The truth seemed to be, however, that the mother-forest, and these wild things which it nourished, all recognized a kindred wildness in a human child. And she was gentler here than in the quarry-margined streets of the settlement, or in her mother's cottage. The flowers appeared to know it; and one and another whispered as she passed, "Adorn thyself with me! thou beautiful child, adorn thyself with me!" — and, to please them, Pearl gathered the violets, and anemones, and columbines, and some twigs of the freshest green, which the old trees held down before her eyes. With these she decorated her hair, and her young waist, and became a nymph-child, or an infant dryad, or whatever else was in closest sympathy with the antique wood. In such guise had Pearl adorned herself, when she heard her mother's voice, and came slowly back.

NATHANIEL HAWTHORNE

HAWTHORNE'S STUDY AT "WAYSIDE."

HENRY W. LONGFELLOW.

FOOTPRINTS OF ANGELS.

AND now the sun was growing high and warm. A little chapel, whose door stood open, seemed to invite Flemming to enter and enjoy the grateful coolness. He went in. There was no one there. The walls were covered with paintings and sculpture of the rudest kind, and with a few funeral tablets. There was nothing there to move the heart to devotion; but in that hour the heart of Flemming was weak — weak as a child's. He bowed his stubborn knees, and wept. And, O, how many disappointed hopes, how many bitter recollections, how much of wounded pride and unrequited love, were in those tears through which he read, on a marble tablet in the chapel wall opposite, this singular inscription : —

"Look not mournfully into the Past. It comes not back. Wisely improve the

Present. It is thine. Go forth to meet the shadowy Future, without fear, and with a manly heart."

It seemed to him as if the unknown tenant of that grave had opened his lips of dust, and spoken to him the words of consolation which his soul needed, and which no friend had yet spoken. In a moment the anguish of his thoughts was still. The stone was rolled away from the door of his heart; death was no longer there, but an angel clothed in white. He stood up, and his eyes were no more bleared with tears; and, looking into the bright morning heaven, he said: —

"I will be strong!"

Men sometimes go down into tombs, with painful longings to behold once more the faces of their departed friends; and as they gaze upon them, lying there so peacefully with the semblance that they wore on earth, the sweet breath of heaven touches them, and the features crumble and fall together, and are but dust. So did his soul then descend for the last time unto the great tomb of the Past, with painful longings to behold once more the dear faces of those he had loved; and the sweet breath of heaven touched them, and they would not stay, but crumbled away and perished as he gazed. They, too, were dust. And thus, far-sounding, he heard the gate of the Past shut behind him, as the divine poet did the gate of Paradise, when the angel pointed him the way up the Holy Mountain; and to him likewise was it forbidden to look back.

In the life of every man there are sudden transitions of feeling, which seem almost miraculous. At once, as if some magician had touched the heavens and the earth, the dark clouds melt into the air, the wind falls, and serenity succeeds the storm. The causes which produce these sudden changes may have been long at work within us; but the changes themselves are instantaneous, and apparently without sufficient cause. It was so with Flemming; and from that hour forth he resolved that he would no longer veer with every shifting wind of circumstance, — no longer be a child's plaything in the hands of Fate, which we ourselves do make or mar. He resolved henceforward not to lean on others; but to walk self-confident and self-possessed, — no longer to waste his years in vain regrets, nor wait the fulfillment of boundless hopes and indiscreet desires; but to live in the Present wisely, alike forgetful of the Past, and careless of what the mysterious Future might bring. And from that moment he was calm and strong; he was reconciled with himself. His thoughts turned to his distant home beyond the sea. An indescribable sweet feeling rose within him.

"Thither will I turn my wandering footsteps," said he, "and be a man among men, and no longer a dreamer among shadows. Henceforth be mine a life of action and reality! I will work in my own sphere, nor wish it other than it is. This alone is health and happiness."

HENRY WADSWORTH LONGFELLOW.

THE GOOD MAN.

A GOOD man lives to his own heart. He thinks it not good manners to slight the world's opinion; though he will regard it only in the second place.

A good man will look upon every accession of power to do good as a new trial to the integrity of his heart.

A good man, though he will value his own countrymen, yet will think as highly of the worthy men of every nation under the sun.

A good man is a prince of the Almighty's creation.

A good man will not engage even in a national cause, without examining the justice of it.

How much more glorious a character is that of the friend of mankind, than that of the conqueror of nations?

The heart of a worthy man is ever on his lips; he will be pained when he cannot speak all that is in it.

An impartial spirit will admire goodness or greatness wherever he meets it, and whether it makes for or against him.

SAMUEL RICHARDSON.

THE GOOD WOMAN.

A GOOD woman is one of the greatest glories of the creation.

How do the duties of a good wife, a good mother, and a worthy matron, well performed, dignify a woman!

A good woman reflects honor on all those who had any hand in her education, and on the company she has kept.

A woman of virtue and of good understanding, skilled in, and delighting to perform the duties of domestic life, needs not fortune to recommend her to the choice of the greatest and richest man, who wishes his own happiness.

SAMUEL RICHARDSON.

JOHN AND LORNA.

AFTER long or short — I know not, yet ere I was weary, ere I yet began to think or wish for any answer — Lorna slowly raised her eyelids, with a gleam of dew below them, and looked at me doubtfully. Any look with so much in it never met my gaze before.

"Darling, do you love me?" was all that I could say to her.

"Yes, I like you very much," she answered, with her eyes gone from me, and her dark hair falling over, so as not to show me things.

"But do you love me, Lorna, Lorna — do you love me more than all the world?"

"No, to be sure not. Why should I?"

"In truth, I know not why you should. Only I hoped that you did, Lorna. Either love me not at all, or as I love you, forever."

"John, I love you very much; and I would not grieve you. You are the bravest, and the kindest, and the simplest of all men — I mean of all people. I like you very much, Master Ridd, and I think of you almost every day."

"That will not do for me, Lorna. Not almost every day I think, but every instant of my life, of you. For you I would give up my home, my love of all the world beside, my duty to my dearest ones; for you I would give up my life, and hope of life beyond it." . . .

With the large tears in her eyes — tears which seemed to me to rise partly from her want to love me with the power of my love — she put her pure bright lips, half-smiling, half-prone to reply to tears, against my forehead, lined with trouble, doubt, and eager longing. And then she drew my ring from off that snowy twig her finger, and held it out to me; and then, seeing how my face was falling, thrice she touched it with her lips, and sweetly gave it back to me. "John, I dare not take it now; else I should be cheating you. I will try to love you dearly, even as you deserve and wish. Keep it for me just till then. Something tells me I shall earn it in a very little time. Perhaps you will be sorry then, sorry, when it is all too late, to be loved by such as I am."

What could I do, at her mournful tone, but kiss a thousand times the hand which she put up to warn me; and vow that I would rather die with one assurance of her love, than without it live forever with all beside that the world could give? Upon this she looked so lovely, with her dark eyelashes trembling, and her soft eyes full of light, and the color of clear sunrise mounting on her cheeks and brow, that I was forced to turn away, being overcome with beauty.

"Dearest darling, love of my life," I whispered, through her clouds of hair; "how long must I wait to know — how long must I linger doubting whether you can ever stoop from your birth and wondrous beauty to a poor, coarse hind like me — an ignorant, unlettered yeoman?"

"I will not have you revile yourself," said Lorna, very tenderly — just as I had meant to make her. "You are not rude and unlettered, John. You know a great deal more than I do; you have learned both Greek and Latin, as you told me long ago, and you have been at the very best school in the West of England. None of us but my grandfather and the Counsellor (who is a great scholar) can compare with you in this. And though I have laughed at your manner of speech, I only laughed in fun, John; I never meant to vex you by it, nor knew that I had done so."

"Naught you say can vex me, dear," I answered, as she leaned toward me, in her generous sorrow; "unless you say, 'Begone, John Ridd; I love another more than you.'" . . .

"Master John Ridd, it is high time for you to go home to your mother. I love your mother very much from what you have told me about her, and I will not have her cheated."

"If you truly love my mother," said I, very craftily, "the only way to show it is by truly loving me."

Upon that, she laughed at me in the sweetest manner, and with such provoking ways, and such come-and-go of glances, and beginning of quick blushes, which she tried to laugh away, that I knew, as well as if she herself had told me, by some knowledge (void of reasoning, and the surer for it), I knew quite well, while all my heart was burning hot within me, and mine eyes were shy of hers, and her eyes were shy of mine — for certain and forever, this I knew, as in a glory, that Lorna Doone had now begun and would go on to love me.

R. D. BLACKMORE.

CARLYLE TO HIS MOTHER

This letter may operate as a spur on the diligence of my beloved and valuable correspondents at Mainhill. There is a small blank made in the sheet for a purpose which you will notice.* I beg you to accept the little picture which fills it without any murmuring. It is a poor testimonial of the grateful love I should ever bear you. If I hope to get a moderate command of money in the course of my life's operations, I long for it chiefly that I may testify to those dear to me what affection I entertain for them. In the meantime we ought to be thankful that we have never known what it was to be in fear of want, but have always had wherewith to gratify one another by these little acts of kindness, which are worth more than millions unblest by a true feeling between the giver and receiver. You must buy yourself any little odd things you want, and think I enjoy it along with you, if it add to your comfort. I do indeed enjoy it with you. I should be a dog if I did not. I am grateful to you for kindness and true affection such as no other heart will ever feel for me. I am proud of my mother, though she is neither rich nor

* Half a page is cut off, and contains evidently a check for a small sum of money. — J. A. FROUDE.

learned. If I ever forget to love and reverence her, I must cease to be a creature myself worth remembering. Often, my dear mother, in solitary, pensive moments does it come across me like the cold shadow of death that we two must part in the course of time. I shudder at the thought, and find no refuge except in humbly trusting that the great God will surely appoint us a meeting in that far country to which we are tending. May he bless you forever, my good mother, and keep up in your heart those sublime hopes which at present serve as a pillar of cloud by day and a pillar of fire by night to guide your footsteps through the wilderness of life. We are in his hands. He will not utterly forsake us. Let us trust in him.

THOMAS CARLYLE.

ON THE MIDDLE STATION OF LIFE.

THE moral of the following fable will easily discover itself without my explaining it. One rivulet meeting another, with whom he had been long united in strictest amity, with noisy haughtiness and disdain thus bespoke him: — "What, brother! still in the same state! still low and creeping! Are you not ashamed when you behold me, who, though lately in a like condition with you, am now become a great river, and shall shortly be able to rival the Danube or the Rhine, provided those friendly rains continue which have favored my banks, but neglected yours?" "Very true," replies the humble rivulet, "you are now, indeed, swollen to a great size; but methinks you are become withal somewhat turbulent and muddy. I am contented with my low condition and my purity."

Instead of commenting upon this fable, I shall take occasion from it to compare the different stations of life, and to persuade such of my readers as are placed in the middle station to be satisfied with it, as the most eligible of all others. These form the most numerous rank of men that can be supposed susceptible of philosophy, and therefore all discourses of morality ought principally to be addressed to them. The great are too much immersed in pleasure, and the poor too much occupied in providing for the necessities of life, to hearken to the calm voice of reason. The middle station, as it is most happy in many respects, so particularly in this, that a man placed in it can, with the greatest leisure, consider his own happiness, and reap a new enjoyment, from comparing his situation with that of persons above or below him.

Agur's prayer is sufficiently noted — "Two things have I required of thee; deny me them not before I die: Remove far from me vanity and lies; give me neither poverty nor riches ; feed me with food convenient for me, lest I be full and deny thee, and say, who is the Lord? or lest I be poor, and steal, and take the name of my God in vain." The middle station is here justly recommended, as affording the fullest security for virtue; and I may also add, that it gives opportunity for the most ample exercise of it, and furnishes employment for every good quality which we can possibly be possessed of. Those who are placed among the lower ranks of men have little opportunity of exerting any other virtue besides those of patience, resignation, industry and integrity. Those who are advanced into the higher stations, have full employment for their generosity, humanity, affability and charity. When a man lies betwixt these two extremes, he can exert the former virtues towards his superiors, and the latter towards his inferiors. Every moral quality which the human soul is susceptible of, may have its turn, and be called up to action; and a man may, after this manner, be much more certain of his progress in virtue, than where his good qualities lie dormant and without employment.

But there is another virtue that seems principally to lie among equals; and is, for that reason, chiefly calculated for the middle station of life. This virtue is friendship. I believe most men of generous tempers are apt to envy the great, when they consider the large opportunities such persons have of doing good to their fellow-creatures, and of acquiring the friendship and esteem of men of merit. They make no advances in vain, and are not obliged to associate with those whom they have little kindness for, like people of inferior stations, who are subject to have their proffers of friendship rejected even where they would be most fond of placing their affections. But though the great have more facility in acquiring friendships, they cannot be so certain of the sincerity of them as men of a lower rank, since the favors they bestow may acquire them flattery, instead of good-will and kindness. It has been very judiciously remarked, that we attach ourselves more by the services we perform than by those we receive, and that a man is in danger of losing his friends by obliging them too far. I should therefore choose to lie in the middle way, and to have my commerce with my friend varied both by obligations given and received. I have too much pride to be willing that all the obligations should lie on my side, and should be afraid that, if they all lay on his, he would also have too much pride to be entirely easy under them, or have a perfect complacency in my company.

DAVID HUME.

MELONS.

I was engaged in filling a void in the Literature of the Pacific Coast. As this void was a pretty large one, and as I was informed that the Pacific Coast languished under it, I set apart two hours each day to this work of filling in. It was necessary that I should adopt a methodical system, so I retired from the world and locked myself in my room at a certain hour each day, after coming from my office. I then carefully drew out my portfolio and read what I had written the day before. This would suggest some alterations, and I would carefully rewrite it. During this operation I would turn to consult a book of reference, which invariably proved extremely interesting and attractive. It would generally suggest another and better method of "filling in." Turning this method over reflectively in my mind, I would finally commence the new method which I eventually abandoned for the original plan. At this time I would become convinced that my exhausted faculties demanded a cigar. The operation of lighting a cigar usually suggested that a little quiet reflection and meditation would be of service to me, and I always allowed myself to be guided by prudential instincts. Eventually, seated by my window, as before stated, Melons asserted himself. Though our conversation rarely went further than "Hello, Mister!" and "Ah, Melons!" a vagabond instinct we felt in common implied a communion deeper than words. In this spiritual commingling the time passed, often beguiled by gymnastics on the fence or line (always with an eye to my window) until dinner was announced and I found a more practical void required my attention. An unlooked-for incident drew us in closer relation.

A sea-faring friend just from a tropical voyage had presented me with a bunch of bananas. They were not quite ripe, and I hung them before my window to mature in the sun of McGinnis's Court, whose forcing qualities were remarkable. In the mysteriously mingled odors of ship and shore which they diffused throughout my room, there was lingering reminiscence of low latitudes. But even that joy was fleeting and evanescent: they never reached maturity.

Coming home one day, as I turned the corner of that fashionable thoroughfare before alluded to, I met a small boy eating a banana. There was nothing remarkable in that, but as I neared McGinnis's Court I presently met another small boy, also eating a banana. A third small boy engaged in a like occupation obtruded a painful coincidence upon my mind. I leave the psychological reader to determine the exact co-relation between the circumstance and the sickening sense of loss that overcame me on witnessing it. I reached my room — and found the bunch of bananas was gone.

There was but one that knew of their existence, but one who frequented my window, but one capable of gymnastic effort to procure them, and that was — I blush to say it — Melons. Melons the depredator — Melons, despoiled by larger boys of his ill-gotten booty, or reckless and indiscreetly liberal; Melons — now a fugitive on some neighborhood house-top. I lit a cigar, and, drawing my chair to the window, sought surcease of sorrow in the contemplation of the fish-geranium. In a few moments something white passed my window at about the level of the edge. There was no mistaking that hoary head, which now represented to me only aged iniquity. It was Melons, that venerable, juvenile hypocrite.

He affected not to observe me, and would have withdrawn quietly, but that horrible fascination which causes the murderer to revisit the scene of his crime, impelled him toward my window. I smoked calmly, and gazed at him without speaking. He walked several times up and down the court with a half-rigid, half-belligerent expression of eye and shoulders, intended to represent the carelessness of innocence.

Once or twice he stopped, and putting his arms their whole length into his capacious trousers, gazed with some interest at the additional width they thus acquired. Then he whistled. The singular conflicting conditions of John Brown's body and soul were at that time beginning to attract the attention of youth, and Melon's performance of that melody was always remarkable. But to-day he whistled falsely and shrilly between his teeth. At last he met my eye. He winced slightly, but recovered himself, and going to the fence, stood for a few moments on his hands, with his bare feet quivering in the air. Then he turned toward me and threw out a conversational preliminary.

"They is a cirkis"—said Melons gravely, hanging with his back to the fence and his arms twisted around the palings — "a cirkis over yonder!"—indicating the locality with his foot

MELONS.

— "with hosses, and hossback riders. They is a man wot rides six hosses to onct — six hosses to onct — and nary saddle" — and he paused in expectation.

Even this equestrian novelty did not affect me. I still kept a fixed gaze on Melon's eye, and he began to tremble and visibly shrink in his capacious garment. Some other desperate means — conversation with Melons was always a desperate means — must be resorted to. He recommenced more artfully.

"Do you know Carrots?"

I had a faint remembrance of a boy of that euphonious name, with scarlet hair, who was a playmate and persecutor of Melons. But I said nothing.

"Carrots is a bad boy. Killed a policeman onct. Wears a dirk knife in his boots, saw him to-day looking in your windy."

I felt that this must end here. I rose sternly and addressed Melons.

"Melons, this is all irrelevant and impertinent to the case. *You* took those bananas. Your proposition regarding Carrots, even if I were inclined to accept it as credible information, does not alter the material issue. You took those bananas. The offense under the Statutes of California is felony. How far Carrots may have been accessory to the fact either before or after, is not my intention at present to discuss. The act is complete. Your present conduct shows the *animo furandi* to have been equally clear."

By the time I had finished this exordium, Melons had disappeared, as I fully expected.

He never reappeared. The remorse that I have experienced for the part I had taken in what I fear may have resulted in his utter and complete extermination, alas, he may not know, except through these pages. For I have never seen him since. Whether he ran away and went to sea to reappear at some future day as the most ancient of mariners, or whether he buried himself completely in his trousers, I never shall know. I have read the papers anxiously for accounts of him. I have gone to the Police Office in the vain attempt of identifying him as a lost child. But I never saw him or heard of him since. Strange fears have sometimes crossed my mind that his venerable appearance may have been actually the result of senility, and that he may have been gathered peacefully to his fathers in a green old age. I have even had doubts of his existence, and have sometimes thought that he was providentially and mysteriously offered to fill the void I have before alluded to. In that hope I have written these pages.

BRET HARTE.

ON REFUSAL TO NEGOTIATE WITH NAPOLEON.

WHAT, then, was the nature of this system? Was it anything but what I have stated it to be? an insatiable love of aggrandizement, an implacable spirit of destruction against all the civil and religious institutions of every country. This is the first moving and acting spirit of the French Revolution; this is the spirit which animated it at its birth, and this is the spirit which will not desert it till the moment of its dissolution, "which grew with its growth, which strengthened with its strength," but which has not abated under its misfortunes, nor declined in its decay. It has been invariably the same in every period, operating more or less, according as accident or circumstance might assist it; but it has been inherent in the Revolution in all its stages; it has equally belonged to Brissot, to Robespierre, to Tallien, to Reubell, to Barras, and to every one of the leaders of the Directory, but to none more than to Bonaparte, in whom now all their powers are united. What are its characters? Can it be accident that produced them?

No, it is only from the alliance of the most horrid principles, with the most horrid means, that such miseries could have been brought upon Europe. It is this paradox which we must always keep in mind when we are discussing any question relative to the effects of the French Revolution. Groaning under every degree of misery, the victim of its own crimes, and as I once before expressed in this House, asking pardon of God and of man for the miseries which it has brought upon itself and others, France still retains (while it has neither left means of comfort nor almost of subsistence to its own inhabitants) new and unexampled means of annoyance and destruction against all the other powers of Europe.

Its first fundamental principle was to bribe the poor against the rich by proposing to transfer into new hands, on the delusive notion of equality, and in breach of every principle of justice, the whole property of the country. The practical application of this principle was to devote the whole of that property to indiscriminate plunder, and to make it the foundation of a revolutionary system of finance, productive in proportion to the misery and desolation which it created. It has been accompanied by an unwearied spirit of proselytism, diffusing itself over all the nations of the earth; a spirit which can apply itself to all circumstances and all situations, which can furnish a list of grievances and hold out a promise of redress equally to all nations; which inspired the teachers of French liberty with the hope of alike recommending themselves to those who live under the feudal code of the German Empire; to the various states of Italy, under all their different institutions; to the old republicans of Holland, and to the new republicans of America; to the Catholic of Ireland, whom it was to deliver from Protestant usurpation; to the Protestant of Switzerland, whom it was to deliver from Popish superstition; and to the Mussulman of Egypt, whom it was to deliver from Christian persecution, to the remote Indian, blindly bigoted to his ancient institutions; and to the natives of Great Britain, enjoying the perfection of practical freedom, and justly attached to their Constitution, from the joint result of habit, of reason, and of experience. The last and distinguishing feature is a perfidy which nothing can bind, which no tie of treaty, no sense of the principles generally received among nations, no obligation, human or divine, can restrain. Thus qualified, thus armed for destruction, the genius of the French Revolution marched forth, the terror and dismay of the world. Every nation has in its turn been the witness, many have been the victims of its principles; and it is left for us to decide whether we will compromise with such a danger, while we have yet resources to supply the sinews of war, while the heart and spirit of the country is yet unbroken, and while we have the means of calling forth and supporting a powerful co-operation in Europe.

WILLIAM PITT.

PARODY ON THE SPEECHES OF CHARLES II.

My lords and gentlemen,

I told you, at our last meeting, the Winter was the fittest time for business, and truly I thought so, till my lord-treasurer assured me the Spring was the best season for salads and subsidies. I hope, therefore, that April will not prove so unnatural a month, as not to afford some kind showers on my parched exchequer, which gapes for want of them. Some of you, perhaps, will think it dangerous to make me too rich; but I do not fear it; for I promise you faithfully, whatever you give me I will always want; and although in other things my word may be thought a slender authority, yet in that, you may rely on me, I will never break it.

My lords and gentlemen,

I can bear my straits with patience; but my lord-treasurer does protest to me, that the revenue, as it now stands, will not serve him and me too. One of us must pinch for it, if you do not help me. I must speak freely to you; I am under bad circumstances. Here is my lord-treasurer can tell, that all the money designed for next Summer's guards must of necessity be applied to the next year's cradles and swaddling clothes. What shall we do for ships then? I hint this only to you, it being your business, not mine. I know, by experience, I can live without ships. I lived ten years abroad without, and never had my health better in my life; but how you will be without, I leave to yourselves to judge, and therefore hint this only by the bye: I do not insist upon it. There is another thing I must press more earnestly, and that is this: it seems a good part of my revenue will expire in two or three years, except you will be pleased to continue it. I have to say for it; pray, why did you give me so much as you have done, unless you resolve to give on as fast as I call for it? The nation hates you already for giving so much, and I will hate you too, if you do not give me more. So that, if you stick not to me, you must not have a friend in England. On the other hand, if you will give me the revenue I desire, I shall be able to do those things for your religion and liberty, that I have had long in my thoughts, but cannot effect them without a little more money to carry me through. Therefore look to't, and take notice, that if you do not make me rich enough to undo you, it shall lie at your doors. For my part, I wash my hands on it.

If you desire more instances of my zeal, I have them for you. For example, I have converted my sons from popery, and I may say, without vanity, it was my own work. 'Twould do one's heart good to hear how prettily George can read already in the psalter. They are all fine children, God bless 'em, and so like me in their understandings.

I must now acquaint you, that, by my lord-treasurer's advice, I have made a considerable retrenchment upon my expenses in candles and charcoal, and do not intend to stop, but will, with your help, look into the late embezzlements of my dripping-pans and kitchen-stuff.

ANDREW MARVELL.

THE DESTINY OF THE REPUBLIC.

We are a young republic, just entering upon the arena of nations; we will be the architects of our own fortunes. Our destiny, under Providence, is in our own hands. With wisdom, prudence, and statesmanship on the part of our public men, and intelligence, virtue, and patriotism on the part of the people, success to the full measure of our most sanguine hopes may be looked for. But, if unwise counsels prevail, if we become divided, if schisms arise, if dissensions spring up, if factions are engendered, if party spirit, nourished by unholy personal ambition, shall rear its hydra head, I have no good to prophesy for you. Without intelligence, virtue, integrity, and patriotism on the part of the people, no republic or representative government can be durable or stable.

ALEXANDER HAMILTON STEPHENS.

ALEXANDER HAMILTON STEPHENS.

NEWS FROM THE FRONT.

THERE came the unwonted sound of horses' feet trampling along the lane. Two men rode up to the gate and dismounted. As they came in we saw that they wore blue uniforms, and a moment later we recognized the latter one whom we had seen several times while the *crevasse* was open and the Yankee soldiers had come down to help rebuild the levee. He was the Colonel of the regiment camped above the bend of the river.

The little boys ran through the gap in the hedge to Aunt Rose, and I shrank back leaving Uncle Joshua to go forward and meet them. The tall Colonel stopped when he came up to Uncle Joshua and said something to him in a low tone. Uncle Joshua's voice in reply sounded sharp and unnatural, though I could not hear what he said. The officer spoke again and seemed to be urging something; and then

Uncle Joshua fell upon his knees and began sobbing and rocking himself to and fro. "Oh," I heard him cry as if half beside himself, "*who* gwine ter tell her! I *cyant* tell her! Oh, Lord, whar give an' whar tek away, hab mussy on her! An' on de chillun! Oh, *my* Marster! *my* Marster!"

The Colonel stood for a moment as if irresolute and perplexed and then walked on slowly toward the house followed by his orderly. He carried in his hand a sheathed sword which I had seen him take from the soldier as they came in the gateway. Once he turned as if to go back. A quick exclamation broke from him as he faced around again. For there in the walk before him and barring his way stood mother. She had come through the gap in the hedge followed by the little boys who were all huddled about her. She was deadly pale and her great eyes were fixed upon the officer's face with a look of terror. I had never seen fear on her brave face before and I shivered at it while I wondered what it meant.

The Colonel uncovered his head and the soldier after a moment's hesitation took off his cap. The weather-beaten faces of the men were almost as pale as mother's!

There was a short silence; the officer seemed to be trying to find a way to begin what he had to say.

"Madame," he said at last, "a messenger coming from the other side of the river, and bearing letters and — other messages for this neighborhood has been captured by some of my men. A number of the letters he carried were old — some of them had been drifting about for months. But one among them was of late date and contained the news of" —

He broke off abruptly and turned away as if unable to bear the look in the eyes gazing into his. His glance fell upon little Percy. He stooped and bent one knee to the ground and drew the child gently to him. "My son," he said, putting the sword into the small hands and closing them upon it, "give this to your mother and tell her that it was the sword of a brave and honorable man who died a gallant death on the battle-field." The empty tray she was holding dropped from mother's hand and a low cry escaped from her blanched lips. "Tell her — " but a tear splashed down upon the little upturned face. He laid a hand caressingly upon the yellow curls and rose to his feet. He thrust a letter into the hands of one of the other children and without another word he hurried off down the walk; the soldier followed, and a moment later they were galloping along the lane toward the river.

I think none of us really understood until little Percy went up to mother and began in his childish way to repeat what the officer had said. But when with one great sob she stooped and lifted him in her arms with father's sword hugged to his breast — oh, then, we all knew!

Father had been killed ten days before at the head of his men while leading a charge; and he had been buried on the battle-field.

M. E. M. DAVIS.

THE GETTYSBURG ADDRESS.

FOURSCORE and seven years ago our fathers brought forth upon this continent a new nation, conceived in liberty, and dedicated to the proposition that all men are created equal. Now we are engaged in a great civil war, testing whether that nation, or any nation so conceived and so dedicated, can long endure. We are met on a great battlefield of that war. We have come to dedicate a portion of that field as a final resting-place for those who here gave their lives that that nation might live. It is altogether fitting and proper that we should do this. But in a larger sense we cannot dedicate, we cannot consecrate, we cannot hallow this ground. The brave men, living and dead, who struggled here, have consecrated it far above our power to add or detract. The world will little note, nor long remember, what we say here, but it can never forget what they did here. It is for us, the living, rather to be dedicated here to the unfinished work which they who fought here have thus far so nobly advanced. It is rather for us to be here dedicated to the great task remaining before us, that from these honored dead we take increased devotion to that cause for which they gave the last full measure of devotion; that we here highly resolve that these dead shall not have died in vain; that this nation, under God, shall have a new birth of freedom, and that government of the people, by the people, and for the people, shall not perish from the earth.

ABRAHAM LINCOLN.

ABRAHAM LINCOLN.

THE MIDNIGHT SUN.

BAYARD TAYLOR.

As we crossed the mouth of the Ulvsfjord that evening, we had an open sea horizon toward the north, a clear sky, and so much sunshine at eleven o'clock that it was evident the Polar day had dawned upon us at last. The illumination of the shores was unearthly in its glory, and the wonderful effects of the orange sunlight, playing upon the dark hues of the island cliffs, can neither be told nor painted. The sun hung low between Fuglöe, rising like a double dome from the sea, and the tall mountains of Arnöe, both of which islands resembled immense masses of transparent purple glass, gradually melting into crimson fire at their bases. The glassy, leaden-colored sea was powdered with a golden bloom, and the tremendous precipices at the mouth of the Lyngen Fjord behind us, were steeped in a dark red mellow, flush, and touched with pencillings of pure, rose-colored light, until their naked ribs seemed to be clothed in imperial velvet. As we turned into the Fjord and ran southward along their bases, a waterfall, struck by the sun, fell in fiery orange foam down the red walls, and the blue ice-pillars of a beautiful glacier filled up the ravine beyond it. We were all on deck; and all faces, excited by the divine splendor of the scene and tinged by the same wonderful aureole, shone as if transfigured. In my whole life I have never seen a spectacle so unearthly beautiful.

Our course brought the sun rapidly toward the ruby cliffs of Arnöe, and it was evident that he would soon be hidden from sight. It was not yet half-past eleven, and an enthusiastic passenger begged the captain to stop the vessel until midnight. "Why," said the latter, "it is midnight now, or very near it; you have Drontheim time, which is almost forty minutes in arrears." True enough, the real time lacked but five minutes of midnight, and those of us who had sharp eyes and strong imaginations saw the sun make his last dip and rise a little, before he vanished in a blaze of glory behind Arnöe. I turned away with my eyes full of dazzling spheres of crimson and gold, which danced before me wherever I looked; and it was a long time before they were blotted out by the semi-oblivion of a daylight sleep.

BAYARD TAYLOR.

THE WHISTLE.

WHEN I was a child, at seven years old, my friends, on a holiday, filled my little pocket with coppers. I went directly to a shop where they sold toys for children; and, being charmed with the sound of a whistle, that I met by the way in the hands of another boy, I voluntarily offered him all my money for one. I then came home, and went whistling all over the house, much pleased with my whistle, but disturbing all the family. My brothers and sisters and cousins, understanding the bargain I had made, told me I had given four times as much for it as it was worth. This put me in mind what good things I might have bought with the rest of my money; and they laughed at me so much for my folly that I cried with vexation; and the reflection gave me more chagrin than the whistle gave me pleasure.

This, however, was afterwards of use to me, the impression continuing on my mind; so that often, when I was tempted to buy some unnecessary thing, I said to myself, don't give too much for the whistle, and so I saved my money.

BENJAMIN FRANKLIN.

As I grew up, came into the world, and observed the actions of men, I thought I met with many, very many, who gave too much for the whistle.

When I saw any one too ambitious of court favor, — sacrificing his time in attendance at levees, his repose, his liberty, his virtue, and perhaps his friends, to attain it, — I have said to myself, this man gives too much for his whistle.

When I saw another fond of popularity, constantly employing himself in political bustles, neglecting his own affairs, and ruining them by that neglect, he pays, indeed, says I, too much for his whistle.

STATUE OF FRANKLIN, INDEPENDENCE HALL, PA.

If I knew a miser who gave up every kind of comfortable living, — all the pleasure of doing good to others, — all the esteem of his fellow-citizens, — and the joys of benevolent friendship, for the sake of accumulating wealth, poor man, says I, you do, indeed, pay too much for your whistle.

When I meet a man of pleasure, sacrificing every laudable improvement of the mind or of his fortune to mere corporeal sensations, — "Mistaken man," says I, "you are providing pain for yourself instead of pleasure, — you give too much for your whistle."

If I see one fond of fine clothes, fine furniture, fine equipages, all above his fortune, for which he contracts debts, and ends his career in prison, — "Alas," says I, "he has paid dear, very dear, for his whistle."

When I see a beautiful, sweet-tempered girl married to an ill-natured brute of a husband, — "What a pity it is," says I, "that she has paid so much for a whistle."

In short, I conceived that a great part of the miseries of mankind were brought upon them by the false estimates they had made of the value of things, and by their giving too much for their whistles.

BENJAMIN FRANKLIN.

INAUGURAL ADDRESS.

Such being the impression under which I have, in obedience to the public summons, repaired to the present station, it would be peculiarly improper to omit in this first official act, my fervent supplications to that Almighty Being who rules over the universe — who presides in the councils of nations — and whose providential aids can supply every human defect, that his benediction may consecrate to the liberties and happiness of the people of the United States, a government instituted by themselves for these essential purposes; and may enable every instrument employed in its administration to execute, with success, the functions allotted to his charge. In tendering this homage to the great author of every public and private good, I assure myself that it expresses your sentiments not less than my own, nor those of my fellow-citizens at large less than either. No people can be bound to acknowledge and adore the invisible hand, which conducts the affairs of men, more than the people of the United States. Every step by which they have advanced to the character of an independent nation, seems to have been distinguished by some token of providential agency; and in the important revolution just accomplished in the system of their united government, the tranquil deliberations and voluntary consent of so many distinct communities, from which the event has resulted, cannot be compared with the means by which most governments have been established, without some return of pious gratitude along with an humble anticipation of the future blessings which the past seems to presage. These reflections, arising out of the present crisis, have forced themselves too strongly on my mind to be suppressed. You will join with me, I trust, in thinking that there are none under the influence of which the proceedings of a new and free government can more auspiciously commence.

GEORGE WASHINGTON.

THE ESSENTIAL PRINCIPLES OF GOVERNMENT.

About to enter, fellow citizens, upon the exercise of duties which comprehend everything dear and valuable to you, it is proper you should understand what I deem the essential principles of our government, and consequently those which ought to shape its administration. I will compress them within the narrowest compass they will bear, stating the general principle, but not all its limitations. Equal and exact justice to all men, of whatever state or persuasion, religious or political; peace, commerce, and honest friendship with all nations, entangling alliances with none; the support of the State governments in all their rights, as the most competent administrations for our domestic concerns, and the surest bulwarks against anti-republican tendencies; the preservation of the general government in its whole constitutional vigor, as the sheet anchor of our peace at home and safety abroad; a jealous care of the right of election by the people, a mild and safe corrective of abuses which are lopped by the sword of revolution where peaceable remedies are unprovided; absolute acquiescence in the decisions of the majority, the vital principle of republics, from which there is no appeal but to force, the vital principle and immediate parent of despotism; a well-disciplined militia, our best reliance in peace, and for the first moments of war, till regulars may relieve them; the supremacy of the civil over the military authority; economy in the public expense, that labor may be lightly burdened; the honest payment of our debts, and sacred preservation of the public faith; encouragement of agriculture, and of commerce as its handmaid; the diffusion of information and arraignment of all abuses at the bar of the public reason; freedom of religion, freedom of the press, and freedom of person, under the protection of the *habeas corpus*, and trial by juries impartially selected. These principles form the bright constellation which has gone before us and guided our steps through an age of revolution and reformation. The wisdom of our sages and blood of our heroes have been devoted to their attainment; they should be the creed of our political faith, the text of civic instruction, the touchstone by which to try the services of those we trust; and should we wander from them in moments of error or of alarm, let us hasten to retrace our steps, and to regain the road which alone leads to peace, liberty, and safety.

THOMAS JEFFERSON.

THOMAS JEFFERSON.

CONCILIATION.

But there is still behind a third consideration concerning this object, which serves to determine my opinion on the sort of policy which ought to be pursued in the management of America, even more than its population and its commerce — I mean its temper and character. In this character of the Americans, a love of freedom is the predominating feature, which marks and distinguishes the whole; and, as an ardent is always a jealous affection, your colonies become suspicious, restive, and untractable, whenever they see the least attempt to wrest from them by force, or shuffle from them by chicane, what they think the only advantage worth living for.

This fierce spirit of liberty is stronger in the English colonies, probably, than in any other people of the earth, and this from a variety of powerful causes, which, to understand the true temper of their minds, and the direction which this spirit takes, it will not be amiss to lay open somewhat more largely.

First, the people of the colonies are descendants of Englishmen. England, sir, is a nation which still, I hope, respects, and formerly adored her freedom. The colonists emigrated from you when this part of your character was most predominant; and they took this bias and direction the moment they parted from your hands. They are, therefore, not only devoted to liberty, but to liberty according to English ideas and on English principles. Abstract liberty, like other mere abstractions, is not to be found. Liberty inheres in some sensible object; and every nation has formed to itself some favorite point which, by way of eminence, becomes the criterion of their happiness. It happened, you know, sir, that the great contests for freedom in this country were, from the earliest times, chiefly upon the question of taxing. Most of the contests in the ancient commonwealths turned primarily on the right of election of magistrates, or on the balance among the several orders of the State. The question of money was not with them so immediate. But in England it was otherwise. On this point of taxes the ablest pens and most eloquent tongues have been exercised; the greatest spirits have acted and suffered. In order to give the fullest satisfaction concerning the importance of this point, it was not only necessary for those who in argument defended the excellence of the English Constitution, to insist on this privilege of granting money as a dry point of fact, and to prove that the right had been acknowledged in ancient parchments and blind usages to reside in a certain body called the House of Commons. They went much farther: they attempted to prove (and they succeeded) that in theory it ought to be so, from the particular nature of a House of Commons, as an immediate representative of the people, whether the old records had delivered this oracle or not. They took infinite pains to inculcate, as a fundamental principle, that, in all monarchies, the people must, in effect, themselves, mediately or immediately, possess the power of granting their own money, or no shadow of liberty could subsist. The colonies draw from you, as with their life-blood, those ideas and principles. Their love of liberty, as with you, fixed and attached on this specific point of taxing. Liberty might be safe or might be endangered in twenty other particulars, without their being much pleased or alarmed. Here they felt its pulse; and, as they found that beat, they thought themselves sick or sound. I do not say whether they were right or wrong in applying your general arguments to their own case. It is not easy, indeed, to make a monopoly of theorems and corollaries. The fact is, that they did thus apply those general arguments; and your mode of governing them, whether through lenity or indolence, through wisdom or mistake, confirmed them in the imagination that they, as well as you, had an interest in these common principles.

They were further confirmed in these pleasing errors by the form of their provincial legislative assemblies. Their governments are popular in a high degree; some are merely popular; in all, the popular representative is the most weighty; and this share of the people in their ordinary government never fails to inspire them with lofty sentiments, and with a strong aversion from whatever tends to deprive them of their chief importance.

If any thing were wanting to this necessary operation of the form of government, religion would have given it a complete effect. Religion, always a principle of energy, in this new people is no way worn out or impaired; and their mode of professing it is also one main cause of this free spirit. The people are Protestants; and of that kind which is the most averse to all implicit submission of mind and opinion. This is a persuasion not only favorable to liberty, but built upon it. I do not think, sir, that the reason of this averseness in the dissenting churches from all that looks like absolute government, is so much to be sought in their religious tenets as in their history. Every one knows that the Roman Catholic religion is at least coeval with most of the governments where it prevails; that it has generally gone hand in hand with them; and received great favor and every kind of support from authority. The Church of England, too, was formed from her cradle under the nursing care of regular government. But the dissenting interests have sprung up in direct opposition to all the ordinary powers of the world, and could justify that opposition only on a strong claim to natural liberty. Their very existence depended on the powerful and unremitted assertion of that claim. All Protestantism, even the most cold and passive, is a kind of dissent. But the religion most prevalent in our northern colonies is a refinement on the principle of resistance; it is the dissidence of dissent; and the Protestantism of the Protestant religion.

This religion, under a variety of denominations, agreeing in nothing but in the communion of the spirit of liberty, is predominant in most of the northern provinces; where the Church of England, notwithstanding its legal rights, is in reality no more than a sort of private sect, not composing, most probably, the tenth of the people. The colonists left England when this spirit was high, and in the emigrants was the highest of all; and even that stream of foreigners, which has been constantly flowing into these colonies, has, for the greatest part, been composed of dissenters from the establishments of their several countries, and have brought with them a temper and character far from alien to that of the people with whom they mixed.

EDMUND BURKE.

SUMNER'S STUDY.

ON THE KANSAS-NEBRASKA BILL.

FROM these expressions, and other tokens which daily greet us, it is evident that at least the religious sentiment of the country is touched, and, under this sentiment, I rejoice to believe that the whole North will be quickened with the true life of freedom. Sir Philip Sidney, speaking to Queen Elizabeth of the spirit which animated every man, woman, and child in the Netherlands against the Spanish power, exclaimed: " It is the spirit of the Lord, and is invincible." A similar spirit is now animating the free States against the slave power, breathing everywhere its precious inspiration, and forbidding repose under the attempted usurpation. The threat of disunion, so often sounded in our ears, will be disregarded by an aroused and indignant people. Ah, sir, Senators vainly expect peace. Not in this way can peace come. In passing this bill you scatter, broadcast through the land, dragon's teeth, and though they may not, as in ancient fable, spring up armed men, yet will they fructify in civil strife and feud.

ON THE KANSAS-NEBRASKA BILL.

From the depths of my soul, as a loyal citizen and as a Senator, I plead, remonstrate, protest against the passage of this bill. I struggle against it as against death; but, as in death itself corruption puts on incorruption, and this mortal body puts on immortality, so from the sting of this hour I find assurance of that triumph by which freedom will be restored to her immortal birthright in the Republic.

Sir, the bill which you are now about to pass is at once the worst and the best bill on which Congress ever acted.

It is the worst bill, inasmuch as it is a present victory of slavery. In a Christian land, and in an age of civilization, a time-honored statute of freedom is struck down, opening the way to all the countless woes and wrongs of human bondage. Among the crimes of history a new one is about to be recorded, which, in better days, will be read with universal shame. The tea tax and stamp act, which aroused the patriotic rage of our fathers, were virtues by the side of this enormity; nor would it be easy to imagine, at this day, any measure which more openly defied every sentiment of justice, humanity, and Christianity. Am I not right, then, in calling it the worst bill on which Congress ever acted?

But there is another side to which I gladly turn. Sir, it is the best bill on which Congress ever acted; for it prepares the way for that "All hail hereafter," when slavery must disappear. It annuls all past compromises with slavery, and makes all future compromises impossible. Thus it puts freedom and slavery face to face, and bids them grapple. Who can doubt the result? It opens wide the door of the future, when, at last, there will really be a North, and the slave power will be broken; when this wretched despotism will cease to dominate over our Government, no longer impressing itself upon all that it does, at home and abroad; when the National Government shall be divorced in every way from slavery, and, according to the true intention of our fathers, freedom shall be established by Congress everywhere, at least beyond the local limits of the States.

Slavery will then be driven from its usurped foothold here in the District of Columbia; in the national territories, and elsewhere beneath the national flag; the fugitive-slave bill, as odious as it is unconstitutional, will become a dead letter; and the domestic slave-trade, so far as it can be reached, but especially on the high seas, will be blasted by Congressional prohibition. Everywhere within the sphere of Congress, the great Northern Hammer will descend to smite the wrong; and the irresistible cry will break forth, "No more slave States!"

Thus, sir, now standing at the very grave of freedom in Kansas and Nebraska, I find assurances of that happy resurrection, by which freedom will be secured hereafter, not only in these territories, but everywhere under the National Government. More clearly than ever before, I now see "the beginning of the end" of slavery. Am I not right, then, in calling this measure the best bill on which Congress ever acted?

Sorrowfully I bend before the wrong you are about to perpetrate. Joyfully I welcome all the promises of the future.

CHARLES SUMNER.

ALREADY IN THE FIELD.

SPEECH BEFORE THE VIRGINIA CONVENTION.

Mr. President, it is natural to man to indulge in the illusions of hope. We are apt to shut our eyes against a painful truth, and listen to the song of that syren till she transforms us into beasts. Is this the part of wise men engaged in a great and arduous struggle for liberty? Are we disposed to be of the number of those who, having eyes, see not, and having ears, hear not, the things which so nearly concern their temporal salvation? For my part, whatever anguish of spirit it may cost, I am willing to know the whole truth; to know the worst and to provide for it.

I have but one lamp by which my feet are guided; and that is the lamp of experience. I know of no way of judging of the future but by the past. And judging by the past, I wish to know what there has been in the conduct of the British ministry for the last ten years, to justify those hopes with which gentlemen have been pleased to solace themselves and the House? Is it that insidious smile with which our petition has been lately received? Trust it not, sir; it will prove a snare to your feet. Suffer not yourselves to be betrayed with a kiss. Ask yourselves how this gracious reception of our petition comports with these war-like preparations which cover our waters and darken our land. Are fleets and armies necessary to a work of love and reconciliation? Have we shown ourselves so unwilling to be reconciled that force must be called in to win back our love? Let us not deceive ourselves, sir. These are the implements of war and subjugation; the last arguments to which kings resort. I ask gentlemen, sir, what means this martial array, if its purpose be not to force us to submission? Can gentlemen assign any other possible motives for it? Has Great Britain any enemy in this quarter of the world, to call for all this accumulation of navies and armies? No, sir, she has none. They are meant for us; they can be meant for no other. They are sent over to bind and rivet upon us those chains which the British ministry have been so long forging. And what have we to oppose to them? Shall we try argument? Sir, we have been trying that for the last ten years. Have we anything new to offer on the subject? Nothing. We have held the subject up in every light of which it is capable; but it has been all in vain. Shall we resort to entreaty and humble supplication? What terms shall we find which have not been already exhausted? Let us not, I beseech you, sir, deceive ourselves longer. Sir, we have done everything that could be done to avert the storm which is now coming on. We have petitioned; we have remonstrated; we have supplicated; we have prostrated ourselves before the throne, and have implored its interposition to arrest the tyrannical hands of the ministry and parliament. Our petitions have been slighted; our remonstrances have produced additional violence and insult; our supplications have been disregarded; and we have been spurned with contempt from the foot of the throne. In vain, after these things, may we indulge the fond hope of peace and reconciliation. There is no longer any room for hope if we wish to be free — if we mean to preserve inviolate those inestimable privileges for which we have been so long contending — if we mean not basely to abandon the noble struggle in which we have been so long engaged, and which we have pledged ourselves never to abandon until the glorious object of our contest shall be obtained, we must fight! I repeat it, sir, we must fight! An appeal to arms and to the God of Hosts is all that is left us!

They tell us, sir, that we are weak; unable to cope with so formidable an adversary. But when shall we be stronger? Will it be the next week, or the next year? Will it be when we are totally disarmed, and when a British guard shall be stationed in every house? Shall we gather strength by irresolution and inaction? Shall we acquire the means of effectual resistance by lying supinely on our backs, and hugging the delusive phantom of hope, until our enemies shall have bound us hand and foot? Sir, we are not weak, if we make a proper use of the means which the God of nature hath placed in our power. Three millions of people, armed in the holy cause of liberty, and in such a country as that which we possess, are invincible by any force which our enemy can send against us. Besides, sir, we shall not fight our battles alone. There is a just God who presides over the destinies of nations; and who will raise up friends to fight our battles for us. The battle, sir, is not to the strong alone; it is to the vigilant, the active, the brave. Besides, sir, we have no election. If we were base enough to desire it, it is now too late to retire from the contest. There is no retreat but in submission and slavery! Our chains are forged! Their clanking may be heard on the plains of Boston! The war is inevitable — and let it come! I repeat it, sir, let it come!

It is in vain, sir, to extenuate the matter. Gentlemen may cry peace, peace — but there is no peace. The war is actually begun! The next gale that sweeps from the north will bring to our ears the clash of resounding arms! Our brethren are already in the field! Why stand we here idle? What is it that gentlemen wish? What would they have? Is life so dear, or peace so sweet, as to be purchased at the price of chains and slavery? Forbid it, Almighty God! I know not what course others may take; but as for me, give me liberty, or give me death!

PATRICK HENRY.

THE LAST TRAIN NORTH.

THE jest and the laugh ran to and fro everywhere. It seemed very strange to Mary to find it so. There were two or three convalescent wounded men in the car, going home on leave, and they appeared never to weary of the threadbare joke of calling their wounds "furloughs." There was one little slip of a fellow — he could hardly have been seventeen — wounded in the hand, whom they kept teased to the point of exasperation by urging him to confess that he had shot himself for a furlough, and of whom they said, later, when he had got off at a flag-station, that he was the bravest soldier in his company. No one on the train seemed to feel that he had got all that was coming to him until the conductor had exchanged a jest with him. The land laughed. On the right hand and on the left it dimpled and wrinkled in gentle depressions and ridges, and rolled away in fields of young corn and cotton. The train skipped and clattered along at a happy-go-lucky, twelve-miles-an-hour gait, over trestles and stock-pits, through flowery cuts and along slender, rain-washed embankments where dewberries were ripening, and whence cattle ran down and galloped off across the meadows on this side and that, tails up and heads down, throwing their horns about, making light of the screaming destruction, in their dumb way, as the people made light of the war. At stations where the train stopped — and it stopped on the faintest excuse — a long line of heads and gray shoulders were thrust out of the windows of the soldiers' car, in front, with all manner of masculine head-coverings, even bloody handkerchiefs; and woe to the negro or negress or "citizen" who, by any conspicuous demerit or excellence of dress, form, stature, speech, or bearing, drew the fire of that line! No human power of face or tongue could stand the incessant volley of stale quips and mouldy jokes, affirmative, interrogative, and exclamatory that fell about their victim.

At one spot, in a lovely natural grove, where the air was spiced with the gentle pungency of the young hickory foliage, the train paused a moment to let off a man in fine gray cloth, whose yellow stripes and one golden star on the coat-collar indicated a major of cavalry. It seemed as though pandemonium had opened. Mules braying, negroes yodling, axes ringing, teamsters singing, men shouting and howling, and all at nothing; mess-fires smoking all about in the same hap-hazard, but roomy, disorder in which the trees of the grove had grown; the railroad side lined with a motley crowd of jolly fellows in spurs, and the atmosphere between them and the line of heads in the car-windows murky with the interchange of compliments that flew back and forth from the "web-foots" [*] to the "critter company," and from the "critter company" to the "web-foots." As the train moved off, "I say, boys," drawled a lank, coatless giant on the roadside, with but one suspender and one spur, "tha-at's right! Gen'l Beerygyard told you to strike fo' yo' homes, an' I see you a-doin' it ez fass as you kin git thah." And the "citizens" in the

[*] Infantry.

rear car-windows giggled even at that; while the "web-foots" he-hawed their derision, and the train went on, as one might say, with its hands in its pockets, whooping and whistling over the fields — after the cows; for the day was declining.

Mary was awed. As she had been forewarned to do, she tried not to seem unaccustomed to, or out of harmony with, all this exuberance. But there was something so brave in it, coming from a people who were playing a losing game, with their lives and fortunes for their stakes; something so gallant in it, laughing and gibing in the sight of blood, and smell of fire, and shortness of food and raiment, that she feared she had betrayed a stranger's wonder and admiration every time the train stopped and the idlers of the station platform lingered about her window and silently paid their ungraceful but complimentary tribute of simulated casual glances.

For, with all this jest, it was very plain there was but little joy. It was not gladness; it was bravery. It was the humor of an invincible spirit — the gayety of defiance. She could easily see the grim earnestness beneath the jocund temper, and beneath the unrepining smile the privation and the apprehension. What joy there was, was a martial joy. The people were confident of victory at last, — a victorious end, whatever might lie between; and of even what lay between they would confess no fear. Richmond was safe, Memphis safer, New Orleans safest. Yea, notwithstanding Porter and Farragut were pelting away at Forts Jackson and St. Philip. Indeed, if the rumor be true, if Farragut's ships had passed those forts, leaving Porter behind, then the Yankee sea-serpent was cut in two, and there was an end of him in that direction. Ha! ha!

"Is to-day the twenty-sixth?" asked Mary, at last, of one of the ladies in real ribbons, leaning over toward her.

"Yes, ma'am."

It was the younger one who replied. As she did so she came over and sat by Mary.

"I judge, from what I heard your little girl asking you, that you are going beyond Jackson."

"I am going to New Orleans."

"Do you live there?" The lady's interest seemed genuine and kind.

"Yes. I am going to join my husband there."

Mary saw by the reflection in the lady's face that a sudden gladness must have overspread her own.

"He'll be mighty glad, I'm sure," said the pleasant stranger, patting Alice's cheek, and looking, with a pretty fellow-feeling, first into the child's face and then into Mary's.

"Yes, he will," said Mary, looking down upon the curling locks at her elbow with a mother's happiness.

"Is he in the army?" asked the lady.

Mary's face fell.

"His health is bad," she replied.

"I know some nice people down in New Orleans," said the lady again.

"We haven't many acquaintances," rejoined Mary, with a timidity that was almost trepidation. Her eyes dropped, and she began softly to smooth Alice's collar and hair.

"I didn't know," said the lady, "but you might know some of them. For instance, there is Dr. Sevier."

Mary gave a start and smiled.

"Why, is he your friend too?" she asked. She looked up into the lady's quiet, brown eyes and down again into her own lap, where her hands had suddenly knit together, and then again into the lady's face. "We have no friend like Dr. Sevier."

"Mother," called the lady softly, and beckoned. The senior lady leaned toward her. "Mother, this lady is from New Orleans, and is an intimate friend of Dr. Sevier."

The mother was pleased.

"What might one call your name?" she asked, taking a seat behind Mary and continuing to show her pleasure.

"Richling."

The mother and daughter looked at each other. They had never heard the name before.

Yet only a little while later the mother was saying to Mary, — they were expecting at any moment to hear the whistle for the terminus of the route, the central Mississippi town of Canton:

"My dear child, no! I couldn't sleep to-night if I thought you was all alone in one o' them old hotels in Canton. No, you must come home with us. We're barely two mile' from town, and we'll have the carriage ready for you bright and early in the morning, and our coachman will put you on the cars just as nice — Trouble?" She laughed at the idea. "No; I tell you what would trouble me, — that is, if we'd allow it; that'd be for you to stop in one o' them hotels all alone, child, and like as not some careless servant not wake you in time for the cars to-morrow." At this word she saw capitulation in Mary's eyes. "Come, now, my child, we're not going to take no for an answer."

Nor did they.

But what was the result? The next morning when Mary and Alice stood ready for the carriage, and it was high time they were gone, the carriage was not ready; the horses had got astray in the night. And while the black coachman was on one horse, which he had found and caught, and was scouring the neighboring fields and lanes and meadows in search of the other, there came out from townward upon the still, country air, the long whistle of the departing train; and then the distant rattle and roar of its far southern journey began, and then its warning notes to the scattering colts and cattle.

"Look away!" — it seemed to sing — "Look away!" — the notes fading, failing on the ear, — "away — away — away down south in Dixie," — the last train that left for New Orleans until the war was over.

GEORGE WASHINGTON CABLE.

THE GRASSHOPPER AND THE ANT.

A FRIVOLOUS Grasshopper, having spent the Summer in Mirth and Revelry, went on the approach of the inclement winter to the Ant, and implored it of its charity to stake him. "You had better go to your Uncle," replied the prudent Ant; "had you imitated my Forethought and deposited your Funds in a Savings Bank, you would not now be compelled to regard your Duster in the light of an Ulster." Thus saying, the virtuous Ant retired, and read in the Papers next morning that the Savings Bank where he had deposited his Funds had suspended.
Moral. — *Dum vivimus, vivamus.*

GEORGE T. LANIGAN.

THE LITTLE WOMEN'S ROMANCE.

SOMETHING in his resolute tone made Jo look up quickly to find him looking down at her with an expression that assured her the dreaded moment had come, and made her put out her hand with an imploring, —

"No, Teddy, please don't!"

"I will, and you *must* hear me. It's no use, Jo; we've got to have it out, and the sooner the better for both of us," he answered, getting flushed and excited all at once.

"Say what you like, then; I'll listen," said Jo, with a desperate sort of patience.

Laurie was a young lover, but he was in earnest, and meant to "have it out," if he died in the attempt; so he plunged into the subject with characteristic impetuosity, saying in a voice that *would* get choky now and then, in spite of manful efforts to keep it steady, —

"I've loved you ever since I've known you, Jo; couldn't help it, you've been so good to me. I've tried to show it, but you wouldn't let me; now I'm going to make you hear, and give me an answer, for I *can't* go on so any longer."

"I wanted to save you this; I thought you'd understand," began Jo, finding it a great deal harder than she expected.

"I know you did; but girls are so queer you never know what they mean. They say No when they mean Yes, and drive a man out of his wits just for the fun of it," returned Laurie, entrenching himself behind an undeniable fact.

"*I* don't. I never wanted to make you care for me so, and I went away to keep you from it if I could."

"I thought so; it was like you, but it was no use. I only loved you all the more, and I worked hard to please you, and I gave up billiards and everything you didn't like, and waited and never complained, for I hoped you'd love me, though I'm not half good enough "—here there was a choke that couldn't be controlled, so he decapitated buttercups while he cleared his "confounded throat."

"Yes, you are; you're a great deal too good for me, and I'm so grateful to you, and so proud and fond of you, I don't see why I can't love you as you want me to. I've tried, but I can't change the feeling, and it would be a lie to say I do when I don't."

"Really, truly, Jo?"

He stopped short, and caught both her hands as he put his question with a look which she did not soon forget.

"Really, truly, dear."

They were in the grove now, close by the stile; and when the last words fell reluctantly from Jo's lips, Laurie dropped her hands, and turned as if to go on, but for once in his life that fence was too much for him; so he just laid his head down on the mossy post, and stood so still that Jo was frightened.

"Oh, Teddy, I'm so sorry, so desperately sorry, I could kill myself if it would do any good! I wish you wouldn't take it so hard. I can't help it; you know it's impossible for people to make themselves love other people if they don't," cried Jo, inelegantly but remorsefully, as she softly patted his shoulder, remembering the time when he had comforted her so long ago.

THE ALCOTT HOME.

"They do sometimes," said a muffled voice from the post.

"I don't believe it's the right sort of love, and I'd rather not try it," was the decided answer.

There was a long pause, while a blackbird sung blithely on the willow by the river, and the tall grass rustled in the wind. Presently Jo said very soberly, as she sat down on the step of the stile,—

"Laurie, I want to tell you something."

He started as if he had been shot, threw up his head, and cried out in a fierce tone, —

"*Don't* tell me that, Jo; I can't bear it now!"

"Tell what?" she asked, wondering at his violence.

"That you love that old man."

"What old man?" demanded Jo, thinking he must mean his grandfather.

"That devilish Professor you were always writing about. If you say you love him, I know I shall do something desperate;" and he looked as if he would keep his word, as he clenched his hands, with a wrathful sparkle in his eyes.

Jo wanted to laugh, but restrained herself, and said warmly, for she, too, was getting excited with all this, —

"Don't swear, Teddy! He isn't old, nor anything bad, but good and kind, and the best friend I've got, next to you. Pray don't fly into a passion; I want to be kind, but I know I shall get angry if you abuse my Professor. I haven't the least idea of loving him or anybody else."

"But you will, after a while, and then what will become of me?"

"You'll love some one else, too, like a sensible boy, and forget all this trouble."

"I *can't* love any one else; and I'll never forget you, Jo, never! never!" with a stamp to emphasize his passionate words.

"What *shall* I do with him?" sighed Jo, finding that emotions were more unmanageable than she expected. "You haven't heard what I wanted to tell you. Sit down and listen; for indeed I want to do right and make you happy," she said, hoping to soothe him with a little reason, which proved that she knew nothing about love.

Seeing a ray of hope in that last speech, Laurie threw himself down on the grass at her feet, leaned his arm on the lower step of the stile, and looked up at her with an expectant face. Now that arrangement was not conducive to calm speech or clear thought on Jo's part; for how *could* she say hard things to her boy while he watched her with eyes full of love and longing, and lashes still wet with the bitter drop or two her hardness of heart had wrung from him? She gently turned his head away, saying, as she stroked the wavy hair which had been allowed to grow for her sake, — how touching that was to be sure! —

"I agree with mother that you and I are not suited to each other, because our quick tempers and strong wills would probably make us very miserable, if we were so foolish as to" — Jo paused a little over the last word, but Laurie uttered it with a rapturous expression, —

"Marry, — no, we shouldn't! If you loved me, Jo, I should be a perfect saint, for you could make me anything you like."

"No, I can't. I've tried it and failed, and I won't risk our happiness by such a serious experiment. We don't agree, and never shall; so we'll be good friends all our lives, but we won't go and do anything rash."

"Yes, we will if we get the chance," muttered Laurie, rebelliously.

"Now do be reasonable, and take a sensible view of the case," implored Jo, almost at her wit's end.

"I won't be reasonable; I don't want to take what you call 'a sensible view;' it won't help me, and it only makes you harder. I don't believe you've got any heart."

"I wish I hadn't."

There was a little quiver in Jo's voice, and thinking it a good omen, Laurie turned round, bringing all his persuasive powers to bear, as he said, in the wheedlesome tone which had never been so dangerously wheedlesome before, —

"Don't disappoint us, dear! Every one expects it. Grandpa has set his heart upon it, your people like it, and I can't get on without you. Say you will, and let's be happy. Do, do!"

Not until months afterward did Jo understand how she had the strength of mind to hold fast to the resolution she had made when she decided that she did not love her boy, and never could. It was very hard to do, but she did it, knowing that delay was both useless and cruel.

"I can't say 'Yes' truly, so I won't say it at all. You'll see that I'm right, by and by, and thank me for it," — she began solemnly.

"I'll be hanged if I do!" and Laurie bounced up off the grass, burning with indignation at the bare idea.

"Yes, you will!" persisted Jo; "you'll get over this after awhile, and find some lovely, accomplished girl who will adore you, and make a fine mistress for your fine house. I shouldn't; I'm homely and awkward and odd and old, and you'd be ashamed of me, and we should quarrel, — we can't help it even now, you see, — and I shouldn't like elegant society, and you would, and you'd hate my scribbling, and I couldn't get on without it, and we should be unhappy, and wish we hadn't done it, and everything would be horrid!"

"Anything more?" asked Laurie, finding it hard to listen patiently to this prophetic burst.

"Nothing more, except that I don't believe I shall ever marry. I'm happy as I am, and love my liberty too well to be in a hurry to give it up for any mortal man."

"I know better!" broke in Laurie. "You think so now; but there'll come a time when you *will* care for somebody, and you'll love him tremendously, and live and die for him. I know you will, it's your way, and I shall have to stand by and see it;" and the despairing lover cast his hat upon the ground with a gesture that would have seemed comical, if his face had not been so tragical.

"Yes, I *will* live and die for him, if he ever comes and makes me love him in spite of myself, and you must do the best you can!" cried Jo, losing patience with poor Teddy. "I've done my best, but you *won't* be reasonable, and it's selfish of you to keep teasing for what I can't give. I shall always be fond of you, very fond indeed, as a friend, but I'll never marry you; and the sooner you believe it the better for both of us, — so now!"

That speech was like fire to gunpowder. Laurie looked at her a minute as if he did not quite know what to do with himself, then turned sharply away, saying, in a desperate sort of tone, —

"You'll be sorry some day, Jo."

"Oh, where are you going?" she cried, for his face frightened her.

"To the devil!" was the consoling answer.

For a minute Jo's heart stood still, as he swung himself down the bank toward the river; but it takes much folly, sin, or misery to send a young man to a violent death, and Laurie was not one of the weak sort who are conquered by a single failure.

* * * * *

THEY had been floating about all the morning from gloomy St. Gingolf to sunny Montreux, with the Alps of Savoy on one side, Mont St. Bernard and the Dent du Midi on the other, pretty Vevay in the valley, and Lausanne on the hill beyond, a cloudless blue sky overhead, and the bluer lake below, dotted with the picturesque boats that look like white-winged gulls.

They had been talking of Bonnivard, as they glided past Chillon, and of Rosseau, as they looked up at Clarens, where he wrote his "Héloise." Neither had read it, but they knew it was a love-story, and each privately wondered if it was half as interesting as their own. Amy had been dabbling her hand in the water during the little pause that fell between them, and when she looked up, Laurie was leaning on his oars, with an expression in his eyes that made her say hastily, merely for the sake of saying something, —

"You must be tired; rest a little, and let me row; it will do me good; for, since you came, I have been altogether lazy and luxurious."

"I'm not tired; but you may take an oar, if you like. There's room enough, though I have to sit nearly in the middle, else the boat won't trim," returned Laurie, as if he rather liked the arrangement.

Feeling that she had not mended matters much, Amy took the proffered third of a seat, shook her hair over her face, and accepted an oar. She rowed as well as she did many other things; and, though she used both hands, and Laurie but one, the oars kept time, and the boat went smoothly through the water.

"How well we pull together, don't we?" said Amy, who objected to silence just then.

"So well, that I wish we might always pull in the same boat. Will you, Amy?" very tenderly.

"Yes, Laurie," very low.

Then they both stopped rowing, and unconsciously added a pretty little tableau of human love and happiness to the dissolving views reflected in the lake.

LOUISA M. ALCOTT.

THE WONDERFUL TAR-BABY STORY.

"DIDN'T the fox *never* catch the rabbit, Uncle Remus?" asked the little boy the next evening.

"He come mighty nigh it, honey, sho's you bawn — Brer Fox did. One day atter Brer Rabbit fool 'im wid dat calamus root, Brer Fox went ter wuk en got 'im some tar, en mix it wid some turkentime, en fix up a contrapshun wat he call a Tar-Baby, en he tuck dish yer Tar-Baby en he sot 'er in de big road, en den he lay off in de bushes fer ter see wat de news wuz gwineter be. En he didn't hatter wait long, nudder, kaze bimeby here come Brer Rabbit pacin' down de road — lippity-clippity, clippity-lippity — dez ez sassy ez a jay bird. Brer Fox, he lay low. Brer Rabbit come prancin' long twel he spy de Tar-Baby, en den he fotch up on his behime legs like he was 'stonished. De Tar-Baby, she sot dar, she did, en Brer Fox, he lay low.

"'Mawnin'!' sez Brer Rabbit, sezee — 'nice wedder dis mawnin',' sezee.

"Tar-Baby ain't sayin' nuthin', en Brer Fox, he lay low.

"'How duz yo' sym'tums seem ter segashuate?' sez Brer Rabbit, sezee.

"Brer Fox, he wink his eye slow, en lay low, en de Tar-Baby, she ain't sayin' nuthin'.

"'How you come on, den? Is you deaf?' sez Brer Rabbit, sezee. 'Kaze if you is, I kin holler louder,' sezee.

"Tar-Baby stay still, en Brer Fox, he lay low.

"'Youer stuck up, dat's w'at you is,' says Brer Rabbit, sezee, 'en I'm gwineter kyore you, dat's w'at I'm a gwineter do,' sezee.

"Brer Fox, he sorter chuckle in his stummuck, he did, but Tar-Baby ain't sayin' nuthin'.

"'I'm gwineter larn you howter talk ter 'specttubble fokes ef hits de las' ack,' sez Brer Rabbit, sezee. 'Ef you don't take off dat hat en tell me howdy, I'm gwinter bus' you wide open,' sezee.

"Tar-Baby stay still, en Brer Fox, he lay low.

"Brer Rabbit keep on axin' 'im, en de Tar-Baby, she keep on sayin' nuthin', twel present'y Brer Rabbit draw back wid his fis', he did, en blip he tuck er side er de head. Right dar's whar he broke his merlasses jug. His fis' stuck, en he can't pull loose. De tar hilt him. But Tar-Baby, she stay still, en Brer Fox, he lay low.

"'Ef you don't lemme loose, I'll knock you agin,' sez Brer Rabbit, sezee, en wid dat he fotch 'er a wipe wid de udder han', en dat stuck. Tar-Baby, she ain't sayin' nuthin', en Brer Fox, he lay low.

"'Tu'n me loose, fo' I kick de natal stuffin' outen you,' sez Brer Rabbit, sezee, but de Tar-Baby, she ain't sayin' nuthin'. She des hilt on, en den Brer Rabbit lose de use er his feet in de same way. Brer Fox, he lay low. Den Brer Rabbit squall out dat ef de Tar-Baby don't tu'n 'im loose he butt 'er cranksided. En den he butted, en his head got stuck. Den Brer Fox, he sa'ntered fort', lookin' des ez innercent ez wunner yo' mammy's mockin'-birds.

"'Howdy, Brer Rabbit,' sez Brer Fox, sezee. 'You look sorter stuck up dis mawnin',' sezee, en den he rolled on de groun', en laft en laft twel he couldn't laff no mo'. 'I speck you'll take dinner wid me dis time, Brer Rabbit. I done laid in some calamus root, en I ain't gwineter take no skuse,' sez Brer Fox, sezee."

Here Uncle Remus paused, and drew a two-pound yam out of the ashes.

"Did the fox eat the rabbit?" asked the little boy to whom the story had been told.

"Dat's all de fur de tale goes," replied the old man. "He mout, en den agin he moutent. Some say Jedge B'ar come 'long en loosed 'im — some say he didn't. I hear Miss Sally callin'. You better run 'long."

JOEL CHANDLER HARRIS.

HOW MR. RABBIT WAS TOO SHARP FOR MR. FOX.

"UNCLE REMUS," said the little boy one evening, when he had found the old man with little or nothing to do, "did the fox kill and eat the rabbit when he caught him with the Tar-Baby?"

"Law, honey, ain't I tell you 'bout dat?" replied the old darky, chuckling slyly. "I 'clar ter grashus I ought er tole you dat, but old man Nod wuz ridin' on my eyelids 'twel a leetle mo'n I'd a dis'member'd my own name, en den on to dat here come yo' mammy hollerin' atter you.

"W'at I tell you w'en I fus' begin? I tole you Brer Rabbit wuz a monstus soon beas'; leas'ways dat's w'at I laid out fer ter tell you. Well, den, honey, don't you go en make no udder kalkalashuns, kaze in dem days Brer Rabbit en his family wuz at de head er de gang w'en enny racket wuz on han', en dar dey stayed. 'Fo' you begins fer ter wipe yo' eyes 'bout Brer Rabbit, you wait en see whar'bouts Brer Rabbit gwineter fetch up at. But dat's needer yer ner dar.

"W'en Brer Fox fine Brer Rabbit mixt up wid de Tar-Baby, he feel mighty good, en he roll on de groun' en laff. Bimeby he up'n say, sezee:

"'Well, I speck I got you dis time, Brer Rabbit,' sezee; 'maybe I ain't, but I speck I is. You been runnin' roun' here sassin' atter me a mighty long time, but I spec' you done come ter de een' er de row. You bin cuttin' up yo' capers en bouncin' roun' in dis naberhood ontwel you come ter b'leeve yo'se'f de boss er de whole gang. En den youer allers some'rs whar you got no bizness,' sez Brer Fox, sezee. 'Who ax you fer ter come en strike up a 'quaintence wid dish yer Tar-Baby? En who stuck you up dar whar you iz? Nobody in de roun' worril. You des tuck en jam yo'se'f on dat Tar-Baby widout waitin' fer enny invite,' sez Brer

Fox, sezee, 'en dar you is, en dar you'll stay twel I fixes up a bresh-pile and fires her up, kaze I'm gwineter bobbycue you dis day, sho,' sez Brer Fox, sezee.

"Den Brer Rabbit talk mighty 'umble.

"'I don't keer w'at you do wid me, Brer Fox,' sezee, 'so you don't fling me in dat brier-patch. Roas' me, Brer Fox,' sezee, 'but don't fling me in that brier-patch,' sezee.

"'Hit's so much trouble fer ter kindle a fier,' sez Brer Fox, sezee, 'dat I speck I'll hatter hang you,' sezee.

"'Hang me des ez high as you please, Brer Fox,' sez Brer Rabbit, sezee, ' but do fer de Lord's sake don't fling me in dat brier-patch,' sezee.

"'I ain't got no string,' sez Brer Fox, sezee, 'en now I speck I'll hatter drown you,' sezee.

"'Drown me des ez deep ez you please, Brer Fox,' sez Brer Rabbit, sezee, 'but do don't fling me in dat brier-patch,' sezee.

"'Dey ain't no water nigh,' sez Brer Fox, sezee, 'en now I speck I'll hatter skin you,' sezee.

"'Skin me, Brer Fox,' sez Brer Rabbit, sezee, 'snatch out my eyeballs, t'ar out my years by de roots, en cut off my legs,' sezee, 'but do please, Brer Fox, don't fling me in dat brier-patch,' sezee.

"Co'se Brer Fox wanter hurt Brer Rabbit bad ez he kin, so he kotch 'im by de behine legs en slung 'im right in the middle er de brier-patch. Dar wuz a considerbul flutter whar Brer Rabbit struck de bushes, en Brer Fox sorter hang 'roun' fer ter see w'at wuz gwineter happen. Bimeby he hear somebody call 'im, en way up de hill he see Brer Rabbit settin' cross-legged on a chinkapin log koamin' de pitch outen his har wid a chip. Den Brer Fox know dat he bin swop off mighty bad. Brer Rabbit was bleedzed fer ter fling back some er his sass, en he holler out:

"'Bred en bawn in a brier-patch, Brer Fox — bred en bawn in a brier-patch!' en wid dat he skip out des ez lively ez a cricket in de embers."

JOEL CHANDLER HARRIS.

TORTURE.

I now lay upon my back, and at full length, on a species of low framework of wood. To this I was securely bound by a long strap resembling a surcingle. It passed in many convolutions about my limbs and body, leaving at liberty only my head, and my left arm to such an extent that I could by dint of much exertion supply myself with food from an earthen dish which lay by my side on the floor. I saw to my horror that the pitcher had been removed. I say to my horror, for I was consumed with intolerable thirst. This thirst it appeared to be the design of my persecutors to stimulate, for the food in the dish was meat pungently seasoned.

Looking upward I surveyed the ceiling of my prison. It was some thirty or forty feet overhead, and constructed much as the side walls. In one of its panels a very singular figure riveted my whole attention. It was the painted figure of Time as he is commonly represented, save that in lieu of a scythe he held what at a casual glance I supposed to be the pictured image of a huge pendulum, such as we see on antique clocks. There was something, however, in the appearance of this machine which caused me to regard it more attentively. While I gazed directly upward at it (for its position was immediately over my own), I fancied that I saw it in motion. In an instant afterward the fancy was confirmed. Its sweep was brief, and of course slow. I watched it for some minutes, somewhat in fear but more in wonder. Wearied at length with observing its dull movement, I turned my eyes upon the other objects in the cell.

A slight noise attracted my notice, and looking to the floor, I saw several enormous rats traversing it. They had issued from the well which lay just within view to my right. Even then while I gazed, they came up in troops, hurriedly, with ravenous eyes, allured by the scent of the meat. From this it required much effort and attention to scare them away.

It might have been half an hour, perhaps even an hour (for I could take but imperfect note of time), before I again cast my eyes upward. What I then saw confounded and amazed me. The sweep of the pendulum had increased in extent by nearly a yard. As a natural consequence, its velocity was much greater. But what mainly disturbed me was the idea that it had perceptibly descended. I now observed, with what horror it is needless to say, that its nether extremity was formed of a crescent of glittering steel, about a foot in length from horn to horn; the horns upward, and the under edge evidently as keen as that of a razor. Like a razor also it seemed massy and heavy, tapering from the edge into a solid and broad structure above. It was appended to a weighty rod of brass, and the whole hissed as it swung through the air.

I could no longer doubt the doom prepared for me by monkish ingenuity in torture. My cognizance of the pit had become known to the inquisitorial agents — the pit, whose horrors had been destined for so bold a recusant as myself, the pit, typical of hell, and regarded by rumor as the Ultima Thule of all their punishments.

The plunge into this pit I had avoided by the merest of accidents, and I knew that surprise or entrapment into torment formed an important portion of all the grotesquerie of these dungeon deaths. Having failed to fall, it was no part of the demon plan to hurl me into the abyss, and thus (there being no alternative) a different and a milder destruction awaited me. Milder! I half smiled in my agony as I thought of such application of such a term.

What boots it to tell of the long, long hours of horror more than mortal, during which I counted the rushing oscillations of the steel! Inch by inch — line by line — with a descent only appreciable as intervals that seemed ages — down and still down it came! Days passed — it might have been that many days passed — ere it swept so closely over me as to fan me with its acrid breath. The odor of the sharp steel forced itself into my nostrils. I prayed — I wearied heaven with my prayer for its more speedy descent. I grew frantically mad, and struggled to force myself upward against the sweep of the fearful scimitar. And then I fell suddenly calm, and lay smiling at the glittering death as a child at some rare bauble.

There was another interval of utter insensibility; it was brief, for upon again lapsing into life there had been no perceptible descent in the pendulum. But it might have been long — for I knew there were demons who took note of my swoon, and who could have arrested the vibration at pleasure. Upon my recovery, too, I felt very — oh! inexpressibly — sick and weak, as if through long inanition. Even amid the agonies of that period the human nature craved food. With painful effort I outstretched my left arm as far as my bonds permitted, and took possession of the small remnant which had been spared me by the rats. As I put a portion of it within my lips there rushed to my mind a half-formed thought of joy — of hope. Yet what business had I with hope? It was, as I say, a half-formed thought — man has many such, which are never completed. I felt that it was of joy — of hope; but I felt also that it had perished in its formation. In vain I struggled to perfect — to regain it. Long suffering had nearly annihilated all my ordinary powers of mind. I was an imbecile — an idiot.

The vibration of the pendulum was at right angles to my length. I saw that the crescent was designed to cross the region of the heart. It would fray the serge of my robe; it would return and repeat its operations — again — and again. Notwithstanding its terrifically wide sweep (some thirty feet or more) and the hissing vigor of its descent, sufficient to sunder these very walls of iron, still the fraying of my robe would be all that for several minutes it would accomplish; and at this thought I paused. I dared not go farther than this reflection. I dwelt upon it with a pertinacity of attention — as if, in so dwelling, I could arrest here the descent of the steel.

I forced myself to ponder upon the sound of the crescent as it should pass across the garment — upon the peculiar thrilling sensation which the friction of cloth produces on the nerves. I pondered upon all this frivolity until my teeth were on edge.

Down — steadily down it crept. I took a frenzied pleasure in contrasting its downward with its lateral velocity. To the right — to the left — far and wide — with the shriek of a damned spirit! to my heart with the stealthy pace of a tiger! I alternately laughed and howled, as the one or the other idea grew predominant.

Down — certainly, relentlessly down! It vibrated within three inches of my bosom! I struggled violently — furiously — to free my left arm. This was free only from the elbow to the hand. I could reach the latter from the platter beside me to my mouth with great effort, but no farther. Could I have broken the fastenings above the elbow, I would have seized and attempted to arrest the pendulum. I might as well have attempted to arrest an avalanche!

Down — still unceasingly — still inevitably down! I gasped and struggled at each vibration. I shrunk convulsively at its every sweep. My eyes followed its outward or upward whirls with the eagerness of the most unmeaning despair; they closed themselves spasmodically at the descent, although death would have been a relief, O, how unspeakable! Still I quivered in every nerve to think how slight a sinking of the machinery would precipitate that keen glistening axe upon my bosom. It was hope that prompted the nerve to quiver — the frame to shrink. It was hope — the hope that triumphs on the rack — that whispers to the death-condemned even in the dungeons of the Inquisition.

I saw that some ten or twelve vibrations would bring the steel in actual contact with my robe, and with this observation there suddenly came over my spirit all the keen collected calmness of despair. For the first time during many hours, or perhaps days, I thought. It now occurred to me that the bandage or surcingle which enveloped me was unique. I was tied by no separate cord. The first stroke of the razor-like crescent athwart any portion of the band would so detach it that it might be unwound from my person by means of my left hand. But how fearful in that case the proximity of the steel! The result of the slightest struggle how deadly! Was it likely, moreover, that the minions of the torturer had not foreseen and provided for this possibility? Was it probable that the bandage crossed my bosom in the track of the pendulum? Dreading to find my faint, and, as it seemed, my last hope frustrated, I so far elevated my head as to obtain a distinct view of my breast. The surcingle enveloped my limbs and body close in all directions save in the path of the destroying crescent.

Scarcely had I dropped my head back into its original position when there flashed upon my mind what I cannot better describe than as the unformed half of that idea of deliverance to which I have previously alluded, and of which a moiety only floated indeterminately through my brain when I raised my food to my burning lips. The whole thought was now present — feeble, scarcely sane, scarcely definite, but still entire. I proceeded at once, with the nervous energy of despair, to attempt its execution.

For many hours the immediate vicinity of the low framework upon which I lay had been literally swarming with rats. They were wild, bold, ravenous, their red eyes glaring upon me as if they waited but for motionlessness on my part to make me their prey. "To what food," I thought, "have they been accustomed in the well?"

They had devoured, in spite of all my efforts to prevent them, all but a small remnant of the contents of the dish. I had fallen into an habitual see-saw or wave of the hand about the platter; and at length the unconscious uniformity of the movement deprived it of effect. In their voracity the vermin frequently fastened their sharp fangs in my fingers. With the particles of the oily and spicy viand which now remained, I thoroughly rubbed the bandage wherever I could reach it; then, raising my hand from the floor, I lay breathlessly still.

At first the ravenous animals were startled and terrified at the change — at the cessation of movement. They shrank alarmedly back; many sought the well. But this was only for a moment. I had not counted in vain upon their voracity. Observing that I remained without motion, one or two of the boldest leaped upon the framework and smelt at the surcingle. This seemed a signal for a general rush. Forth from the well they hurried in fresh troops. They clung to the wood, they overran it, and leaped in hundreds upon my person. The measured movement of the pendulum disturbed them not at all. Avoiding its strokes, they busied themselves with the anointed bandage. They pressed, they swarmed upon me in ever accumulating heaps. They writhed upon my throat; their cold lips sought my own; I was half stifled by their thronging pressure; disgust, for which the world has no name, swelled my bosom, and chilled with heavy clamminess my heart. Yet one minute and I felt that the struggle would be over. Plainly I perceived the loosening of the bandage. I knew that in more than one place it must be already severed. With more than human resolution I lay still.

Nor had I erred in my calculations, nor had I endured in vain. I at length felt that I was free. The surcingle hung in ribands from my body. But the stroke of the pendulum already pressed upon my bosom. It had divided the serge of the robe. It had cut through the linen beneath. Twice again it swung and a sharp sense of pain shot through every nerve. But the moment of escape had arrived. At a wave of my hand my deliverers hurried tumultuously away. With a steady movement, cautious, sidelong, shrinking, and slow, I slid from the embrace of the bandage and beyond the reach of the scimitar. For the moment, at least, I was free.

EDGAR ALLAN POE.

TO THOMAS MURRAY.

O, TOM, what a foolish flattering creature thou art! To talk of future eminence in connection with the literary history of the nineteenth century to such a one as me! Alas! my good lad, when I and all my fancies and reveries and speculations shall have been swept over with the besom of oblivion, the literary history of no century will feel itself the worse. Yet think not, because I talk thus, I am careless of literary fame. No; Heaven knows that ever since I have been able to form a wish, the wish of being known has been the foremost.

O Fortune! thou that givest unto each his portion in this dirty planet, bestow (if it shall please thee) coronets, and crowns, and principalities, and purses, and pudding, and powers upon the great and noble and fat ones of the earth. Grant me that, with a heart of independence unyielding to thy favors and unbending to thy frowns, I may attain to literary fame; and though starvation be my lot, I will smile that I have not been born a king.

But alas! my dear Murray, what am I, or what are you, or what is any other poor unfriended stripling in the ranks of learning? . .

THOMAS CARLYLE.

TO HIS MOTHER.

DEAN STANLEY.

I WILL begin my letter in the midst of my agony of expectation and fear. I finished my examination to-day at two o'clock. At eight to-night the decision takes place, so that my next three quarters of an hour will be dreadful. As I do not know how the other schools have done, my hope of success can depend upon nothing, except that I think I have done pretty well, better, perhaps, from comparing notes, than the rest of the Rugby men. O, the joy if I do get it! and the disappointment if I do not. And from two of us trying at once, I fear the blow to the school would be dreadful if none of us get it. We had to work the second day as hard as on the first, on the third and fourth not so hard, nor to-day — Horace to turn into English verse, which was good for me; a divinity and a mathematical paper, in which I hope my copiousness in the first made up for my scantiness in the second. Last night I dined at Magdalen, which is enough of itself to turn one's head upside down, so very magnificent. . . . I will go on now. We all assembled in the hall, and had to wait an hour, the room getting fuller and fuller with Rugby Oxonians crowding in to hear the result. Every time the door opened my heart jumped, but many times it was nothing. At last the dean appeared in his white robes, and moved up to the head of the table. He began a long preamble — that they were well satisfied with all, and that those who were disappointed were many in comparison with those who were successful, etc. All this time every one was listening with the most intense eagerness, and I almost bit my lips off till — The successful candidates are — Mr. Stanley — I gave a great jump, and there was a half shout amongst the Rugby men. The next was Lonsdale from Eton. The dean then took me into the chapel where the Master and all the Fellows were, and there I swore that I would not reveal the secrets, disobey the statutes, or dissipate the wealth of the college. I was then made to kneel on the steps, and admitted to the rank of scholar and Exhibitioner of Balliol College, *nomine, Patris, Fillii, et Spiritus.* I then wrote my name, and it was finished. We start to-day in a chaise and four for the glory of it. You may think of my joy, the honor of Rugby is saved, and I am a Scholar of Balliol!

ARTHUR PENRHYN STANLEY.

NIL NISI BONUM.

IN America the love and regard for Irving was a national sentiment. Party wars are perpetually raging there, and are carried on by the press with a rancor and fierceness against individuals which exceed British, almost Irish, virulence. It seemed to me, during a year's travel in the country, as if no one ever aimed a blow at Irving. All men held their hand from that harmless, friendly peacemaker. I had the good fortune to see him at New York, Philadelphia, Baltimore, and Washington, and remarked how in every place he was honored and welcome. Every large city has its "Irving House." The country takes pride in the fame of its men of letters. The gate of his own charming little domain on the beautiful Hudson River was forever swinging before visitors who came to him. He shut out no one. I had seen many pictures of his house, and read descriptions of it, in both of which it was treated with a not unusual American exaggeration. It was but a pretty little cabin of a place; the gentleman of the press who took notes of the place, whilst his kind old host was sleeping, might have visited the whole house in a couple of minutes. And how came it that this house was so small, when Mr. Irving's books were sold by hundreds of thousands, nay, millions; when his profits were known to be large, and the habits of life of the good old bachelor were notoriously modest and simple? He had loved once in life. The lady he loved died; and he, whom all the world loved, never sought to replace her. I can't say how much the thought of that fidelity has touched me. Does not the very cheerfulness of his after-life add to the pathos of that untold story? To grieve always was not in his nature; or, when he had his sorrow, to bring all the world in to condole with him and bemoan it. Deep and quiet he lays the love of his heart, and buries it; and grass and flowers grow over the scarred ground in due time.

Irving had such a small house and such narrow rooms, because there was a great number of people to occupy them. He could only afford to keep one old horse (wnich, lazy and aged as it was, managed once or twice to run away with that careless old horseman). He could only afford to give plain sherry to that amiable British paragraph-monger from New York, who saw the patriarch asleep over his modest, blameless cup, and fetched the public into his private chamber to look at him. Irving could only live very modestly, because the wifeless, childless man had a number of children to whom he was as a father. He had as many as nine nieces, I am told — I saw two of these ladies at his house — with all of whom the dear old man had shared the produce of his labor and genius.

WILLIAM MAKEPEACE THACKERAY.

LORNA DOONE.

But when the weather changed in earnest, and the frost was gone, and the southwest wind blew softly, and the lambs were at play with the daisies, it was more than I could do to keep from thought of Lorna. For now the fields were spread with growth, and the waters clad with sunshine, and light and shadow, step by step, wandered over the furzy cleves. All the sides of the hilly wood were gathered in and out with green, silver-gray, or russet points, according to the several manner of the trees beginning. And if one stood beneath an elm, with any heart to look at it, lo! all the ground was strewn with flakes (too small to know their meaning), and all the sprays above were rasped and trembling with a redness. And so I stopped beneath the tree, and carved L. D. upon it, and wondered at the buds of thought that seemed to swell inside me.

The upshot of it all was this, that as no Lorna came to me, except in dreams or fancy, and as my life was not worth living without constant sign of her, forth I must again to find her, and say more than a man can tell. Therefore, without waiting longer for the moving of the spring, dressed I was in grand attire (so far as I had gotten it), and thinking my appearance good, although with doubts about it (being forced to dress in the hay-tallat), round the corner of the wood-stack went I very knowingly — for Lizzie's eyes were wondrous sharp — and then I was sure of meeting none who would care or dare to speak of me.

It lay upon my conscience often that I had not made dear Annie secret to this history; although in all things I could trust her, and she loved me like a lamb. Many and many a time I tried, and more than once began the thing; but there came a dryness in my throat, and a knocking under the roof of my mouth, and a longing to put it off again, as perhaps might be the wisest. And then I would remember too that I had no right to speak of Lorna as if she were common property.

This time I longed to take my gun, and was half resolved to do so; because it seemed so hard a thing to be shot at, and have no chance of shooting; but when I came to remember the steepness and the slippery nature of the water-side, there seemed but little likelihood of keeping dry the powder. Therefore I was armed with nothing but a good stout holly staff, seasoned well for many a winter in our back kitchen chimney.

Although my heart was leaping high with the prospect of some adventure, and the fear of meeting Lorna, I could not but be gladdened by the softness of the weather, and the welcome way of everything. There was that power all round, that power and that goodness, which make us come, as it were, outside our bodily selves to share them. Over and beside us breathes the joy of hope and promise; under foot are troubles past; in the distance bowering newness tempts us ever forward. We quicken with largesse of life, and spring with vivid mystery.

And in good sooth, I had to spring, and no mystery about it, ere ever I got to the top of the rift leading into Doone Glade. For the stream was rushing down in strength, and raving at every corner; a mort of rain having fallen last night, and no wind come to wipe it. However, I reached the head ere dark with more difficulty than danger, and sat in a place which comforted my back and legs desirably.

Thereupon I grew so happy at being on dry land again, and come to look for Lorna, with pretty trees around me, that what did I do but fall asleep with the holly-stick in front of me and my best coat sunk in a bed of moss, with wood and water-sorrel. Mayhap I had not done so, nor yet enjoyed the spring so much, if so be I had not taken three parts of a gallon of cider at home, at Plover's Barrows, because of the lowness and sinking ever since I met Mother Melldrum.

There was a little runnel going softly down beside me, falling from the upper rock by the means of moss and grass, as if it feared to make a noise, and had a mother sleeping. Now and then it seemed to stop, in fear of its own dropping, and waiting for some orders; and the blades of grass that straightened to it turned their points a little way, and offered their allegiance to wind instead of water. Yet before their carkled edges bent more than a driven saw, down the water came again, with heavy drops and pats of running, and bright anger at neglect.

This was very pleasant to me, now and then, to gaze at, blinking as the water blinked, and falling back to sleep again. Suddenly my sleep was broken by a shade cast over me; between me and the low sunlight Lorna Doone was standing.

"Master Ridd, are you mad?" she said, and took my hand to move me.

"Not mad, but half asleep," I answered, feigning not to notice her, that so she might keep hold of me.

"Come away, come away, if you care for life. The patrol will be here directly. Be quick, Master Ridd, let me hide thee."

"I will not stir a step," said I, though being in the greatest fright that might be well imagined, "unless you call me 'John.'"

"Well, John, then — Master John Ridd, be quick, if you have any to care for you."

"I have many that care for me," I said, just to let her know; "and I will follow you, Mistress Lorna: albeit without any hurry, unless there be peril to more than me."

Without another word she led me, though with many timid glances toward the upper valley, to, and into, her little bower, where the inlet through the rock was. I am almost sure that I spoke before (though I cannot now go seek for it, and my memory is but a worn-out tub) of a certain deep and perilous pit, in which I was like to drown myself through hurry and fright of boyhood. And even then I wondered greatly, and was vexed with Lorna for sending me in that heedless manner into such an entrance.

But now it was clear that she had been right, and the fault mine own entirely; for the entrance to the pit was only to be found by seeking it. Inside the niche of native stone, the plainest thing of all to see, at any rate by daylight, was the stairway hewn from rock, and leading up the mountain, by means of which I had escaped, as before related. To the right side of this was the mouth of the pit, still looking very formidable; though Lorna laughed at my fear of it, for she drew her water thence. But on the left was a narrow crevice, very difficult to espy, and having a sweep of gray ivy laid, like a slouching beaver, over it. A man here coming from the brightness of the outer air, with eyes dazed by the twilight, would never think of seeing this and following it to its meaning.

Lorna raised the screen for me, but I had much ado to pass, on account of bulk and stature. Instead of being proud of my size (as it seemed to me she ought to be), Lorna laughed so quietly that I was ready to knock my head or elbows against anything, and say no more about it. However, I got through at last without a word of compliment, and broke into the pleasant room, the lone retreat of Lorna.

The chamber was of unhewn rock, round, as near might be, eighteen or twenty feet across, and gay with rich variety of fern and moss and lichen. The fern was in its winter still, or coiling for the spring-tide; but moss was in abundant life, some feathering, and some gobleted, and some with fringe of red to it. Overhead, there was no ceiling but the sky itself, flaked with little clouds of April whitely wandering over it. The floor was made of soft, low grass, mixed with moss and primroses; and in a niche of shelter moved the delicate wood-sorrel. Here and there, around the sides, were "chairs of living stone," as some Latin writer says, whose name has quite escaped me; and in the midst a tiny spring arose, with crystal beads in it, and a soft voice as of a laughing dream, and dimples like a sleeping babe. Then after going round a little, with surprise of daylight, the water overwelled the edge, and softly went through lines of light to shadows and an untold bourne.

While I was gazing at all these things with wonder and some sadness, Lorna turned upon me lightly (as her manner was) and said:

"Where are the new-laid eggs, Master Ridd? Or hath blue hen ceased laying?"

I did not altogether like the way in which she said it, with a sort of a dialect, as if my speech could be laughed at.

"Here be some," I answered, speaking as if in spite of her. "I would have brought thee twice as many, but that I feared to crush them in the narrow ways, Mistress Lorna."

And so I laid her out two dozen upon the moss of the rock ledge, unwinding the wisp of hay from each as it came safe out of my pocket.

Lorna looked with growing wonder, as I added one to one; and when I had placed them side by side, and bidden her now to tell them, to my amazement what did she do but burst into a flood of tears!

"What have I done?" I asked, with shame, scarce daring even to look at her, because her grief was not like Annie's — a thing that could be coaxed away, and left a joy in going — "oh! what have I done to vex you so?"

"It is nothing done by you, Master Ridd," she answered, very proudly, as if naught I did could matter; "it is only something that comes upon me with the scent of the pure, true clover hay. Moreover, you have been too kind; and I am not used to kindness."

Some sort of awkwardness was on me, at her words and weeping, as if I would like to say something, but feared to make things worse, perhaps, than they were already. Therefore I abstained from speech, as I would in my own pain. And as it happened, this was the way to make her tell me more about it. Not that I was curious, beyond what pity urged me and the strange affairs around her; and now I gazed upon the floor, lest I should seem to watch her; but none the less for that I knew all that she was doing.

Lorna went a little way, as if she would not think of me, nor care for one so careless; and all my heart gave a sudden jump, to go like a mad thing after her; until she turned of her own accord, and with a little sigh came back to me. Her eyes were soft with trouble's shadow, and the proud lift of her neck was gone, and beauty's vanity borne down by woman's want of sustenance.

"Master Ridd," she said, in the softest voice that ever flowed between two lips, "have I done aught to offend you?"

Hereupon it went hard with me not to catch her up and kiss her, in the manner in which she was looking; only it smote me suddenly that this would be a low advantage of her trust and helplessness. She seemed to know what I would be at, and to doubt very greatly about it, whether, as a child of old, she might permit the usage. All sorts of things went through my head, as I made myself look away from her, for fear of being tempted beyond what I could bear. And the upshot of it was that I said within my heart and through it: "John Ridd, be on thy very best manners with this lonely maiden."

Lorna liked me all the better for my good forbearance, because she did not love me yet, and had not thought about it; at least so far as I knew. And though her eyes were so beauteous, so very soft and kindly, there was (to my apprehension) some great power in them, as if she would not have a thing, unless her judgment leaped with it.

But now her judgment leaped with me because I had behaved so well; and being of quick, urgent nature — such as I delight in from the change from mine own slowness — she without any let or hindrance, sitting over against me, now raising and now dropping fringe over those sweet eyes that were the road-lights of her tongue, Lorna told me all about everything I wished to know, every little thing she knew, except, indeed, that point of points, how Master Ridd stood with her.

R. D. Blackmore.

THE TYRANNY OF ANDROS.

The general sentiment of the early New England writers was like that of the "Wonder-working Providence," though it did not always find such rhapsodic expression. It has left its impress upon the minds of their children's children down to our own time, and has affected the opinions held about them by other people. It has had something to do with a certain tacit assumption of superiority on the part of New Englanders, upon which the men and women of other communities have been heard to comment in resentful and carping tones. There has probably never existed, in any age or at any spot on the earth's surface, a group of people that did not take for granted its own preëminent excellence. Upon some such assumption, as upon an incontrovertible axiom, all historical narratives, from the chronicles of a parish to the annals of an empire, alike proceed. But in New England it assumed a form especially apt to provoke challenge. One of its unintentional effects was the setting up of an unreal and impossible standard by which to judge the acts and motives of the Puritans of the seventeenth century. We come upon instances of harshness and cruelty, of narrow-minded bigotry, and superstitious frenzy; and feel, perhaps, a little surprised that these men had so much in common with their contemporaries. Hence, the interminable discussion which has been called forth by the history of the Puritans, in which the conclusions of the writer have generally been determined by circumstances of birth or creed, or perhaps of reaction against creed. One critic points to the Boston of 1659, or the Salem of 1692 with such gleeful satisfaction as used to stir the heart of Thomas Paine when he alighted upon an inconsistency in some text of the Bible; while another, in the firm conviction that Puritans could do no wrong, plays fast and loose with arguments that might be made to justify the deeds of a Torquemada.

John Fiske.

THE LONG PATH.

I FELT very weak indeed (though of a tolerably robust habit) as we came opposite the head of this path on that morning. I think I tried to speak twice without making myself distinctly audible. At last I got out the question, "Will you take the long path with me?"

"Certainly," said the school-mistress, "with much pleasure."

"Think," I said, "before you answer; if you take the long path with me now, I shall interpret it that we part no more!"

The school-mistress stepped back with a sudden movement, as if an arrow had struck her.

One of the long granite blocks used as seats was hard by — the one you may still see close by the Gingko-tree. "Pray sit down," I said.

"No, no," she answered softly, "I will walk the *long path* with you!"

O. W. HOLMES.

LUCY AND THE "RAJAH."

"THEN let me say, Lucy, to-day, for perhaps I shall never say that, or anything that is sweet to say, again. Lucy, you know what I came for?"

"Oh, yes! to receive my congratulations."

"More than that — a great deal. To ask you to go halves in the 'Rajah.'"

Lucy's eyebrows demanded an explanation.

"She is worth two thousand a year to her commander, and that is too much for a bachelor."

Lucy colored and smiled.

"Why, it is only just enough for most of them to live on."

"It is too much for me alone, under the circumstances," said David, gravely; and there was a little silence.

"Lucy, I love you. With you the 'Rajah' would be a godsend. She will help me keep you in the company you have been used to, and were made to brighten and adorn; but without you I cannot take her from your hand — and, to speak plain, I won't."

"Oh! Mr. Dodd!"

"No, Lucy, before I knew you, to command a ship was the height of my ambition, the quarter-deck my heaven on earth; and this is a clipper, I own it, I saw her in the docks. But you have taught me to look higher. Share my ship and my heart with me, and she will be all the dearer to me that she came to us from her I love. But don't say to me, 'Me you sha'n't have, you are not good enough for that, but there is a ship for you in my place!' I wouldn't accept a star out of the firmament on those terms."

"How unreasonable! On the contrary, you should say, 'I am doubly fortunate; I escape a weak, foolish companion for life, and I have a beautiful ship.' But friendship such as mine for you was never appreciated; I do you injustice; you only talk like that to tease me, and make me unhappy."

"Oh, Lucy, Lucy! did you ever know me" —

"There, now, forgive me! and own that you are not in earnest."

"This will show you," said David, sadly, and he took out two letters from his pocket. "Here are two letters to the Secretary. In one I accept the ship with thanks, and offer to superintend her when her rigging is being set up; and in this one I decline her altogether, with my humble and sincere thanks."

"Oh, yes, you are very humble, sir," said Lucy. "Now — dear friend — listen to reason. You have others " —

"Excuse my interrupting you, but it is a rule with me never to reason about right and wrong. I notice that whoever does that, ends by choosing wrong. I don't go to my head to find my duty, I go to my heart; and what little manhood there is in me all cries out against me compounding with the woman I love, and taking a ship instead of her." . . .

"See how power hurts people, and brings out their true character. Since you commanded the 'Rajah' you are all changed. You used to be submissive; now you must have your own way entirely; you will fling my poor ship in my face unless I give you — but this is really using force; yes, Mr. Dodd, this is using force. Somebody has told you that my sex yield when downright compulsion is used. It is true. And the more ungenerous to apply it." And she melted into a few placid tears.

David did not know this sign of yielding in a woman, and he groaned at the sight of them, and hung his head.

"Advise me what I had better do."

To this singular proposal, David, listening to the ill-advice of the fiend Generosity, groaned out, "Why should you be tormented and made cry?"

"Why, indeed?"

"Nothing can change me. I advise you to cut it short."

"I *will* cut this short, Mr. Dodd; give me that paper."

"Which?"

"The wicked one where you refuse my 'Rajah.'" . . .

She took it, and with both her supple white hands tore it with insulting precision exactly in half.

"There, sir; and there, sir" (exactly in four); "and there" (in eight, with malicious exactness); "and there;" and, though it seemed impossible to effect another separation, yet the taper fingers and a resolute will reduced it to tiny bits. She then made a gesture to throw them in the fire, but thought better of it and held them.

David looked on, almost amused at this zealous demolition of a thing he could so easily replace. He said, part sadly, part doggedly, part apologetically, "I can write another."

"But you will not. Oh, Mr. Dodd, don't you see?"

He looked up at her eagerly. To his surprise her haughty, eagle look had gone, and she seemed a pitying goddess, all tenderness and benignity; only her mantling, burning cheek showed her to be a woman.

She faltered, in answer to his wild, eager look, "Was I ever so rude before? What right have I to tear your letter, unless I " —

The characteristic full stop, and above all, the heaving bosom, the melting eye, and the red cheek were enough even for poor simple David. Heaven seemed to open on him. His burning kisses fell on the sweet hands that had torn his death-warrant. . . . David drew her closer and closer to him, till she hid her forehead and wet eyelashes on his shoulder and murmured —

"How could I let *you* be unhappy?"

Neither spoke for a while. Each felt the other's heart beat; and David drank that ecstasy of silent, delirious bliss which comes to great hearts once in a life.

CHARLES READE.

TWENTY-THREE!

THE supposed Evrémonde descends, and the seamstress is lifted out next after him. He has not relinquished her patient hand in getting out, but still holds it as he promised. He gently places her with her back to the crashing engine that constantly whirrs up and falls, and she looks into his face and thanks him.

"But for you, dear stranger, I should not be so composed, for I am naturally a poor little thing, faint of heart; nor should I have been able to raise my thoughts

GADSHILL. — THE HOME OF CHARLES DICKENS.

to Him who was put to death, that we might have hope and comfort here to-day. I think you were sent to me by Heaven."

"Or you to me," says Sydney Carton. "Keep your eyes upon me, dear child, and mind no other object."

"I mind nothing while I hold your hand. I shall mind nothing when I let it go, if they are rapid."

"They will be rapid. Fear not!"

The two stand in the fast-thinning throng of victims, but they speak as if they were alone. Eye to eye, voice to voice, hand to hand, heart to heart, these two children of the Universal Mother, else so wide apart and differing, have come together on the dark highway, to repair home together, and to rest in her bosom.

"Brave and generous friend, will you let me ask you one last question? I am very ignorant, and it troubles me — just a little."

"Tell me what it is."

"I have a cousin, an only relative and an orphan, like myself, whom I love very dearly. She is five years younger than I, and she lives in a farmer's house in the south country. Poverty parted us, and she knows nothing of my fate — for I cannot write — and if I could, how should I tell her! It is better as it is."

"Yes, yes; better as it is."

"What I have been thinking as we came along, and what I am still thinking now, as I look into your kind strong face which gives me so much support, is this: — If the Republic really does good to the poor, and they come to be less hungry, and in all ways to suffer less, she may live a long time; she may even live to be old."

"What then, my gentle sister?"

"Do you think:" the uncomplaining eyes in which there is so much endurance, fill with tears, and the lips part a little more and tremble: "that it will seem long to me, while I wait for her in the better land where I trust both you and I will be mercifully sheltered?"

"It cannot be, my child; there is no Time there, and no trouble there."

"You comfort me so much! I am so ignorant. Am I to kiss you now? Is the moment come?"

"Yes."

She kisses his lips; he kisses hers; they solemnly bless each other. The spare hand does not tremble as he releases it; nothing worse than a sweet, bright constancy is in the patient face. She goes next before him — is gone; the knitting women count Twenty-Two.

"I am the Resurrection and the Life, saith the Lord: he that believeth in me, though he were dead, yet shall he live: and whosoever liveth and believeth in me, shall never die."

The murmuring of many voices, the upturning of many faces, the pressing on of many footsteps in the outskirts of the crowd, so that it swells forward in a mass, like one great heave of water, all flashes away. Twenty-Three.

CHARLES DICKENS.

"DE BAPTIZIN' IN ELKHORN CREEK."

CONSIDERING THE NEXT TEXT.

"Hit wuz long time," he continued, "'fo' Phillis come to heah me preach any mo'. But 'long 'bout de nex' fall we had big meetin', en heap mo' 'um j'ined. But Phillis, she ain't nuver j'ined yit. I preached mighty nigh all roun' my coat-tails till I say to myse'f, D' ain't but one tex' lef', en I jes got to fetch 'er wid dat! De tex' wuz on de *right* tail o' my coat : 'Come unto me, all ye dat labor en is heavy laden.' Hit wuz a ve'y momentous sermon, en all 'long I jes see Phillis wras'lin' wid 'erse'f, en I say, 'She *got* to come *dis* night, de Lohd he'pin' me.' En I had n' mo' 'n said de word, 'fo' she jes walked down en guv me 'er han'. Den we had de baptizin' in Elkhorn Creek, en de watter wuz deep en de curren' tol'ble swif'. Hit look to me like dere wuz five hundred uv 'um on de creek side. By en by I stood on de edge o' de watter, en Phillis she come down to let me baptize 'er. En me en 'er j'ined han's en waded out in de creek, mighty slow, caze Phillis didn't have no shot roun' de bottom uv 'er dress, en it kep' floatin' on top de watter till I pushed it down. But by en by we got 'way out in de creek, en bof of us wuz tremblin'. En I says to 'er ve'y kindly, 'When I put you un'er de watter, Phillis, you mus' try en hole yo'se'f stiff, so I can lif' you up easy.' But I hadn't mo' 'n jes got 'er laid back over de watter ready to souze 'er un'er when 'er feet flew up off de bottom uv de creek, en when I retched out to fetch 'er up, I stepped in a hole ; en 'fo' I knowed it, we wuz flounderin' roun' in de watter, en de hymn dey wuz singin' on de bank sounded mighty confused-like. En Phillis she swallowed some watter, en all 't oncet she jes grab me right tight roun' de neck, en said mighty quick, says she, 'I gwine marry whoever gits me out 'n dis yere watter !'

"En by en by, when me en 'er wuz walkin' up de bank o' de creek, drippin' all over, I says to 'er, says I :

"'Does you 'member whut you said back yon'er in de watter, Phillis ?'

"'I ain't out'n no watter yit,' says she, ve'y contemptuous.

"'When does you consider yo'se'f out 'n de watter?' says I, ve'y humble.
"'When I get dese soakin' clo'es off 'n my back,' says she.
"Hit wuz good dark when we got home, en atter a while I crope up to de doah o' Phillis's cabin en put my eye down to de keyhole, en see Phillis jes settin' 'fo' dem blazin' walnut logs dressed up in 'er new red linsey dress, en 'er eyes shinin'. En I shuk so I mos' faint. Den I tap easy on de doah, en say in a mighty tremblin' tone, says I:
"'Is you out 'n de watter yit, Phillis?'
"'I got on dry dress,' says she.
"'Does you 'member what you said back yon'er in de watter, Phillis?' says I.
"'De latch-string on de outside de doah,' says she, mighty sof'.
"En I walked in."

JAMES LANE ALLEN.

SCOTCHMEN.

. . . Shall I tire you with a description of this unfruitful country, where I must lead you over their hills all brown with heath, or their valleys scarcely able to feed a rabbit? Man alone seems to be the only creature who has arrived to the natural size in this poor soil. Every part of the country presents the same dismal landscape. No grove nor brook lend their music to cheer the stranger, or make the inhabitants forget their poverty. Yet with all these disadvantages, enough to call him down to humility, a Scotchman is one of the proudest things alive. The poor have pride ever ready to relieve them. If mankind should happen to despise them they are masters of their own admiration, and that they can plentifully bestow upon themselves.

From their pride and poverty, as I take it, results one advantage this country enjoys: namely, the gentlemen here are much better bred than among us. No such character here as our fox-hunter; and they have expressed great surprise when I informed them that some men in Ireland of one thousand pounds a year spend their whole lives in running after a hare, drinking to be drunk, and . . . Truly, if such a being, equipped in his hunting dress, came among a circle of Scotch gentry, they would behold him with the same astonishment that a countryman does King George on horseback.

The men here have generally high cheek-bones, and are lean and swarthy, fond of action, dancing in particular. Now that I have mentioned dancing, let me say something of their balls, which are very frequent here. When a stranger enters the dancing-hall, he sees one end of the room taken up by the ladies, who sit dismally in a group by themselves; in the other end stand their pensive partners that are to be; but no more intercourse between the sexes than there is between two countries at war. The ladies indeed may ogle, and the gentlemen sigh ; but an embargo is laid on any closer commerce. At length, to interrupt hostilities, the lady directress, or intendant, or what you will, pitches upon a lady and gentleman to walk a minuet ; which they perform with a formality that approaches to despondence. After five or six couples have thus walked the gauntlet, all stand up to country dances, each gentleman furnished with a partner from the aforesaid lady directress ; so they dance much, say nothing, and thus concludes our assembly. I told a Scotch gentleman that such profound silence resembled the ancient procession of the Roman matrons in honor of Ceres ; and the Scotch gentleman told me (and, faith, I believe he was right) that I was a very great pedant for my pains. . . .

OLIVER GOLDSMITH.

ON ENGLAND'S FOREIGN POLICY.

I BELIEVE there is no permanent greatness to a nation except it be based upon morality. I do not care for military greatness or military renown. I care for the condition of the people among whom I live. There is no man in England who is less likely to speak irreverently of the crown and monarchy of England than I am ; but crowns, coronets, mitres, military display, the pomp of war, wide colonies, and a huge empire are, in my view, all trifles light as air, and not worth considering, unless with them you can have a fair share of comfort, contentment, and happiness among the great body of the people. Palaces, baronial castles, great halls, stately mansions, do not make a nation. The nation in every country dwells in the cottage ; and unless the light of your constitution can shine there, unless the beauty of your legislation and the excellence of your statesmanship are impressed there on the feelings and condition of the people, rely upon it you have yet to learn the duties of government.

I have not, as you have observed, pleaded that this country should remain without adequate and scientific means of defence. I acknowledge it to be the duty of your statesman, acting upon the known opinions and principles of ninety-nine out of every hundred persons in the country, at all times, with all possible moderation, but with all possible efficiency, to take steps which shall preserve order within and on the confines of your kingdom. But I shall repudiate and denounce the expenditure of every shilling, the engagement of every man, the employment of every ship, which has no object but intermeddling in the affairs of other countries, and endeavoring to extend the boundaries of an empire which is already large enough to satisfy the greatest ambition, and I fear is much too large for the highest statesmanship to which any man has yet attained.

The most ancient of profane historians has told us that the Scythians of his time were a very warlike people, and that they elevated an old cimeter upon a platform as a symbol of Mars, for to Mars alone, I believe, they built altars and offered sacrifices. To this cimeter they offered sacrifices of horses and cattle, the main wealth of the country, and more costly sacrifices than to all the rest of their gods. I often ask myself whether we are at all advanced in one respect beyond those Scythians. What are our contributions to charity, to education, to morality, to religion, to justice, and to civil government, when compared with the wealth we expend in sacrifices to the old cimeter? Two nights ago I addressed in this hall a vast assembly composed to a great extent of your countrymen who have no political power, who are at work from the dawn of the day to the evening, and who have therefore limited means of informing themselves on these great subjects. Now I am privileged to speak to a somewhat different audience. You represent those of your great community who have a more complete education, who have on some points greater intelligence, and in whose hands reside the power and influence of the district. I am speaking, too, within the hearing of those whose gentle nature, whose finer instincts, whose purer minds, have not suffered as some of us have suffered in the turmoil and strife of life. You can mould opinion, you can create political power; — you cannot think a good thought on this subject and communicate it to your neighbors, — you cannot make these points topics of discussion in your social circles and more general meetings, without affecting sensibly and speedily the course which the government of your country will pursue.

May I ask you, then, to believe, as I do most devoutly believe, that the moral law was not written for men alone in their individual character, but that it was written as well for nations, and for nations great as this of which we are citizens. If nations reject and deride that moral law, there is a penalty which will inevitably follow. It may not come at once, it may not come in our lifetime; but rely upon it, the great Italian is not a poet only, but a prophet, when he says:

"The sword of heaven is not in haste to smite,
Nor yet doth linger."

We have experience, we have beacons, we have landmarks enough. We know what the past has cost us, we know how much and how far we have wandered, but we are not left without a guide. It is true we have not, as an ancient people had, Urim and Thummim — those oraculous gems on Aaron's breast, — from which to take counsel, but we have the unchangeable and eternal principles of the moral law to guide us, and only so far as we walk by that guidance can we be permanently a great nation, or our people a happy people.

JOHN BRIGHT.

VIRTUE ALONE BEAUTIFUL.

JOHN G WHITTIER.

"HANDSOME is that handsome does, — hold up your heads, girls," is the language of Primrose in the play, when addressing her daughters. The worthy matron was right. Would that all my female readers, who are sorrowing foolishly because they are not in all respects like Dubufe's Eve, or that statue of Venus which enchants the world, could be persuaded to listen to her. What is good-looking, as Horace Smith remarks, but looking good? Be good, be womanly, be gentle, — generous in your sympathies, heedful of the well-being of those around you, and, my word for it, you will not lack kind words or admiration. Loving and pleasant associations will gather about you. Never mind the ugly reflection which your glass may give you. That mirror has no heart. But quite another picture is given you on the retina of human sympathy. There the beauty of holiness, of purity, of that inward grace "which passeth show," rests over it, softening and mellowing its features, just as the full, calm moonlight melts those of a rough landscape into harmonious loveliness.

"Hold up your heads, girls," I repeat after Primrose. Why should you not? Every mother's daughter of you can be beautiful. You can envelop yourselves in an atmosphere of moral and intellectual beauty, through which your otherwise plain faces will look forth like those of angels. Beautiful to Ledyard, stiffening in the cold of a northern winter, seemed the diminutive, smoke-stained women of Lapland, who wrapped him in their furs, and ministered to his necessities with kind and gentle words of compassion. Lovely to the homesick Park seemed the dark maids of Sigo, as they sung their low and simple songs of welcome beside his bed, and sought to comfort the white stranger who had "no mother to bring him milk, and no wife to grind him corn."

Oh! talk as you may of beauty, as a thing to be chiselled upon marble or wrought on canvas, — speculate as you may upon its colors and outline, — what is it but an intellectual abstraction after all? The heart feels a beauty of another kind, — looking through outward environments, discovers a deeper and more real loveliness.

This was well understood by the old painters. In their pictures of Mary, the virgin mother, the beauty which melts and subdues the gazer is that of the soul and the affections, — uniting the awe and the mystery of the mother's miraculous allotment with the inexpressible love, the unutterable tenderness, of young maternity, — Heaven's crowning miracle with nature's sweetest and holiest instinct. And their pale Magdalens, holy with the look of sins forgiven, — how the divine beauty of their penitence sinks into the heart! Do we not feel that the only real deformity is sin, and that goodness evermore hallows and sanctifies its dwelling-place?

JOHN G. WHITTIER.

CUVIER.

CUVIER has performed for the kingdoms of animated nature the work which Newton wrought for the mechanism of the heavens. His generalizations now seem final and complete. They bind together all tribes of being in one vast and beautiful system, pervaded by analogies and equivalent provisions; and reveal, in the structure and adaptations of the animal economy, numberless mysteries of divine wisdom which had been hidden from the foundation of the world. He reached these sublime results because his religious nature prompted him to look for unity and harmony in the works of God, — to search everywhere for traces of the all-pervading and all-perfect mind, — to seek in the humblest zoophyte the expression of an idea of God, — the not unworthy type of the Infinite Archetype. He wrought in glowing faith. He served at the altar of science as a priest of the Most High. Infidelity went from his presence rebuked and humbled. His soul was kindled, his lips were touched ever more and more with the fire of heaven, as, with waning strength and under the burden of bereavement, he still drew bolder, fuller harmonies, unheard before, from the lyre of universal nature. Says one who was present at the lecture from which he went home to die, "In the whole of this lecture there was an omnipresence of the Omnipotent and Supreme Cause. The examination of the visible world seemed to touch upon the invisible. The search into creation invoked the presence of the Creator. It seemed as if the veil were to be torn from before us, and science was about to reveal eternal wisdom."

ANDREW P. PEABODY.

IS GARDENING A PLEASURE?

"THE country," exclaimed Mr. Bluff, with an air of candor and impartiality, "is, I admit, a very necessary and sometimes a very charming place. I thank Heaven for the country when I eat my first green peas, when the lettuce is crisp, when the potatoes are delicate and mealy, when the well-fed poultry comes to town, when the ruddy peach and the purple grape salute me at the fruit-stands. I love the country when I think of a mountain ramble; when I am disposed to wander with rod and reel along the forest-shadowed brook; when the apple-orchards are in blossom; when the hills blaze with autumn foliage. But I protest against the dogmatism of rural people, who claim all the cardinal and all the remaining virtues for their rose-beds and cabbage-patches. The town, sir, bestows felicities higher in character than the country does; for men and women, and the works of men and women, are always worthier our love and concern than the rocks and the hills. . . .

. . . "Oh, yes! I have heard before of the pleasures of the garden. Poets have sung, enthusiasts have written, and old men have dreamed of them since History began her chronicles. But have the *pains* of the garden ever been dwelt upon? Have people, now, been entirely honest in what they have said and written on this theme? When enthusiasts have told us of their prize pears, their early peas of supernatural tenderness, their asparagus, and their roses, and their strawberries, have they not hidden a good deal about their worm-eaten plums — about their cherries that were carried off by armies of burglarious birds; about their potatoes that proved watery and unpalatable; about their melons that fell victims to their neighbors' fowls; about their peaches that succumbed to the unexpected raid of Jack Frost; about their grapes that fell under the blight of mildew; about their green corn that withered in the hill; about the mighty host of failures that, if all were told, would tower in high proportion above the few much-blazoned successes?

"Who is it that says a garden is a standing source of pleasure? Amend this, I say, by asserting that a garden is a standing source of discomfort and vexation. . . . A hopeless restlessness, according to my observation, takes possession of every amateur gardener. Discontent abides in his soul. There is, indeed, so much to be done, changed, re-arranged, watched, nursed, that the amateur gardener is really entitled to praise and generous congratulations when one of his thousand schemes comes to fruition. We ought in pity to rejoice with him over his big Lawton blackberries, and say nothing of the cherries, and the pears, and the peaches, that once were budding hopes, but have gone the way of Moore's 'dear gazelle.' Then the large expenditures which were needed to bring about his triumph of the Lawtons. 'Those potatoes,' said an enthusiastic amateur gardener to me once, 'cost twenty-five cents apiece!' And they were very good potatoes, too — almost equal to those that could be bought in market at a dollar a bushel.

"And then, amateur gardeners are feverishly addicted to early rising. Men with gardens are like those hard drinkers whose susceptibilities are hopelessly blunted. Who but a man diverted from the paths of honest feeling and natural enjoyment, possessed of a demoniac mania, lost to the peace and serenity of the virtuous and the blessed, could find pleasure amid the damps, and dews, and chills, and rawedgedness of a garden in the early morning, absolutely find pleasure in saturated trousers, in shoes swathed in moisture, in skies that are gray and gloomy, in flowers that are, as Mantalini would put it, 'demnition moist'? The thing is incredible! Now, a garden, after the sun has dried the paths, warmed the air, absorbed the dew, is admissible. But a possession that compels an early turning out into fogs and discomforts deserves for this fact alone the anathema of all rational beings.

"I really believe, sir, that the literature of the garden, so abundant everywhere, is written in the interest of suburban land-owners. The inviting one-sided picture so persistently held up is only a covert bit of advertising, intended to seduce away happy cockneys of the town — men supremely contented with their attics, their promenades in Fifth Avenue, their visits to Central Park, where all is arranged for them without their labor or concern, their evenings at the music gardens, their soft morning slumbers, which know no dreadful chills and dews! How could a back-ache over the pea-bed compensate for these felicities? How could sour cherries, or half-ripe strawberries, or wet rosebuds, even if they do come from one's own garden, reward him for the lose of the ease and the serene conscience of one who sings merrily in the streets, and cares not whether worms burrow, whether suns burn, whether birds steal, whether winds overturn, whether droughts destroy, whether floods drown, whether gardens flourish, or not?"

Oliver Bell Bunce.

THE ROSE OF GLENGARY.

"Shall I sing you one of our old songs?"

The soft, pure voice sounded in his ears like some fine melody of olden poets; her frank, kind eyes, as she looked at him, soothed and quieted him. Again she was the little laughing star of his childhood, as when they wandered about over the fields — little children — that period so recent, yet which seemed so far away, because the opening heart lives long in a brief space of time. Again she was to him Little Redbud, he to her was the boy playmate, Verty. She had done all by a word — a look, a kind, frank smile, a single glance of confiding eyes. He loved her more than ever — yes, a thousand times more strongly, and was calm.

He followed her to the harpsicord, and watched her in every movement with quiet happiness; he seemed to be under the influence of a charm.

"I think I will try and sing the 'Rose of Glengary,'" she said, smiling. "You know, Verty, it is one of the old songs you loved so much; and it will make us think of old times — in childhood, you know. Though that is not such old, old time — at least, for me," added Redbud, with a smile more soft and confiding than before.

"Shall I sing it? Well, give me the book — the brown-backed one."

The old volume — such as we find to-day in ancient country houses, was opened, and Redbud commenced singing. The girl sang the sweet ditty with much expression; and her kind, touching voice filled the old homestead with a tender melody, such as the autumn time would utter, could its spirit become vocal. The clear, tender carol made the place fairyland for Verty long years afterwards; and always he seemed to hear her singing when he visited the room.

Redbud sang, afterwards, more than one of those old ditties — "Jock o' Hazeldean," and "Flowers of the Forest," and many others — ditties which, for us to-day, seem like so many utterances of the fine old days in the far past.

For, who does not hear them floating above those sweet fields of the olden time — those bright Hesperian gardens, where, for us at least, the fruits are all golden, and the airs all happy?

Beautiful, sad ditties of the brilliant past! not he who writes would have you lost from memory, for all the modern world of music. Kind madrigals! which have an aroma of the former day in all your cadences, and dear old-fashioned trills — from whose dim ghosts now, in the faded volumes stored away in garrets and on upper shelves, we gather what you were in the old immemorial years! Soft melodies of another age, that sound still in the present with such moving sweetness, one heart at least knows what a golden treasure you clasp, and listens thankfully when you deign to issue out from silence; for he finds in you alone — in your gracious cadences, your gay or stately voices — what he seeks; the life, and joy, and splendor of the antique day sacred to love and memory!

And Verty felt the nameless charm of the good old songs, warbled by the young girl's sympathetic voice ; and more than once his wild-wood nature stirred within him, and his eyes grew moist. And when she ceased, and the soft carol went away to the realm of silence, and was heard no more, the young man was a child again, and Redbud's hand was in his own, and all his heart was still.

JOHN ESTEN COOKE.

THE FISHWIFE.

"THERE is poetry in everything." True, quite true, Emerson — thou true man, poet of the backwoods ! But there is not poetry in a fishwife, surely ? Surely there is ; lots of it. Her creel has more than all Dugald Moore's tomes. Why, there was one — I mean a fishwife — this moment in the lobby. She had a hooked nose. It seemed to be the type, nay the ancestor, of a cod-hook. Her mouth was a skate or turbot humanized ; her teeth, selected from the finest oyster pearl ; her eyes, whelks with the bonnets on — bait for old fish on sea or land ; her hands and fingers in redness and toughness rivalled the crab, barring him of the Zodiac. Yet she was all poetry. I had been fagging, reading, and writing since 6 A. M. (on honor !) — had dived into Owen, was drowned in Edwards, and wrecked on Newman — my brain was wearied, when suddenly I heard the sound of " Flukes ! " followed by " Had — dies ! " (a name to which Haidee was as prose). I descended and gazed into the mysterious creel, and then came a gush of sunlight upon my spirit — visions of sunny mornings with winding shores, and clean, sandy, pearly beaches, and rippling waves glancing and glittering over white shells and polished stones, and breezy headlands ; and fishing-boats moving like shadows onward from the great deep ; and lobsters, and crabs, and spoutfish, and oysters, crawling, and chirping, and spouting out sea-water, the old " ocean gleaming like a silver shield." The fishwife was a Claude Lorraine ; her presence painted what did my soul good, and as her reward I gave her what I'll wager never during her life had been given her before — all that she asked for her fish ! And why, you ask, have I sat down to write to you, beloved John, all this — to spend a sheet of paper, to pay one penny, to abuse ten tickings of my watch to write myself, like Dogberry, an ass ? Why ? "Nature," quoth d'Alembert, "puts questions which Nature cannot answer." And shall I beat Nature, and be able to answer questions put to me by John — Nature's own child ? Be silent, and let neither of us shame our parent. Modesty forbids me to attempt any solution of thy question, dear John. Now for work. My pipe is out !

NORMAN MACLEOD.

AN ENGLISH SUNSET.

. . . Imagine us on our evening walk out upon the East Cliff, a mile and a half from our present abode. We have passed a rough pathway, and, weary of a long, low hedge, the very symbol of sameness and almost of nothingness, have struck in by a beach which the sailors, who sit there with their observatory telescopes, have made upon the grassy cliff, and are looking upon the sea and sky and straggling town of Herne Bay. The ruddy ball is sinking; over it is a large feathery mass of cloudage that *was* swan's-down but now, thrilled through with rosy light, resembles pinky crimson flames, and the dark waters below are tinged with rose color. In the distance appears the straggling town, with its tall watch, or rather clock tower, and its long pier, like a leviathan centipede, walking out into the waves. This time we are home before dark; another evening we set out later, and by the time we descend the cliff it is dark, and as we are pacing down the velvet path, as we call the smooth, grassy descent which leads to the town, there is Nurse in her black cloak waving in the wind, moving toward us through the dusk like a magnified bat. . . .

Mrs. Sara Coleridge.

SECESSION.

I would not exaggerate the fearful consequences of dissolution. It is the breaking up of a federative Union, but it is not like the breaking up of society. It is not anarchy. A link may fall from the chain, and the link may still be perfect, though the chain have lost its length and its strength. In the uniformity of commercial regulations, in matters of war and peace, postal arrangements, foreign relations, coinage, copyrights, tariff, and other Federal and national affairs, this great government may be broken; but in most of the essential liberties and rights which government is the agent to establish and protect, the seceding State has no revolution, and the remaining States can have none. This arises from that refinement of our polity which makes the States the basis of our instituted labor. Greece was broken by the Persian power, but her municipal institutions remained. Hungary lost her national crown, but her home institutions remain. South Carolina may preserve her constituted domestic authority, but she must be content to glimmer obscurely remote rather than shine and revolve in a constellated band. She even goes out by the ordinance of a so-called sovereign convention, content to lose by her isolation that youthful, vehement, exultant, progressive life, which is our Nationality! She foregoes the hopes, the boasts, the flag, the music, all the

A FISHER LAD.

emotions, all the traits, and all the energies which, when combined in our United States, have won our victories in war and our miracles of national advancement. Her Governor, Colonel Pickens, in his inaugural, regretfully "looks back upon the inheritance South Carolina had in the common glories and triumphant power of this wonderful confederacy, and fails to find language to express the feelings of the human heart as he turns from the contemplation." The ties of brotherhood, interest, lineage, and history are all to be severed. No longer are we to salute a South Carolinian with the "*idem sententiam de republica,*" which makes unity and nationality. What a prestige and glory are here dimmed and lost in the contaminated reason of man!

Can we realize it? Is it a masquerade, to last for a night, or a reality to be dealt with, with the world's rough passionate handling? It is sad and bad enough; but let us not overtax our anxieties about it as yet. It is not the sanguinary regimen of the French revolution; not the rule of assignats and guillotine; not the cry of "*Vivent les Rouges! A mort les gendarmes!*" but as yet, I hope I may say, the peaceful attempt to withdraw from the burdens and benefits of the Republic. Thus it is unlike every other revolution. Still it is revolution. It may, according as it is managed, involve consequences more terrific than any revolution since government began.

If the Federal Government is to be maintained, its strength must not be frittered away by conceding the theory of secession. To concede secession as a right, is to make its pathway one of roses and not of thorns. I would not make its pathway so easy. If the government has any strength for its own preservation, the people demand it should be put forth in its civil and moral forces. Dealing, however, with a sensitive public sentiment, in which this strength reposes, it must not be rudely exercised. It should be the iron hand in the glove of velvet. Firmness should be allied with kindness. Power should assert its own prerogative, but in the name of law and love. If these elements are not thus blended in our policy, as the Executive proposes, our government will prove either a garment of shreds or a coat of mail. We want neither. . . .

Before we enter upon a career of force, let us exhaust every effort at peace. Let us seek to excite love in others by the signs of love in ourselves. Let there be no needless provocation and strife. Let every reasonable attempt at compromise be considered. Otherwise we have a terrible alternative. War, in this age and in this country, sir, should be the *ultima ratio.* Indeed, it may well be questioned whether there is any reason in it for war. What a war! Endless in its hate, without truce and without mercy. If it ended ever, it would only be after a fearful struggle; and then with a heritage of hate which would forever forbid harmony. . . .

Small States and great States; new States and old States; slave States and free States; Atlantic States and Pacific States; gold and silver States; iron and copper States; grain States and lumber States; river States and lake States;—all having varied interests and advantages, would seek superiority in armed strength. Pride, animosity, and glory would inspire every movement. God shield our country from such a fulfilment of the prophecy of the revered founders of the Union! Our struggle would be no short, sharp struggle. Law, and even religion herself, would become false to their divine purpose. Their voice would no longer be the voice of God, but of his enemy. Poverty, ignorance, oppression, and its handmaid, cowardice, breaking out into merciless cruelty; slaves false; freemen slaves, and society itself poisoned at the cradle and dishonored at the grave;— its life, now so full of blessings, would be gone with the life of a fraternal and united Statehood. What sacrifice is too great to prevent such a calamity? Is such a picture overdrawn? Already its outlines appear. What means the inaugural of Governor Pickens, when he says: "From the position we may occupy toward the Northern States, as well as from our own internal structure of society, the government may, from necessity, become strongly military in its organization"? What mean the minute-men of Governor Wise? What the Southern boast that they have a rifle or shot-gun to each family? What means the Pittsburgh mob? What this alacrity to save Forts Moultrie and Pinckney? What means the boast of the Southern men of being the best-armed people in the world, not counting the two hundred thousand stand of United States arms stored in Southern arsenals? Already Georgia has her arsenals, with eighty thousand muskets. What mean these lavish grants of money by Southern Legislatures to buy more arms? What mean these rumors of arms and force on the Mississippi? . . .

Mr. Speaker, he alone is just to his country; he alone has a mind unwarped by section, and a memory unparalyzed by fear, who warns against precipitancy. He who could hurry this nation to the rash wager of battle is not fit to hold the seat of legislation. What can justify the breaking up of our institutions into belligerent fractions? Better this marble Capitol were levelled to the dust; better were this Congress struck dead in its deliberations; better an immolation of every ambition and passion which here have met to shake the foundations of society than the hazard of these consequences! . . . I appeal to Southern men, who contemplate a step so fraught with hazard and strife, to pause. Clouds are about us! There is lightning in their frown! Cannot we direct it harmlessly to the earth? The morning and evening prayer of the people I speak for in such weakness rises in strength to that Supreme Ruler who, in noticing the fall of a sparrow, cannot disregard the fall of a nation, that our States may continue to be as they have been —*one;* one in the unreserve of a mingled national being; one as the thought of God is one!

SAMUEL SULLIVAN COX.

COVETOUSNESS.

Of covetousness we may truly say, that it makes both the Alpha and Omega in the devil's alphabet, and that it is the first vice in corrupt nature which moves, and the last which dies. For look upon any infant, and as soon as it can but move a hand, we shall see it reaching out after something or other which it should not have; and he who does not know it to be the proper and peculiar sin of old age, seems himself to have the dotage of that age upon him, whether he has the years or no.

The covetous person lives as if the world were made altogether for him, and not he for the world, to take in every thing, and to part with nothing. Charity is accounted no grace with him, nor gratitude any virtue. The cries of the poor never enter into his ears; or if they do, he has always one ear readier to let them out than the other to take them in. In a word, by his rapines and extortions, he is always for making as many poor as he can, but for relieving none whom he either finds or makes so. So that it is a question, whether his heart be harder, or his fist closer. In a word, he is a pest and a monster: greedier than the sea, and barrener than the shore.

Robert South.

BERGERSON AND MOE.

The firm of Bergerson & Moe, cabinet-makers, hired a tumble-down shanty in an out-of-the-way street in a flourishing Western city, and hung out a big sign, which served the double purpose of hiding the insignificance of the shanty and inviting custom. The sign was Moe's idea; the money that paid for it was Bergerson's. As they were contrasts in everything, so also in this: Bergerson had a little capital, but no ideas; Moe had an abundance of ideas, but no capital. He was so handsome, however, so overflowing with life and activity, that his impecuniosity did not trouble him. The streets delighted him; the enormous drays and trucks, loaded with merchandise, gave him the keenest enjoyment; even the swinging bridges, which tried the souls and provoked the profanity of good citizens, exhilarated him. He swam like a dexterous eel through the labyrinthine turmoil, and noted the unlimited possibilities for advancement which this seething industrial democracy afforded. He saw himself in spirit as one of the pillars of the city, commanding multitudes of men, signing subscription papers with a grand flourish, and making speeches at public dinners with the proud feeling of a representative citizen. He saw himself vividly in all these situations, and felt his bosom expand with the anticipated triumph.

In the meanwhile Bergerson was making chairs and tables, which no one bought. Moe was not fond of making chairs, but he made some clever and tasteful designs, which, after much discussion, he induced his partner to copy. He also got up an ingenious puzzle with polished sticks and rings, and, after having peddled this invention for a few days on the street, he sold it to a large firm for three hundred dollars. He gained immensely in Bergerson's esteem by this enterprise; but lost again more than he had gained by investing his surplus in a tall hat and a fine suit of clothes of the latest fashion. Bergerson was on the point of dissolving the partnership when he saw him enter the shop in this inappropriate attire; but he only growled, and worked on with fiercer energy. Talking was always a serious business with him, and not to be engaged in except on severe provocation. And he had reason to congratulate himself in this instance that he did not act on his first impulse. For during the next days he was dumfounded by a sudden rush of customers, who bought everything he had to sell at prices which he himself regarded as exorbitant.

It turned out that Moe, dressed in his modish costume, had marched through the most populous streets with a chair on his head, and on his back an enormous placard, on which the following verse was painted in big letters: —

> "Ho! ho! ho!
> For Bergerson and Moe!
> They make chairs that never break, sir!
> Of the latest style and make, sir!
> Speed on nimble toe
> To Bergerson and Moe."

This jingle had a kind of captivating rhythm to it which made men unconsciously march to it, hum it, curse it, and lay awake repeating it in the small hours of the night. One tormented man recited it to his neighbor in the hope of getting rid of it, and the neighbor, finding all other remedies unavailing, took the hint and sped to Bergerson & Moe.

With the proceeds of their unexpected popularity Bergerson & Moe hired a larger shop, and engaged a couple of journeymen. As it happened, their chairs were equal to their poetic reputation, for Truls Bergerson knew but one way to work, and that was the solid Norwegian way, which had a view both to time and eternity. You might sit on his chairs, or stand on them, ride horseback on them with your children, or fling them at inconvenient visitors — they bore it all with perfect equanimity; they scarcely changed their complexion, and they never broke. These qualities came to be remarked upon, and Moe took pains that no one should remain in ignorance of them. At the same time he visited, in the guise of a critical customer, every furniture dealer in town, and took note of prices, designs and workmanship. To the factories, too, he gained access as a workman out of employment, and made everywhere profitable observations. He had a natural knack, also, at designing, and kept Bergerson and the journeymen busy executing his brilliant ideas. Within a year a second removal became necessary, and a dozen journeymen scarcely sufficed to satisfy the public craving for the furniture of Bergerson & Moe.

If Bergerson had been capable of any such violent emotion as surprise, he would, no doubt, have indulged in vague wonder at his own prosperity. But Bergerson was not at all emotional. He pocketed his money stolidly, and with no reflection except where he had better keep it. And after having carried some twelve hundred dollars on his person for several months, he began to make cautious inquiries, and ended by investing his surplus in two building lots. The ground, he reasoned, could not run away, nor could any one run away with it. For more than a week he entertained himself, every evening, by reading the deed (with the aid of a pocket dictionary), and gazing at the seals and signatures with quiet satisfaction. Like all his countrymen, he had the earth hunger.

HJALMAR HJORTH BOYESEN.

PALM SUNDAY.

I WRITE to you, Madam, from a place, the name of which is, I fancy, hardly known to you. It is a little town on the borders of Wales, which I have hurried to from the circuit in order to pass a week with my sister. She has lately come hither for the sake of her children's breathing the pure air which blows from the Welsh mountains, and enjoying the pleasures which this beautiful country affords. It is the most beautiful country that I have seen in England, or any where else, except in Switzerland; indeed, it very much resembles some parts of Switzerland, but every thing is on a smaller scale; the mountains are less high, the rocks less craggy, and the torrents less rapid. The valleys are perfectly Swiss, and are enchanting: scattered over with villages and farm-houses, and portioned out into a multitude of small fields, they bespeak a happy equality of property, and transport one back in idea to the infancy of society. . . . But the most beautiful objects in this country, and which are in a great degree independent of the season, are the health, the cheerfulness, and the contentment which appear on the countenances of the inhabitants.

The poor people here have a custom which I never knew observed anywhere else, and which is very poetical, and very affecting. Once a year (on Palm Sunday) they get up early in the morning, and gather the violets and primroses, and the few other flowers which at this season are to be found in the fields, and with their little harvest they hasten to the churchyard and strew the flowers over the graves of their nearest relations. Some arrange their humble tribute of affection in different forms with a great deal of taste. The young girls who are so fortunate as never to have lost any near relation or any friend, exert themselves that the tombs of the strangers who have died in the village, at a distance from all who knew them, may not be left unhonored; and hardly a grave appears without some of these affectionate ornaments. I came here soon after this ceremony had been observed and was surprised, on walking through a churchyard, to find in it the appearance of a garden; and to see the flowers withering, each in the place in which it had been fixed. . . .

Sir Samuel Romilly.

MR. BARKIS.

MR. BARKIS.

As this was a great deal for the carrier (whose name was Mr. Barkis) to say — he being, as I observed in a former chapter, of a phlegmatic temperament, and not at all conversational — I offered him a cake as a mark of attention, which he ate at one gulp, exactly like an elephant, and which made no more impression on his big face than it would have done on an elephant's.

"Did *she* make 'em, now?" said Mr. Barkis, always leaning forward, in his slouching way, on the foot-board of the cart, with an arm on each knee.

"Peggotty, do you mean, sir?"

"Ah!" said Mr. Barkis. "Her."

"Yes. She makes all our pastry, and does all our cooking."

"Do she, though?" said Mr. Barkis.

He made up his mouth as if to whistle, but he didn't whistle. He sat looking at the horse's ears as if he saw something new there; and sat so for a considerable time. By-and-by he said:

"No sweethearts, I b'lieve?"

"Sweetmeats did you say, Mr. Barkis?" For I thought he wanted something else to eat, and had pointedly alluded to that description of refreshment.

"Hearts," said Mr. Barkis. "Sweethearts; no person walks with her?"

"With Peggotty?"

"Ah!" he said. "Her."

"Oh, no. She never had a sweetheart."

"Didn't she, though?" said Mr. Barkis.

Again he made up his mouth to whistle, and again he didn't whistle, but sat looking at the horse's ears.

"So she makes," said Mr. Barkis, after a long interval of reflection, "all the apple parsties, and does all the cooking, do she?"

I replied that such was the fact.

"Well, I'll tell you what," said Mr. Barkis. "P'raps you might be writin' to her?"

"I shall certainly write to her," I rejoined.

"Ah!" he said, slowly turning his eyes toward me. "Well! if you was writin' to her, p'raps you'd recollect to say that Barkis was willin'; would you?"

"That Barkis is willing," I repeated innocently. "Is that all the message?"

"Ye-es," he said, considering. "Ye-es. Barkis is willin'."

"But you will be at Blunderstone again to-morrow, Mr. Barkis," I said, faltering a little at the idea of my being far away from it then, "and could give your own message so much better."

As he repudiated this suggestion, however, with a jerk of his head, and once more confirmed his previous request by saying with profound gravity, "Barkis is willin'. That's the message," I readily undertook its transmission. While I was waiting for the coach in the hotel at Yarmouth that very afternoon, I procured a sheet of paper and an inkstand, and wrote a note to Peggotty which ran thus: "My dear Peggotty. I have come here safe. Barkis is willing. My love to mamma. Yours affectionately. — P. S. He says he particularly wants you to know — *Barkis is willing.*"

. . . .

Mr. Barkis was to call for me in the morning at nine o'clock. I got up at eight, a little giddy from the shortness of my night's rest, and was ready for him before the appointed time. He received me exactly as if not five minutes had elapsed since we were last together, and I had only been into the hotel to get change for sixpence, or something of that sort.

As soon as I and my box were in the cart, and the carrier seated, the lazy horse walked away with us all at his accustomed pace.

"You look very well, Mr. Barkis," I said, thinking he would like to know it.

Mr. Barkis rubbed his cheek with his cuff, and then looked at his cuff as if he expected to find some of the bloom upon it; but made no other acknowledgement of the compliment.

"I gave your message, Mr. Barkis," I said; "I wrote to Peggotty."

"Ah!" said Mr. Barkis. Mr. Barkis seemed gruff, and answered dryly.

"Wasn't it right, Mr. Barkis?" I asked, after a little hesitation.

"Why, no," said Mr. Barkis.

"Not the message?"

"The message was right enough, perhaps," said Mr. Barkis. "But it come to an end there."

Not understanding what he meant, I repeated inquisitively, "Came to an end, Mr. Barkis?"

"Nothing come of it," he explained, looking at me sideways. "No answer."

"There was an answer expected, was there, Mr. Barkis?" said I, opening my eyes. For this was a new light to me.

"When a man says he's willin'," said Mr. Barkis, turning his glance slowly on me again, "it's as much as to say, that man's a-waitin' for a answer."

"Well, Mr. Barkis?"

"Well," said Mr. Barkis, carrying his eyes back to his horse's ears; "that man's been a-waitin' for a answer ever since."

"Have you told her so, Mr. Barkis?"

"N-no," growled Mr. Barkis, reflecting about it. "I a'n't got no call to go and tell her so. I never said six words to her myself. *I* a'n't a-goin' to tell her so."

"Would you like me to do it, Mr. Barkis?" said I, doubtfully.

"You might tell her, if you would," said Mr. Barkis, with another slow look at me, "that Barkis was a-waitin' for a answer. Says you — what name is it?"

"Her name?"

"Ah!" said Mr. Barkis, with a nod of his head.

"Peggotty."

"Chrisen name, or nat'ral name?" said Mr. Barkis.

"Oh, it's not her Christian name. Her Christian name is Clara."

"Is it though!" said Mr. Barkis.

He seemed to find an immense fund of reflection in this circumstance, and sat pondering and inwardly whistling for some time.

"Well!" he resumed at length. "Says you, 'Peggotty! Barkis is a-waitin' for a answer.' Says she, perhaps, 'Answer to what?' Says you, 'To what I told you.' 'What is that?' says she. 'Barkis is willin',' says you."

This extremely artful suggestion Mr. Barkis accompanied with a nudge of his elbow that gave me quite a stitch in my side. After that, he slouched over his horse in his usual manner; and made no other reference to the subject except, half an hour afterward, taking a piece of chalk from his pocket, and writing up inside the tilt of the cart, "Clara Peggotty," — apparently as a private memorandum.

CHARLES DICKENS.

IN A SEA OF GLORY.

SPRING IN NEW ENGLAND.

In our methodical New England life, we still recognize some magic in summer. Most persons at least resign themselves to being decently happy in June. They accept June. They compliment its weather. They complain of the earlier months as cold, and so spend them in the city; and they complain of the later months as hot, and so refrigerate themselves on some barren sea-coast. God offers us yearly a necklace of twelve pearls; most men choose the fairest, label it June, and cast the rest away. It is time to chant a hymn of more liberal gratitude.

There are no days in the whole round year more delicious than those which often come to us in the latter half of April.

On these days one goes forth in the morning, and finds an Italian warmth brooding over all the hills; taking visible shape in a glistening mist of silvered azure, with which mingles the smoke from many bonfires. The sun trembles in his own soft rays, till one understands the old English tradition, that he dances on Easter-Day. Swimming in a sea of glory, the tops of the hills look nearer than their bases, and their glistening water-courses seem close to the eye, as is their liberated murmur to the ear. All across this broad intervale the teams are ploughing. The grass in the meadow seems all to have grown green since yesterday. The blackbirds jangle in the oak, the robin is perched upon the elm, the songsparrow on the hazel, and the bluebird on the apple-tree. There rises a hawk and sails slowly, the stateliest of airy things, a floating dream of long and languid summer-hours. But as yet, though there is warmth enough for a sense of luxury, there is coolness enough for exertion. No tropics can offer such a burst of joy; indeed, no zone much warmer than our Northern States can offer a genuine spring. There can be none where there is no winter, and the monotone of the season is broken only by wearisome rains. Vegetation and birds being distributed over the year, there is no burst of verdure nor of song.

But with us, as the buds are swelling, the birds are arriving; they are building their nests almost simultaneously; and in all the Southern year there is no such rapture of beauty and of melody as here marks every morning from the last of April onward.

But days even earlier than those in April have a charm; — even days that seem raw and rainy, when the sky is dull and a bequest of March-wind lingers, chasing the squirrel from the tree and the children from the meadows. There is a fascination in walking through these bare early woods — there is such a pause of preparation, winter's work is so cleanly and thoroughly done. Everything is taken down and put away; throughout the leafy arcades the branches show no remnant of last year, save a few twisted leaves of oak and beech, a few empty seed-vessels of the tardy witch-hazel, and a few gnawed nutshells dropped coquettishly by the squirrels into the crevices of the bark. All else is bare, but prophetic; buds everywhere, the whole splendor of the coming summer concentrated in those hard little knobs on every bough, and clinging here and there among them, a brown, papery chrysalis, from which shall yet wave the superb wings of the Luna moth.

An occasional shower patters on the dry leaves, but it does not silence the robin on the outskirts of the wood; indeed, he sings louder than ever during rain, though the song-sparrow and the bluebird are silent.

THOMAS WENTWORTH HIGGINSON.

CONTINENTAL CONGRESS.

THE dying embers of the Continental Congress, barely kept alive for some months by the occasional attendance of one or two delegates, as the day approached* for the new system to be organized, quietly went out, without note or observation. History knows few bodies so remarkable. The Long Parliament of Charles I., and the French National Assembly, are alone to be compared with it. Coming together, in the first instance, a mere collection of consulting delegates, the Continental Congress had boldly seized the reins of power, assumed the leader-

* March 3, 1789.

A PROMISE OF THE SPRING.

ship of the insurgent States, issued bills of credit, raised armies, declared independence, negotiated foreign treaties, carried the nation through an eight years' war; finally, had extorted from the proud and powerful mother-country an acknowledgment of the sovereign authority so daringly assumed and so indomitably maintained. But this brilliant career had been as short as it was glorious. The decline had commenced even in the midst of the war. Exhausted by such extraordinary efforts, — smitten with the curse of poverty, their paper money first depreciating and then repudiated, overwhelmed with debts which they could not pay, pensioners on the bounty of France, insulted by mutineers, scouted at by the public creditors, unable to fulfill the treaties they had made, bearded and encroached upon by the State authorities, issuing fruitless requisitions which they had no power to enforce, vainly begging for additional authority which the States refused to grant, thrown more and more into the shade by the very contrast of former power — the Continental Congress sunk fast into decrepitude and contempt. Feeble is the sentiment of political gratitude! Debts of that sort are commonly left for posterity to pay. While all eyes were turned — some with doubt and some with apprehension, but the greater part with hope and confidence — towards the ample authority vested in the new government now about to be organized, not one respectful word seems to have been uttered, not a single reverential regret to have been dropped over the fallen greatness of the exhausted and expiring Continental Congress.

RICHARD HILDRETH.

THE SIEGE OF LEYDEN.

MEANTIME, the besieged city was at its last gasp. The burghers had been in a state of uncertainty for many days; being aware that the fleet had set forth for their relief, but knowing full well the thousand obstacles which it had to surmount. They had guessed its progress by the illumination from the blazing villages; they had heard its salvos of artillery on its arrival at North Aa; but since then, all had been dark and mournful again, hope and fear, in sickening alternation, distracting every breast. They knew that the wind was unfavorable, and at the dawn of each day every eye was turned wistfully to the vanes of the steeples. So long as the easterly breeze prevailed, they felt, as they anxiously stood on towers and housetops, that they must look in vain for the welcome ocean. Yet, while thus patiently waiting, they were literally starving; for even the misery endured at Harlem had not reached that depth and intensity of agony to which Leyden was now reduced. Bread, malt-cake, horse-flesh, had entirely disappeared; dogs, cats, rats, and other vermin, were esteemed luxuries. A small number of cows, kept as long as possible, for their milk, still remained; but a few were killed from day to day, and distributed in minute proportions, hardly sufficient to support life among the famishing population. Starving wretches swarmed daily around the shambles where these cattle were slaughtered, contending for any morsel which might fall, and lapping eagerly the blood as it ran along the pavement; while the hides, chopped and boiled, were greedily devoured. Women and children, all day long, were seen searching gutters and dunghills for morsels of food, which they disputed fiercely with the famishing dogs. The green leaves were stripped from the trees, every living herb was converted into human food; but these expedients could not avert starvation. The daily mortality was frightful: infants starved to death on the maternal breasts which famine had parched and withered; mothers dropped dead in the streets, with their dead children in their arms. In many a house the watchmen, in their rounds, found a whole family of corpses — father, mother, children, side by side; for a disorder called the plague, naturally engendered of hardship and famine, now came, as if in kindness, to abridge the agony of the people. The pestilence stalked at noonday through the city, and the doomed inhabitants fell like grass beneath its scythe. From six thousand to eight thousand human beings sank before this scourge alone; yet the people resolutely held out — women and men mutually encouraging each other to resist the entrance of their foreign foe — an evil more horrible than pest or famine.

Leyden was sublime in its despair. A few murmurs were, however, occasionally heard at the steadfastness of the magistrates, and a dead body was placed at the door of the burgomaster, as a silent witness against his inflexibility. A party of the more faint-hearted even assailed the heroic Adrian Vander Werf with threats and reproaches as he passed through the streets. A crowd had gathered around him as he reached a triangular place in the center of the town, into which many of the principal streets emptied themselves, and upon one side of which stood the church of St. Pancras. There stood the burgomaster, a tall, haggard, imposing figure, with dark visage and a tranquil but commanding eye. He waved his broad-leaved felt hat for silence, and then exclaimed, in language which has been almost literally preserved, "What would ye, my friends? Why do ye murmur that we do not break our vows and surrender the city to the Spaniards? — a fate more horrible than the agony which she now endures. I tell you I have made an oath to hold the city; and may God give me strength to keep my oath! I can die but once, whether by your hands, the enemy's, or by the hand of God. My own fate is indifferent to me; not so that of the city intrusted to my care. I know that we shall starve if not soon relieved; but starvation is preferable to the dishonored death which is the only alternative. Your menaces move me not; my life is at your disposal; here is my sword, plunge it into my breast, and divide my flesh among you. Take my body to appease your hunger, but expect no surrender so long as I remain alive." . . .

On the 28th of September, a dove flew into the city, bringing a letter from Admiral Boisot. In this despatch, the position of the fleet at North Aa was described in encouraging terms, and the inhabitants were assured that, in a very few days at furthest, the long-expected relief would enter their gates. The tempest came to their relief. A violent equinoctial gale, on the night of the 1st and 2d of October, came storming from the northwest, shifting after a few hours full eight points, and then blowing still more violently from the southwest. The waters of the North Sea were piled in vast masses upon the southern coast of Holland, and then dashed furiously landward, the ocean rising over the earth and sweeping with unrestrained power across the ruined dykes. In the course of twenty-four hours, the fleet at North Aa, instead of nine inches, had more than two feet of water. . . . On it went, sweeping over the broad waters which lay between Zoeterwoude and Zwieten; as they approached some shallows which led into the great mere, the Zealanders dashed into the sea, and with sheer strength shouldered every vessel through. . . . On again the fleet of Boisot still went, and, overcoming every obstacle, entered the city on the morning of the 3d of October. Leyden was relieved.

JOHN LOTHROP MOTLEY.

POPULAR CULTURE.

. . . It is true that the old world moves tardily on its arduous way, but even if the results of all our efforts in the cause of education were smaller than they are, there are still two considerations that ought to weigh with us and encourage us.

There is another thought to encourage us, still more direct, and still more positive. The boisterous old notion of hero-worship, which has been preached by so eloquent a voice in our age, is, after all, now seen to be a half-truth, and to contain the less edifying and the less profitable half of the truth. The world will never be able to spare its hero, and the man with the rare and inexplicable gift of genius will always be as commanding a figure as he has ever been. What we see every day with increasing clearness is that not only the well-being of the many, but the chances of exceptional genius, moral or intellectual, in the gifted few, are highest in a society where the *average* interest, curiosity, capacity, are all highest. The moral of this for you and for me is plain. We cannot, like Beethoven or Handel, lift the soul by the magic of divine melody into the seventh heaven of ineffable vision and hope incommensurable; we cannot, like Newton, weigh the far-off stars in a balance, and measure the heavings of the eternal flood; we cannot, like Voltaire, scorch up what is cruel and false by a word as a flame; nor, like Milton or Burke, awaken men's hearts with the note of an organ-trumpet; we cannot, like the great saints of the churches and the great sages of the schools, add to those acquisitions of spiritual beauty and intellectual mastery which have, one by one, and little by little, raised man from being no higher than the brute to be only a little lower than the angels. But what we can do — the humblest of us in this great hall — is by diligently using our own minds and diligently seeking to extend our own opportunities to others, to help to swell that common tide, on the force and the set of whose currents depends the prosperous voyaging of humanity. When our names are blotted out, and our place knows us no more, the energy of each social service will remain, and so, too, let us not forget, will each social disservice remain, like the unending stream of one of nature's forces. The thought that this is so, may well lighten the poor perplexities of our daily life, and even soothe the pang of its calamities ; it lifts us from our feet as on wings, opening a larger meaning to our private toil and a higher purpose to our public endeavor ; it makes the morning as we awake to its welcome, and the evening like a soft garment as it wraps us about ; it nerves our arm with boldness against oppression and injustice, and strengthens our voice with deeper accents against falsehood, while we are yet in the full noon of our days — yes, and perhaps it will shed some ray of consolation, when our eyes are growing dim to it all, and we go down into the Valley of Darkness.

JOHN MORLEY.

A QUESTION OF SUPREMACY.

I HAVE received a present of a pair of Cochin-Chinas, a superb cock and a duncolored hen. I put them with my other fowls in the cellar, to protect them for a short time from the severity of the weather. My Shanghai rooster had for several nights been housed up ; for on one occasion, when the cold was snapping, he was discovered under the lee of a stone wall, standing on one leg, taking no notice of the approach of any one, and nearly gone. When brought in, he backed up against the red-hot kitchen stove, and burnt his tail off. Before this he had no feathers in the rear to speak of, and now he is bob-tailed indeed. Anne sewed upon him a jacket of carpet, and put him in a tea-box for the night ; and it was ludicrous on the next morning to see him lifting up his head above the square prison-box, and crowing lustily to greet the day. But before breakfast-time he had a dreadful fit. He retreated against the wall, he fell upon his side, he kicked, and he "carried on " ; but when the carpet was taken off, he came to himself, and ate corn with a voracious appetite. His indisposition was, no doubt, occasioned by a rush of blood to the head from the tightness of the bandages. When Shanghai and Cochin met together in the cellar, they enacted in that dusky hole all the barbarities of a profane cockpit. I heard a sound as if from the tumbling of barrels, followed by a dull, thumping noise, like spirit-rappings, and went below, where the first object which met my eye was a mouse creeping along the beam out of an excavation in my pine-apple cheese. As for the fowls, instead of salutation after the respectful manner of their country — which is expressed thus : Shang knocks knees to Cochin, bows three times, touches the ground, and makes obeisance — they were engaged in a bloody fight, unworthy of celestial poultry. With their heads down, eyes flashing, and red as vipers, and with a feathery frill or ruffle about their necks, they were leaping at each other, to see who should hold dominion over the ash-heap. It put me exactly in mind of two Scythians or two Greeks in America, where each wished to be considered the only Scythian or only Greek in the country. A contest or emulation is at all times highly animating and full of zest, whether two scholars write, two athletes strive, two boilers strain, or two cocks fight. Every lazy dog in the vicinity is immediately at hand. I looked on until I saw the Shanghai's peepers darkened, and his comb streaming with blood. These birds contended for some days after for pre-eminence, on the lawn, and no flinching could be observed on either part, although the Shanghai was by one third the smaller of the two. At last the latter was thoroughly mortified; his eyes wavered and wandered vaguely, as he stood opposite the foe; he turned tail and ran. From that moment he became the veriest coward, and submitted to every indignity without attempting to resist. He suffered himself to be chased about the lawn, fled from the Indian meal, and was almost starved. Such submission on his part at last resulted in peace, and the two rivals walked side by side without fighting, and ate together, with a mutual concession, of the corn. This, in turn, engendered a degree of presumption on the part of the Shanghai cock; and one day, when the dew sparkled and the sun shone peculiarly bright, he so far forgot himself as to ascend a hillock and venture on a tolerably triumphant crow. It showed a lack of judgment; his cock-a-doodle-doo proved fatal. Scarcely had he done so, when Cochin-China rushed upon him, tore out his feathers, and flogged him so severely that it was doubtful whether he would remain with us. Now, alas! he presents a sad spectacle: his comb frozen off, his tail burnt off, and his head knocked to a jelly. While the corn jingles in the throats of his compeers when they eagerly snap it, as if they were eating from a pile of shilling pieces or fi'-penny bits, he stands aloof and grubs in the ground. How changed!

Frederick William Shelton.

ON THE ART OF LIVING WITH OTHERS.

We come now to the consideration of temper, which might have been expected to be treated first. But to cut off the means and causes of bad temper is, perhaps, of as much importance as any direct dealing with the temper itself. Besides, it is probable that in small social circles there is more suffering from unkindness than ill-temper. Anger is a thing that those who live under us suffer more from than those who live with us. But all the forms of ill-humor and sour-sensitiveness, which especially belong to equal intimacy (though indeed they are common to all), are best to be met by impassiveness. When two sensitive persons are shut up together, they go on vexing each other with a reproductive irritability. But sensitive and hard people get on well together. The supply of temper is not altogether out of the usual laws of supply and demand.

Intimate friends and relations should be careful when they go out into the world together, or admit others to their own circle, that they do not make a bad use of the knowledge which they have gained of each other by their intimacy. Nothing is more common than this, and did it not mostly proceed from mere carelessness it would be superlatively ungenerous. You seldom need wait for the written life of a man to hear about his weaknesses, or what are supposed to be such, if you know his intimate friends, or meet him in company with them. . . .

Lastly, in conciliating those we live with, it is most surely done, not by consulting their interests, nor by giving away to their opinions, so much as by not offending their tastes. The most refined part of us lies in this region of taste, which is perhaps a result of our whole being rather than a part of our nature, and at any rate is the region of our most subtle sympathies and antipathies. . . .

It may be said that if the great principles of Christianity were attended to, all such rules, suggestions, and observations as the above would be needless. True enough! Great principles are at the bottom of all things; but to apply them to daily life, many little rules, precautions, and insights are needed. Such things hold a middle place between real life and principles, as form does between matter and spirit; moulding the one and expressing the other.

ARTHUR HELPS.

THE INVENTION OF GUNPOWDER.

THE only hope of salvation for the Greek empire and the adjacent kingdoms, would have been some more powerful weapon, some discovery in the art of war, that should give them a decisive superiority over their Turkish foes. Such a weapon was in their hands; such a discovery had been made in the critical moment of their fate. The chemists of China or Europe had found, by casual or elaborate experiments, that a mixture of saltpetre, sulphur, and charcoal, produces, with a spark of fire, a tremendous explosion. It was soon observed, that if the expansive force were compressed in a strong tube, a ball of stone or iron might be expelled with irresistible and destructive velocity. The precise era of the invention and application of gunpowder is involved in doubtful traditions and equivocal language; yet we may clearly discern that it was known before the middle of the fourteenth century; and that before the end of the same, the use of artillery in battles and sieges, by sea and land, was familiar to the states of Germany, Italy, Spain, France, and England. The priority of nations is of small account; none could derive any

THE INVENTION OF GUNPOWDER.

exclusive benefit from their previous or superior knowledge; and in the common improvement, they stood on the same level of relative power and military science. Nor was it possible to circumscribe the secret within the pale of the church; it was disclosed to the Turks by the treachery of apostates and the selfish policy of rivals; and the sultans had sense to adopt, and wealth to reward, the talents of a Christian engineer. The Genoese, who transported Amurath into Europe, must be accused as his preceptors; and it was probably by their hands that his cannon was cast and directed at the siege of Constantinople. The first attempt was indeed unsuccessful; but in the general warfare of the age, the advantage was on their side who were most commonly the assailants; for a while the proportion of the attack and defence was suspended; and this thundering artillery was pointed against the walls and towers which had been erected only to resist the less potent engines of antiquity. By the Venetians, the use of gunpowder was communicated without reproach to the sultans of Egypt and Persia, their allies against the Ottoman power; the secret was soon propagated to the extremities of Asia; and the advantage of the European was confined to his easy victories over the savages of the New World. If we contrast the rapid progress of this mischievous discovery with the slow and laborious advances of reason, science, and the arts of peace, a philosopher, according to his temper, will laugh or weep at the folly of mankind.

EDWARD GIBBON.

EXPERTS IN THE ART OF WAR.

A LESSON IN PATRIOTISM.

I FIRST came to understand anything about "the man without a country" one day when we overhauled a dirty little schooner which had slaves on board. An officer was sent to take charge of her, and, after a few minutes, he sent back his boat to ask that some one might be sent him who could speak Portuguese. We were all looking over the rail when the message came, and we all wished we could interpret when the captain asked who spoke Portuguese. But none of the officers did; and just as the captain was sending forward to ask if any of the people could, Nolan stepped out and said he should be glad to interpret, if the captain wished, as he understood the language. The captain thanked him, fitted out another boat with him, and in this boat it was my luck to go.

When we got there, it was such a scene as you seldom see, and never want to. Nastiness beyond account, and chaos run loose in the midst of the nastiness. There were not a great many of the negroes; but by way of making what there were understand that they were free, Vaughan had had their hand-cuffs and ankle-cuffs knocked off, and, for convenience' sake, was putting them upon the rascals of the schooner's crew. The negroes were, most of them, out of the hold, and swarming all round the dirty deck, with a central throng surrounding Vaughan and addressing him in every dialect and patois of a dialect, from the Zulu click up to the Parisian of Beledeljereed.

As we came on deck, Vaughan looked down from a hogshead, on which he had mounted in desperation, and said, —

"For God's love, is there anybody who can make these wretches understand something? The men gave them rum, and that did not quiet them. I knocked that big fellow down twice, and that did not soothe him. And then I talked Choctaw to all of them together; and I'll be hanged if they understood that as well as they understood the English."

Nolan said he could speak Portuguese, and one or two fine-looking Kroomen were dragged out who, as it had been found already, had worked for the Portuguese on the coast at Fernando Po.

"Tell them they are free," said Vaughan; "and tell them that these rascals are to be hanged as soon as we can get rope enough."

Nolan "put that into Spanish"; that is, he explained it in such Portuguese as the Kroomen could understand, and they in turn to such of the negroes as could understand them. Then there was such a yell of delight, clinching of fists, leaping and dancing, kissing of Nolan's feet, and a general rush made to the hogshead by way of spontaneous worship of Vaughan, as the *deus ex machina* of the occasion.

"Tell them," said Vaughan, well pleased, "that I will take them all to Cape Palmas."

This did not answer so well. Cape Palmas was practically as far from the homes of most of them as New Orleans or Rio Janeiro was; that is, they would be eternally separated from home there. And their interpreters, as we could understand, instantly said, "Ah, non Palmas," and began to propose infinite other expedients in most voluble language. Vaughan was rather disappointed at this result of his liberality, and asked Nolan eagerly what they said. The drops stood on poor Nolan's white forehead, as he hushed the men down, and said, "He says, 'Not Palmas.' He says, 'Take us home, take us to our own country, take us to our own house, take us to our own pickaninnies and our own women.' He says he has an old father and mother who will die if they do not see him. And this one says he left his people all sick, and paddled down to Fernando to beg the white doctor to come and help them, and that these devils caught him in the bay just in sight of home, and that he has never seen anybody from home since then. And this one says," choked out Nolan, "that he has not heard a word from his home in six months, while he has been locked up in an infernal barracoon."

Vaughan always said he grew gray himself while Nolan struggled through this interpretation. I, who did not understand anything of the passion involved in it, saw that the very elements were melting with fervent heat, and that something was to pay somewhere. Even the negroes themselves stopped howling, as they saw Nolan's agony, and Vaughan's almost equal agony of sympathy. As quick as he could get words, he said: "Tell them yes, yes, yes; tell them they shall go to the Mountains of the Moon if they will. If I sail the schooner through the Great White Desert, they shall go home!"

And after some fashion Nolan said so. And then they all fell to kissing him again, and wanted to rub his nose with theirs.

But he could not stand it long; and getting Vaughan to say he might go back, he beckoned me down into our boat. As we lay back in the stern-sheets and the men gave way, he said to me, "Youngster, let that show you what it is to be without a family, without a home, and without a country. And if you are ever tempted to say a word or to do a thing that shall put a bar between you and your family, your home, and your country, pray God in his mercy to take you that instant home to his own in heaven. Stick by your family, boy; forget you have a self, while you do everything for them. Think of your home, boy; write and send, and talk about it. Let it be nearer and nearer to your thought, the farther you have to travel from it; and rush back to it, when you are free, as that poor black slave is doing now. And for your country, boy," and the words rattled in his throat, "and for that flag," and he pointed to the ship, "never dream a dream but of serving her as she bids you, though the service carry you through a thousand hells. No matter what happens to you, no matter who flatters you or who abuses you, never look at another flag, never let a night pass but you pray God to bless that flag. Remember, boy, that behind all these men you have to do with, behind officers, and government, and people even, there is the Country Herself, your Country, and that you belong to Her as you belong to your own mother. Stand by Her, boy, as you would stand by your mother, if those devils there had got hold of her to-day!"

EDWARD EVERETT HALE.

TO HIS DAUGHTER.

. . . So you were very sorry, old girl, when we left you that day? You thought you would not care. Hem! I knew better.

And so the poor lassie cried, and was so lonely the first night, and would have given worlds to be at home again! And your old dad was not a bit sorry to leave you, not he — cruel-hearted man that he is! Nor was your mother, wretched old woman that she is! And yet "you would wonder" how sorry we both were, and how often the old man said "Poor dear darling!" But no tear filled our eye. Are you sure of that? I'm not. And the old father said: "I'm not afraid of my girl. I'm sure she will prove herself good, kind, loving, and obedient, and won't be lazy, but do her work like a heroine, and remember all her old dad told her!" and her mammy said the same. And then the mammy would cry, and the old dad would call her a fool (respectfully). And so we reached London, and then we got your letter, which made us very happy, and then the old man said: "Never fear! she will do right well, and will be very happy, and Miss —— will like her, and she will like Miss ——!" and "We shall soon meet again!" chimed in the mammy. "If it be God's will we shall," said the dad, "and won't we be happy!"

God bless you, my darling! May you love your own Father in heaven far more than you love your own father on earth, and I know how truly you love me, and you know how truly I love you; but He loves you infinitely more than I can possibly do, though I give you my whole heart.

Will you write a line to the old man? And, remember, he won't criticise it, but be glad to hear all your chatter.

NORMAN MACLEOD.

ON HISTORY.

The effect of historical reading is analogous, in many respects, to that produced by foreign travel. The student, like the tourist, is transported into a new state of society. He sees new fashions. He hears new modes of expression. His mind is enlarged by contemplating the wide diversities of laws, of morals, and of manners. But men may travel far, and return with minds as contracted as if they had never stirred from their own market-town. In the same manner, men may know the dates of many battles, and the genealogies of many royal houses, and yet be no wiser. Most people look at past times, as princes look at foreign countries. More than one illustrious stranger has landed on our island amidst the shouts of a mob, has dined with the king, has hunted with the master of the stag-hounds, has seen the Guards reviewed, and a knight of the garter installed; has cantered along Regent Street; has visited St. Paul's, and noted down its dimensions, and has then departed, thinking he has seen England. He has, in fact, seen a few public buildings, public men, and public ceremonies. But of the vast and complex system of society, of the fine shades of national character, of the practical operation of government and laws, he knows nothing. He who would understand these things rightly, must not confine his observations to palaces and solemn days. He must see ordinary men as they appear in their ordinary business and in their ordinary pleasures. He must mingle in the crowds of the exchange and the coffee-house. He must obtain admittance to the convivial table and the domestic hearth. He must bear with vulgar expressions. He must not shrink from exploring even the retreats of misery. He who wishes to understand the condition of mankind in former ages, must proceed on the same principle. If he attends only to public transactions, to wars, congresses, and debates, his studies will be as unprofitable as the travels of those imperial, royal, and serene sovereigns, who form their judgment of our island from having gone in state to a few fine sights, and from having held formal conference with a few great officers.

The perfect historian is he in whose work the character and spirit of an age are exhibited in miniature. He relates no fact, he attributes no expression to his characters, which is not authenticated by sufficient testimony. But by judicious selection, rejection, and arrangement, he gives to truth those attractions which have been usurped by fiction. In his narrative a due subordination is observed; some transactions are prominent, others retire. But the scale on which he represents them is increased or diminished, not according to the dignity of the persons concerned in them, but according to the degree in which they elucidate the condition of society and the nature of man. He shows us the court, the camp, and the senate. But he shows us also the nation. He considers no anecdote, no peculiarity of manner, no familiar saying, as too insignificant for his notice, which is not too insignificant to illustrate the operation of laws, of religion, and of education, and to mark the progress of the human mind. Men will not merely be described, but will be made intimately known to us. The changes of manners will be indicated, not merely by a few general phrases, or a few extracts from statistical documents, but by appropriate images presented in every line.

If a man, such as we are supposing, should write the history of England, he would assuredly not omit the battles, the sieges, the negotiations, the seditions, the ministerial changes. But with these he would intersperse the details which are the charm of historical romances. At Lincoln Cathedral there is a beautiful painted window, which was made by an apprentice out of the pieces of glass which had been rejected by his master. It is so far superior to every other in the church, that, according to the tradition, the vanquished artist killed himself from mortification. Sir Walter Scott, in the same manner, has used those fragments of truth which historians have scornfully thrown behind them, in a manner which may well excite their envy. He has consrtucted out of their gleaning works which, even considered as histories, are scarcely less valuable than theirs. But a truly great historian would reclaim these materials which the novelist has appropriated. The history of the government, and the history of the people, would be exhibited in that mode in which alone they can be exhibited justly, in inseparable conjunction and intermixture. We should not then have to look for the wars and votes of the Puritans in Clarendon, and for their phraseology in "Old Mortality"; for one half of King James in Hume, and for the other half in the "Fortunes of Nigel."

THOMAS BABINGTON MACAULAY.

DEMOCRACY.

He must have been a born leader or misleader of men, or must have been sent into the world unfurnished with that modulating and restraining balance-wheel which we call a sense of humor, who, in old age, has as strong a confidence in his opinions and in the necessity of bringing the universe into conformity with them as he had in youth. In a world the very condition of whose being is that it should be in perpetual flux, where all seems mirage, and the one avoiding thing is the effort to distinguish realities from appearances, the elderly man must be indeed of a singularly tough and valid fiber who is certain that he has any clarified residuum of experience, any assured verdict of reflection, that deserves to be called an opinion, or who, even if he had, feels that he is justified in holding mankind by the button while he is expounding it. And in a world of daily — nay, almost hourly — journalism, where every clever man, every man who 'hinks himself clever, or whom anybody else thinks clever, is called upon to deliver his judgment point-blank and at the word of command on every conceivable subject of human thought, or, on what sometimes seems to him very much the same thing, on every inconceivable display of human want of thought, there is such a spendthrift waste of all those commonplaces which furnish the permitted staple of public discourse that there is little chance of beguiling a new tune out of the one-stringed instrument on which we have been thrumming so long. In this desperate necessity one is often tempted to think that, if all the words of the dictionary were tumbled down in a heap and then all those fortuitous juxtapositions and combinations that made tolerable sense were picked out and pieced together, we might find among them some poignant suggestions toward novelty of thought or expression. But, alas! it is only the great poets who seem to have this unsolicited profusion of unexpected and incalculable phrase, this infinite variety of topic. For everybody else everything has been said before, and said over again after.

JAMES RUSSELL LOWELL.

SELFISHNESS VERSUS NOBILITY.

Now, that which especially distinguishes a high order of man from a low order of man — that which constitutes human goodness, human greatness, human nobleness — is surely not the degree of enlightenment with which men pursue their own advantage; but it is self-forgetfulness, it is self-sacrifice; it is the disregard of personal pleasure, personal indulgence, personal advantages remote or present, because some other line of conduct is more right.

We are sometimes told that this is but another way of expressing the same thing; that, when a man prefers doing what is right, it is only because to do right gives him a higher satisfaction. It appears to me, on the contrary, to be a difference in the very heart and nature of things. The martyr goes to the stake, the patriot to the scaffold, not with a view to any future reward, to themselves, but because it is a glory to fling away their lives for truth and freedom. And so through all the phases of existence, to the smallest details of common life, the beautiful character is the unselfish character. Those whom we most love and admire are those to whom the thought of self seems never to occur; who do simply and with no ulterior aim — with no thought whether it will be pleasant to themselves or unpleasant — that which is good and right and generous.

Is this still selfishness, only more enlightened? I do not think so. The essence of true nobility is neglect of self. Let the thought of self pass in, and the beauty of a great action is gone, like the bloom from a soiled flower. Surely it is a paradox to speak of the self-interest of a martyr who dies for a cause, the triumph of which he will never enjoy; and the greatest of that great company in all ages would have done what they did, had their personal prospects closed with the grave. Nay, there have been those so zealous for some glorious principle as to wish themselves blotted out of the book of Heaven if the cause of Heaven could succeed.

And out of this mysterious quality, whatever it be, arise the higher relations of human life, the higher modes of human obligation. Kant, the philosopher, used to say that there were two things which overwhelmed him with awe as he thought of them. One was the star-sown deep of space, without limit and without end; the other was, right and wrong. Right, the sacrifice of self to good; wrong, the sacrifice of good to self — not graduated objects of desire, to which we are determined by the degrees of our knowledge, but wide asunder as pole and pole, as light and darkness: one the object of infinite love; the other, the object of infinite detestation and scorn. It is in this marvellous power in men to do wrong (it is an old story, but none the less true for that), — it is in this power to do wrong — wrong or right, as it lies somehow with ourselves to choose — that the impossibility stands of forming scientific calculations of what men will do before the fact, or scientific explanations of what they have done after the fact. If men were consistently selfish, you might analyze their motives; if they were consistently noble, they would express in their conduct the laws of the highest perfection. But so long as two natures are mixed together, and the strange creature which results from the combination is now under one influence and now under another, so long you will make nothing of him except from the old-fashioned moral — or, if you please, imaginative — point of view.

Even the laws of political economy itself cease to guide us when they touch moral government. So long as labor is a chattel to be bought and sold, so long, like other commodities, it follows the condition of supply and demand. But if, for his misfortune, an employer considers that he stands in human relations toward his workmen; if he believes, rightly or wrongly, that he is responsible for them; that in return for their labor he is bound to see that their children are decently taught, and they and their families decently fed and clothed and lodged; that he ought to care for them in sickness and in old age, — then political economy will no longer direct him, and the relations between himself and his dependents will have to be arranged on quite other principles.

JAMES ANTHONY FROUDE.

HIERONYMUS AND TIDDLEKINS.

IT seemed to Hieronymus that the climax of his impositions had come, when he was forced to stay at home and mind the baby, while his mother and the rest of them trotted off, gay as larks, to see a man hanged.

It was a hot afternoon, and the unwilling nurse suffered. The baby wouldn't go to sleep. He put it on the bed — a feather-bed — and why it didn't drop off to sleep, as a proper baby should, was more than the tired soul of Hieronymus could tell. He did every thing to sooth Tiddlekins. (The infant had not been named as yet, and by way of affection they addressed it as Tiddlekins.) He even went so far as to wave the flies away from it with a mulberry branch for the space of five or ten minutes. But as it still fretted and tossed, he let it severely alone, and the flies settled on the little black thing as if it had been a licorice stick.

After a while Tiddlekins grew aggressive, and began to yell. Hieronymus, who had almost found consolation in the contemplation of a bloody picture pasted on the wall, cut from the weekly paper of a wicked city, was deprived even of this solace. He picked up "de miserbul little screech-owl," as he called it in his wrath. He trotted it. He sang to it the soothing ditty of —

> "'Tain't never gwine to rain no mo';
> Sun shines down on rich and po'."

But all was vain. Finally, in despair, he undressed Tiddlekins. He had heard his mother say: "Of'en and of'en when a chile is a-screamin' its breff away, 'tain't nothin' ails it 'cep'n pins."

But there were no pins. Plenty of strings and hard knots; but not a pin to account for the antics of the unhappy Tiddlekins.

How it did scream! It lay on the stiffly braced knees of Hieronymus, and puckered up its face so tightly that it looked as if it had come fresh from a wrinkle mould. There were no tears, but sharp regular yells, and rollings of its head, and a distracting monotony in its performances.

"Dis here chile looks's if it's got de measles," muttered Hi, gazing on the squirming atom with calm eyes of despair. Then, running his fingers over the neck and breast of the small Tiddlekins, he cried, with the air of one who makes a discovery, "It's got de heat! Dat's what ails Tiddlekins!"

There was really a little breaking out on the child's body that might account for his restlessness and squalls. And it was such a hot day! Perspiration streamed down Hi's back, while his head was dry. There was not a quiver in the tree-leaves, and the silver poplars showed only their leaden side. The sunflowers were dropping their big heads; the flies seemed to stick to the window-panes, and were too languid to crawl.

HIERONYMUS AND TIDDLEKINS.

Hieronymus had in him the materials of which philosophers are made. He said to himself: "'Tain't nothin' but heat dat's de matter wid dis baby; so uf cose he ought ter be cooled off."

But how to cool him off — that was the great question. Hi knitted his dark brows and thought intently.

It happened that the chiefest treasure of the Pop estate was a deep old well that in the hottest days yielded water as refreshing as iced champagne. The neighbors all made a convenience of the Pop well. And half way down its long cool hollow hung, pretty much all of the time, milk cans, butter pats, fresh meats — all things that needed to be kept cool in summer days.

THE QUICK-WITTED YOUTH.

He looked at the hot, squirming, wretched black baby on his lap; then he looked at the well; and, simple, straightforward lad that he was, he put this and that together.

"If I was ter hang Tiddlekins down de well," he reflected, "'twouldn't be mo' dan three jumps of a flea befo' he's as cool as Christmas."

With this quick-witted youth to think was to act. Before many minutes he had stuffed poor little Tiddlekins into the well bucket, though it must be mentioned to his credit that he tied the baby securely in with his own suspenders.

Warmed up with his exertions, content in this good riddance of such bad rubbish as Tiddlekins, Hieronymus reposed himself on the feather-bed, and dropped off into a sweet slumber. From this he was aroused by the voice of a small boy.

"Hello, Hi! I say, Hi Pop! whar is yer?"

"Here I is!" cried Hi, starting up. "What you want?"

Little Jim Rogers stood in the door-way.

"Towzer's dog," he said, in great excitement, "and daddy's bull-pup is gwine to have a fight dis evenin'. Come on quick, if yer wants ter see de fun."

Up jumped Hi, and the two boys were off like a flash. *Not one thought to Tiddlekins in the well bucket.*

In due time the Pop family got home, and Mother Pop, fanning herself, was indulging in the moral reflections suitable to the occasion, when she checked herself suddenly, exclaiming, "But, land o' Jerusalem! whar's 'Onymus an' de baby?"

"I witnessed Hieronymus," said the elegant Savannah, "as I wandered from

THE POP FAMILY.

school. He was with a multitude of boys, who cheered, without a sign of disapperation, two canine beasts, that tore each other in deadly feud."

"Yer don't mean to say, Sissy, dat 'Onymus Pop is gone ter a dog-fight?"

"Such are my meaning," said Sissy, with dignity.

"Den whar's de baby?"

For answer, a long low wail smote upon their ears, as Savannah would have said.

"Fan me!" cried Mother Pop. "Dat's Tiddlekins's voice."

"Never min' about fannin', mammy," cried Weekly, Savannah's twin, a youth of fifteen, who could read, and was much addicted to gory tales of thunder and blood; "let's fin' de baby. P'r'aps he's been murdered by dat ruffian Hi, an' dat's his ghos' dat we hears a-callin'."

A search was instituted — under the bed, in the bed, in the wash-tub, and the soup-kettle; behind the wood-pile, and in the pea-vines; up the chimney, and in the ash-hopper; but all in vain. No Tiddlekins appeared, though still they heard him cry.

"Shade of Ole Hickory!" cried the father Pop, "whar, whar is dat chile?" Then, with a sudden lighting of the eye. "Unchain de dog," said he; "he'll smell him out."

There was a superannuated bloodhound pertaining to the Pop ménage that they kept tied up all day under a delusion that he was fierce. They unchained this wild animal, and with many kicks endeavored to goad his nostrils to their duty.

It happened that a piece of fresh pork hung in the well, and Lord Percy — so was the dog called — was hungry. So he hurried with vivacity toward the fresh pork.

"De well!" shrieked Mother Pop, tumbling down all in a heap, and looking somehow like Turner's "Slave-Ship," as one stumpy leg protruded from the wreck of red flannel and ruffled petticoats.

"What shall we do?" said Sissy, with a helpless squeak.

A NEIGHBOR'S BOY.

"Why, git him out," said Mr. Pop, who was the practical one of the family.

He began to draw up the well bucket, aided by Weekly, who whispered darkly: "Dar'll be anudder hangin' in town befo' long *and Hi won't miss dat hangin'*."

Soon appeared a little woolly head, then half a black body, the rest of him being securely wedged in the well bucket. He looked like a jack-in-the-box. But he was cool, Tiddlekins was, no doubt of that.

Mother Pop revived at sight of her offspring, still living, and feebly sucking his thumb.

"Ef we had a whiskey bath ter put him in!" she cried.

Into the house flew Father Pop, seized the quart cup, and was over to the white house on the hill in the wink of a cat's eye.

"He stammered forth his piteous tale," said Savannah, telling the story the next day to her schoolmates; "and Judge Chambers himself filled his cup with the best of Bourbon, and Miss Clara came over to see us resusirate the infant."

Mother Pop had Tiddlekins wrapped in hot flannel when he got back; and with a never-to-be-sufficiently-admired economy Mr. Pop moistened a rag with "the best of Bourbon," and said to his wife, "Jes rub him awhile, Cynthy, an' see if dat won't bring him roun'."

As she rubbed, he absent-mindedly raised the quart cup to his lips, and with three deep and grateful gulps the whiskey bath went to refresh the inner man of Tiddlekins's papa.

Then who so valorous and so affectionate as he? Dire were his threats against Hieronymus, deep his lamentations over his child.

"My po' little lammie!" he sobbed. "Work away, Cynthy. Dat chile mus' be saved, even if I should have ter go over ter de judge's for anudder quart o' whiskey. Nuthin' shall be spared to save that preciousest kid o' my ole age."

Miss Clara did not encourage his self-sacrificing proposal; but for all that, it was not long before Tiddlekins grew warm and lively, and winked at his father — so that good old man declared — as he lay on his back, placidly sucking a pig's tail. Savannah had roasted it in the ashes, and it had been cut from the piece of pork that had shared the well with Tiddlekins. The pork belonged to a neighbor, by-the-way; but at such a time the Pop family felt that they might dispense with the vain and useless ceremony of asking for it.

The excitement was over, the baby asleep, Miss Clara gone, and the sun well on its way to China, when a small figure was seen hovering about the gate. It had a limp air of dejection, and seemed to feel some delicacy about coming further.

"The miscreant is got back," remarked Savannah.

"Hieronymus," calls Mrs. Pop, "you may thank yo' heavenly stars dat you ain't a murderer dis summer day" —

"A-waitin' ter be hung nex' wild-grape-time," finished Weekly, pleasantly.

Mr. Pop said nothing. But he reached down from the mantel-shelf a long thin something, shaped like a snake, and quivered it in the air.

Then he walked out to Hi, and taking him by the left ear, led him to the wood-pile.

And here — But I draw a veil.

KATHERINE SHERWOOD BONNER McDOWELL.

OBEDIENCE TO LAW.

A LITTLE consideration of what takes place around us every day, would show us that a higher law than that of our will regulates events; that our painful labors are unnecessary and fruitless, that only in our easy, simple, spontaneous action are we strong, and by contenting ourselves with obedience we become divine. Belief and love, — a believing love will relieve us of a vast load of care. O my brothers, God exists. There is a soul at the center of nature and over the will of every man, so that none of us can wrong the universe. It has so infused its strong enchantment into nature that we prosper when we accept its advice, and when we struggle to wound its creatures our hands are glued to our sides, or they beat our own breasts. The whole course of things goes to teach us faith. We need only obey. There is guidance for each of us, and by lowly listening we shall hear the right word. Why need you choose so painfully your place and occupation and associates and mode of action and of entertainment? Certainly there is a possible right for you that precludes the need of balance and wilful election. For you there is a reality, a fit place and congenial duties. Place yourself in the middle of the stream of power and wisdom which animates all whom it floats, and you are without effort impelled to truth, to right, and a perfect contentment. Then you put all gainsayers in the wrong. Then you are the world, the measure of right, of truth, of beauty. If we would not be marplots with our miserable interferences, the work, the society, letters, arts, science, religion of men would go on far better than now, and the heaven predicted from the beginning of the world, and still predicted from the bottom of the heart, would organize itself, as do now the rose and the air and the sun.

RALPH WALDO EMERSON.

WAR THE DESTROYER.

WE must keep Bonaparte for some time longer at war, as a state of probation. Gracious God, sir! is war a state of probation? Is peace a rash system? Is it dangerous for nations to live in amity with each other? Are your vigilance, your policy, your common powers of observation, to be extinguished by putting an end to the horrors of war? Cannot this state of probation be as well undergone without adding to the catalogue of human sufferings? "But we must pause!" What! must the bowels of Great Britain be torn out — her best blood be spilled — her treasure wasted — that you may make an experiment?

Put yourselves, oh! that you would put yourselves in the field of battle, and learn to judge of the sort of horrors that you excite! In former wars a man might, at least, have some feeling, some interest, that served to balance in his mind the impressions which a scene of carnage and of death must inflict. If a man had been present at the battle of Blenheim, for instance, and had inquired the motive of the battle, there was not a soldier engaged who could not have satisfied his curiosity, and even, perhaps, allayed his feelings. They were fighting, they knew, to repress the uncontrolled ambition of the Grand Monarch. But if a man were present now at a field of slaughter, and were to inquire for what they were fighting — "Fighting!" would be the answer; "they are not fighting; they are pausing." "Why is that man expiring? Why is that other writhing with agony? What means this implacable fury?" The answer must be: "You are quite wrong, sir; you deceive yourself — they are not fighting — do not disturb them — they are merely pausing! This man is not expiring with agony — that man is not dead — he is only pausing! Lord help you, sir! they are not angry with one another; they have now no cause of quarrel; but their country thinks that there should be a pause. All that you see, sir, is nothing like fighting — there is no harm, nor cruelty, nor bloodshed in it whatever; it is nothing more than a political pause! It is merely to try an experiment — to see whether Bonaparte will not behave himself better than heretofore; and in the meantime we have agreed to a pause, in pure friendship!" And is this the way, sir, that you are to show yourselves the advocates of order? You take up a system calculated to uncivilize the world — to destroy order — to trample on religion — to stifle in the heart, not merely the generosity of noble sentiment, but the affections of social nature; and in the prosecution of this system, you spread terror and devastation all around you.

CHARLES JAMES FOX.

THE GIFT OF GOLD.

SHE had set out at an early hour, but had lingered on the road, inclined by her indolence to believe that if she waited under a warm shed the snow would cease to fall. She had waited longer than she knew, and now that she found herself belated in the snow-hidden ruggedness of the long lanes, even the animation of a vindictive purpose could not keep her spirit from failing. It was seven o'clock, and by this time she was not very far from Raveloe, but she was not familiar enough with those monotonous lanes to know how near she was to her journey's end. She needed comfort, and she knew but one comforter — the familiar demon in her bosom; but she hesitated a moment, after drawing out the black remnant, before she raised it to her lips. In that moment the mother's love pleaded for painful

BONAPARTI.

consciousness rather than oblivion — pleaded to be left in aching weariness, rather than to have the encircling arms benumbed so that they could not feel the dear burden. In another moment Molly had flung something away, but it was not the black remnant — it was an empty vial. And she walked on again under the breaking cloud, from which there came now and then the light of a quickly-veiled star, for a freezing wind had sprung up since the snowing had ceased. But she walked always more and more drowsily, and clutched more and more automatically the sleeping child at her bosom.

Slowly the demon was working his will, and cold and weariness were his helpers. Soon she felt nothing but a supreme immediate longing that curtained off all futurity — the longing to lie down and sleep. She had arrived at a spot where her footsteps were no longer checked by a hedgerow, and she had wandered vaguely, unable to distinguish any objects, notwithstanding the wide whiteness around her, and the growing starlight. She sank down against a straggling furze bush, an easy pillow enough; and the bed of snow, too, was soft. She did not feel that the bed was cold, and did not heed whether the child would wake and cry for her. But her arms had not yet relaxed their instinctive clutch; and the little one slumbered on as gently as if it had been rocked in a lace-trimmed cradle.

But the complete torpor came at last: the fingers lost their tension, the arms unbent; then the little head fell away from the bosom, and the blue eyes opened wide on the cold starlight. At first there was a little peevish cry of "mammy," and an effort to regain the pillowing arm and bosom; but mammy's ear was deaf, and the pillow seemed to be slipping away backward. Suddenly, as the child rolled downward on its mother's knees, all wet with snow, its eyes were caught by a bright glancing light on the white ground, and, with the ready transition of infancy, it was immediately absorbed in watching the bright living thing running towards it, yet never arriving. That bright living thing must be caught; and in an instant the child had slipped on all fours, and held out one little hand to catch the gleam. But the gleam would not be caught in that way, and now the head was held up to see where the cunning gleam came from. It came from a very bright place; and the little one, rising on its legs, toddled through the snow, the old grimy shawl in which it was wrapped trailing behind it, and the queer little bonnet dangling at its back — toddled on to the open door of Silas Marner's cottage, and right up to the warm hearth, where there was a bright fire of logs and sticks, which had thoroughly warmed the old sack (Silas's great-coat) spread out on the bricks to dry. The little one, accustomed to be left to itself for long hours without notice from its mother, squatted down on the sack, and spread its tiny hands towards the blaze, in perfect contentment, gurgling and making many inarticulate communications to the cheerful fire, like a new-hatched gosling beginning to find itself comfortable. But presently the warmth had a lulling effect, and the little golden head sank down on the old sack, and the blue eyes were veiled by their delicate half-transparent lids.

But where was Silas Marner while this strange visitor had come to his hearth? He was in the cottage, but he did not see the child. During the last few weeks,

THE GIFT OF GOLD.

since he had lost his money, he had contracted the habit of opening his door and looking out from time to time, as if he thought that his money might be somehow coming back to him, or that some trace, some news of it, might be mysteriously on the road, and be caught by the listening ear or the straining eye. It was chiefly at night, when he was not occupied in his loom, that he fell into this repetition of an act for which he could have assigned no definite purpose, and which can hardly be understood except by those who have undergone a bewildering separation from a supremely loved object. In the evening twilight, and later whenever the night was not dark, Silas looked out on that narrow prospect round the Stonepits, listening and gazing, not with hope, but with mere yearning and unrest.

This morning he had been told by some of his neighbors that it was New Year's Eve, and that he must sit up and hear the old year rung out and the new rung in, because that was good luck, and might bring his money back again. This was only a friendly Raveloe-way of jesting with the half-crazy oddities of a miser, but it had perhaps helped to throw Silas into a more than usually excited state. Since the on-coming of twilight he had opened his door again and again, though only to shut it immediately at seeing all distance veiled by the falling snow. But the last time he opened it the snow had ceased, and the clouds were parting here and there. He stood and listened, and gazed for a long while — there was really something on the road coming towards him then, but he caught no sign of it; and the stillness and the wide trackless snow seemed to narrow his solitude, and touched his yearning with the chill of despair. He went in again, and put his right hand on the latch of the door to close it, — but he did not close it: he was arrested, as he had been already since his loss, by the invisible wand of catalepsy, and stood like a graven image, with wide but sightless eyes, holding open his door, powerless to resist either the good or evil that might enter there.

When Marner's sensibility returned, he continued the action which had been arrested, and closed his door, unaware of the chasm in his consciousness, unaware of any intermediate change, except that the light had grown dim, and that he was chilled and faint. He thought he had been too long standing at the door and looking out. Turning towards the hearth, where the two logs had fallen apart, and sent forth only a red uncertain glimmer, he seated himself on his fireside chair, and was stooping to push his logs together, when, to his blurred vision, it seemed as if there were gold on the floor in front of the hearth. Gold! — his own gold — brought back to him as mysteriously as it had been taken away! He felt his heart begin to beat violently, and for a few moments he was unable to stretch out his hand and grasp the restored treasure. The heap of gold seemed to glow and get large beneath his agitated gaze. He leaned forward at last, and stretched forth his hand; but instead of the hard coin with the familiar resisting outline, his fingers encountered soft warm curls. In utter amazement, Silas fell on his knees and bent his head low to examine the marvel: it was a sleeping child — a round, fair thing, with soft yellow rings all over his head. Could this be his little sister come back to him in a dream — his little sister whom he had carried about in his arms for a year before she died, when he was a small boy without shoes or stockings?

That was the first thought that darted across Silas's blank wonderment. Was it a dream? He rose to his feet again, pushed his logs together, and, throwing on some dried leaves and sticks, raised a flame; but the flame did not disperse the vision — it only lit up more distinctly the little round form of the child, and its shabby clothing. It was very much like his little sister. Silas sank into his chair powerless, under the double presence of an inexplicable surprise and a hurrying influx of memories. How and when had the child come in without his knowledge? He had never been beyond the door. But along with that question, and almost thrusting it away, there was a vision of the old home and the old streets leading to Lantern Yard — and within that vision another, of the thoughts which had been present with him in those far-off scenes. The thoughts were strange to him now, like old friendships impossible to revive; and yet he had a dreamy feeling that this child was somehow a message come to him from that far-off life: it stirred fibers that had never been moved in Raveloe — old quiverings of tenderness — old impressions of awe at the presentiment of some Power presiding over his life; for his imagination had not yet extricated itself from the sense of mystery in the child's sudden presence, and had formed no conjectures of ordinary natural means by which the event could have been brought about.

But there was a cry on the hearth: the child had awaked, and Marner stooped to lift it on his knee. It clung round his neck, and burst louder and louder into that mingling of inarticulate cries with "mammy" by which little children express the bewilderment of waking. Silas pressed it to him, and almost unconsciously uttered sounds of hushing tenderness, while he bethought himself that some of his porridge, which had got cool by the dying fire, would do to feed the child with if it were only warmed up a little.

He had plenty to do through the next hour. The porridge, sweetened with some dry brown sugar from an old store which he had refrained from using for himself, stopped the cries of the little one, and made her lift her blue eyes with a wide quiet gaze at Silas, as he put the spoon into her mouth. Presently she slipped from his knee and began to toddle about, but with a pretty stagger that made Silas jump up and follow her lest she should fall against anything that would hurt her. But she only fell in a sitting posture on the ground, and began to pull at her boots, looking up at him with a crying face as if the boots hurt her. He took her on his knee again, but it was some time before it occurred to Silas's dull bachelor mind that the wet boots were the grievance, pressing on her warm ankles. He got them off with difficulty, and baby was at once happily occupied with the primary mystery of her own toes, inviting Silas, with much chuckling, to consider the mystery too. But the wet boots had at last suggested to Silas, that the child had been walking on the snow, and this roused him from his entire oblivion of any ordinary means by which it could have entered or been brought into his house. Under the prompting of this new idea, and without waiting to form conjectures, he raised the child in his arms, and went to the door. As soon as he had opened it, there was the cry of "mammy" again, which Silas had not heard since the child's first hungry waking. Bending forward, he could just discern the marks made by the little feet on the virgin snow, and he followed their track to the furze bushes. "Mammy!" the little one cried again and again, stretching itself forward so as almost to escape from Silas's arms, before he himself was aware that there was something more than the bush before him—that there was a human body, with the head sunk low in the furze, and half-covered with the shaken snow.

GEORGE ELIOT.

OF KINGS' TREASURIES.

MIGHTY of heart, mighty of mind — magnanimous — to be this, is, indeed, to be great in life ; to become this increasingly, is, indeed, to "advance in life," — in life itself, not in the trappings of it. My friends, do you remember that old Scythian custom, when the head of a house died ? How he was dressed in his finest dress, and set in his chariot, and carried about to his friends' houses ; and each of them placed him at his table's head, and all feasted in his presence ? Suppose it were offered to you in plain words, as it is offered to you in dire facts, that you should gain this Scythian honor gradually, while you yet thought yourself alive. Suppose the offer were this : You shall die slowly ; your blood shall daily grow cold, your flesh petrify, your heart beat at last only as a rusted group of iron valves. Your life shall fade from you, and sink through the earth into the ice of Caina ; but day by day your body shall be dressed more gaily, and set in higher chariots, and have more orders on its breast — crowns on its head, if you will. Men shall bow before it, stare and shout round it, crowd after it up and down the streets ; build palaces for it ; feast with it at their tables' heads all the night long. Your soul shall stay enough within it to know what they do, and feel the weight of the golden dress on its shoulders, and the furrow of the crown-edge on the skull ; — no more. Would you take the offer verbally made by the death-angel ? Would the meanest among us take it, think you ? Yet practically and verily we grasp at it, every one of us, in a measure ; many of us grasp at it in its fulness of horror. Every man accepts it who desires to advance in life without knowing what life is ; who means only that he is to get more horses and more footmen and more fortune and more public honor, and — not more personal soul. He only is advancing in life whose heart is getting softer, whose blood warmer, whose brain quicker, whose spirit is entering into living peace. And the men who have this life in them are the true lords or kings of the earth — they, and they only. All other kingships, so far as they are true, are only the practical issue and expression of theirs ; if less than this, they are either dramatic royalties, — costly shows, set off, indeed, with real jewels instead of tinsel, but still only the toys of nations — or else they are no royalties at all, but tyrannies, or the mere active and practical issue of national folly ; for which reason I have said of them elsewhere, "Visible governments are the toys of some nations, the diseases of others, the harness of some, the burdens of more."

But I have no words for the wonder with which I hear kinghood still spoken of, even among thoughtful men, as if governed nations were a personal property, and might be bought and sold, or otherwise acquired, as sheep, of whose flesh their king was to feed, and whose fleece he was to gather; as if Achilles' indignant epithet of base kings, "people-eating," were the constant and proper title of all monarchs; and enlargement of the king's dominion meant the same thing as the increase of a private man's estate! Kings who think so, however powerful, can no more be the true kings of the nation than gadflies are the kings of a horse; they suck it, and may drive it wild, but do not guide it. They and their courts and their armies are, if one could see clearly, only a large species of marsh mosquito, with bayonet proboscis and melodious, bandmastered trumpeting in the summer air; the twilight being, perhaps, sometimes fairer, but hardly more wholesome, for its glittering mists of midge companies. The true kings, meanwhile, rule quietly, if at all, and hate ruling; too many of them make *il gran rifiuto;* and if they do not, the mob, as soon as they are likely to become useful to it, is pretty sure to make its *gran refiuto* of them.

Yet the visible king may also be a true one some day, if ever day comes when he will estimate his dominion by the force of it, — not the geographical boundaries. It matters very little whether Trent cuts you a cantel out here, or Rhine rounds you a castle less there; but it does matter to you, king of men, whether you can verily say to this man, "Go," and he goeth, and to another, "Come," and he cometh. Whether you can turn your people as you can Trent; and where it is that you bid them come, and where go. It matters to you, king of men, whether your people hate you, and die by you, or love you, and live by you. You may measure your dominion by multitudes, better than by miles; and count degrees of love-latitude, not from, but to, a wonderfully warm and infinite equator.

Measure! — nay, you cannot measure. Who shal measure the difference between the power of those who "do and teach," and who are greatest in the kingdoms of earth, as of heaven, and the power of those who undo and consume, whose power, at the fullest, is only the power of the moth, and the rust? Strange! to think how the Moth-kings lay up treasures for the moth; and the Rust-kings, who are to their people's strength as rust to armor, lay up treasures for the rust; and the Robber-kings, treasures for the robber; but how few kings have ever laid up treasures that needed no guarding — treasures of which the more thieves there were the better! Broidered robe, only to be rent; helm and sword, only to be dimmed; jewel and gold, only to be scattered;— there have been three kinds of kings who have gathered these. Suppose there ever should arise a fourth order of kings who had read in some obscure writing of long ago that there was a fourth kind of treasure which the jewel and gold could not equal, neither should it be valued with pure gold. A web made fair in the weaving by Athena's shuttle; an armor forged in divine fire by Vulcanian force; a gold to be mined in the very sun's red heart, where he sets over the Delphian cliffs, — deep-pictured tissue, impenetrable armor, potable gold, the three great Angels of Conduct, Toil, and Thought, still calling to us, and waiting at the posts of our doors, to lead us with their winged power, and guide us with their unerring eyes, by the path which no fowl knoweth, and which the vulture's eye has not seen! Suppose kings should ever arise who heard and believed this word, and at last gathered and brought forth treasures of Wisdom for their people.

John Ruskin.

A PERILOUS VOYAGE.

"AFTER you are fairly in the water," said Mrs. Aleshine, as she swept along, although without the velocity which that phrase usually implies, "it isn't half so bad as I thought it would be. For one thing, it don't feel a bit salt, although I must say it tasted horribly that way when I first went into it."

"You didn't expect to find pickle-brine, did you?" said Mrs. Lecks. "Though if it was, I suppose we could float on it settin'."

"And as to bein' cold," said Mrs. Aleshine, "the part of me that's in is actually more comfortable than that which is out."

"There's one thing I would have been afraid of," said Mrs. Lecks, "if we hadn't made preparations for it, and that's sharks."

"Preparations!" I exclaimed, "how in the world did you prepare for sharks?"

"Easy enough," said Mrs. Lecks. "When we went down into our room to get ready to go away in the boats we both put on black stockin's. I've read that sharks never bite colored people, although if they see a white man in the water they'll snap him up as quick as lightnin'; and black stockin's was the nearest we could come to it. You see, I thought as like as not we'd have some sort of an upset before we got through."

"It's a great comfort," remarked Mrs. Aleshine, "and I'm very glad you thought of it, Mrs. Lecks. After this I shall make it a rule: Black stockin's for sharks."

"I suppose in your case," said Mrs. Lecks, addressing me, "dark trousers will do as well."

To which I answered that I sincerely hoped they would.

"Another thing I'm thankful for," said Mrs. Aleshine, "is that I thought to put on a flannel skeert."

"And what's the good of it," said Mrs. Lecks, "when it's soppin' wet?"

"Flannel's flannel," replied her friend, "whether it's wet or dry; and if you'd had the rheumatism as much as I have, you'd know it."

To this Mrs. Lecks replied with a sniff, and asked me how soon I thought we would get sight of the ship, for if we were going the wrong way, and had to turn round and go back, it would certainly be very provoking.

I should have been happy indeed to be able to give a satisfactory answer to this question. Every time that we rose upon a swell I threw a rapid glance around the whole circle of the horizon, and at last, not a quarter of an hour after Mrs. Lecks' question, I was rejoiced to see, almost in the direction in which I supposed it ought to be, the dark spot which I had before discovered. I shouted the glad news, and as we rose again my companions strained their eyes in the direction to which I pointed. They both saw it, and were greatly satisfied.

"Now, then," said Mrs. Aleshine, "it seems as if there was somethin' to work for," and she began to sweep her oar with great vigor.

"If you want to tire yourself out before you get there, Barb'ry Aleshine," said Mrs. Lecks, "you'd better go on in that way. Now what I advise is that we stop rowin' altogether and have somethin' to eat, for I'm sure we need it to keep up our strength."

"Eat!" I cried. "What are you going to eat? Do you expect to catch fish?"

"And eat 'em raw?" said Mrs. Lecks. "I should think not. But do you suppose, Mr. Craig, that Mrs. Aleshine and me would go off and leave that ship without takin' somethin' to eat by the way? Let's all gether here in a bunch, and see what sort of a meal we can make. And now, Barb'ry Aleshine, if you lay your oar down there on the water, I recommend you to tie it to one of your bonnet-strings, or it'll be floatin' away, and you won't get it again."

As she said this, Mrs. Lecks put her right hand down into the water, and fumbled about apparently in search of a pocket. I could not but smile as I thought of the condition of food when, for an hour or more, it had been a couple of feet under the surface of the ocean; but my ideas on the subject were entirely changed when I saw Mrs. Lecks hold up in the air two German sausages, and shake the briny drops from their smooth and glittering surfaces.

"There's nothin'," she said, "like sausages for shipwreck and that kind o' thing. They're very sustainin', and bein' covered with a tight skin, water can't get at 'em, no matter how you carry 'em. I wouldn't bring these out in the boat, because havin' the beans we might as well eat them. Have you a knife about you, Mr. Craig?"

I produced a dripping jack-knife, and after the open blade had been waved in the air to dry it a little, Mrs. Lecks proceeded to divide one of the sausages, handing the other to me to hold meanwhile.

"Now don't go eatin' sausages without bread, if you don't want 'em to give you dyspepsy," said Mrs. Aleshine, who was tugging at a submarine pocket.

"I'm very much afraid your bread is all soaked," said Mrs. Lecks.

To which her friend replied that that remained to be seen, and forthwith produced with a splash a glass preserve-jar with a metal top.

"I saw this, nearly empty, as I looked into the ship's pantry, and I stuffed into it all the soft biscuits it would hold. There was some sort of jam left at the bottom, so that the one who gets the last biscuit will have somethin' of a little spread on it. And now, Mrs. Lecks," she continued triumphantly, as she unscrewed the top, "that rubber ring has kept 'em as dry as chips. I'm mighty glad of it, for I had trouble enough gettin' this jar into my pocket, and gettin' it out, too, for that matter."

Floating thus, with our hands and shoulders above the water, we made a very good meal from the sausages and and soft biscuit.

"Barb'ry Aleshine," said Mrs. Lecks, as her friend proceeded to cut the second sausage, "don't you lay that knife down when you've done with it, as if 'twas an oar; for if you do it'll sink, as like as not, about six miles. I've read that the ocean is as deep as that in some places."

"Goodness gracious me!" exclaimed Mrs. Aleshine, "I hope we are not over one of them deep spots."

"There's no knowin'," said Mrs. Lecks, "but if it's more comfortin' to think it's shallerer, we'll make up our minds that way. Now, then," she continued, "we'll finish off this meal with a little somethin' to drink. I'm not given to takin' spirits, but I never travel without a little whiskey, ready mixed with water, to take if it should be needed."

So saying, she produced from one of her pockets a whiskey-flask tightly corked, and of its contents we each took a sip, Mrs. Aleshine remarking that, leaving out being chilled or colicky, we were never likely to need it more than now.

Thus refreshed and strengthened, Mrs. Lecks and Mrs. Aleshine took up their oars while I swam slightly in advance, as before. When, with occasional intermissions of rest, and a good deal of desultory conversation, we had swept and swam for about an hour, Mrs. Lecks suddenly exclaimed: "I can see that thing ever so much plainer now, and I don't believe it's a ship at all. To me it looks like bushes."

"You're mighty long-sighted without your specks," said Mrs. Aleshine, "and I'm not sure but what you're right."

For ten minutes or more I had been puzzling over the shape of the dark spot which was now nearly all the time in sight. Its peculiar form had filled me with a dreadful fear that it was the steamer, bottom upwards, although I knew enough about nautical matters to have no good reason to suppose that this could be the case. I am not far-sighted, but when Mrs. Lecks suggested bushes, I gazed at the distant object with totally different ideas, and soon began to believe that it was not a ship, either right side up or wrong side up, but that it might be an island. This belief I proclaimed to my companions, and for some time we all worked with increased energy in the desire to get near enough to make ourselves certain in regard to this point.

"As true as I'm standin' here," said Mrs. Lecks, who, although she could not read without spectacles, had remarkably good sight at long range, "them is trees and bushes that I see before me, though they do seem to be growin' right out of the water."

"There's an island under them; you may be sure of that!" I cried. "And isn't this ever so much better than a sinking ship?"

"I'm not so sure about that," said Mrs. Aleshine. "I'm used to the ship, and as long as it didn't sink I'd prefer it. There's plenty to eat on board of it, and good beds to sleep on, which is more than can be expected on a little bushy place like that ahead of us. But then, the ship might sink all of a suddint, beds, vittles and all."

"Do you suppose that is the island the other boats went to?" asked Mrs. Lecks.

This question I had already asked of myself. I had been told that the island to which the captain intended to take his boats lay about thirty miles south of the point where we left the steamer. Now I knew very well that we had not come thirty miles, and had reason to believe, moreover, that the greater part of the progress we had made had been towards the north. It was not at all probable that the position of this island was unknown to our captain; and it must, therefore, have been considered by him as an unsuitable place for the landing of his passengers. There might be many reasons for this unsuitableness; the island might be totally barren and desolate; it might be the abode of unpleasant natives; and more important than anything else, it was, in all probability, a spot where steamers never touched.

But, whatever its disadvantages, I was most wildly desirous to reach it; more so, I believe, than either of my companions. I do not mean that they were not sensible of their danger, and desirous to be freed from it; but they were women who had probably had a rough time of it during a great part of their lives, and on emerging from their little circle of rural experiences accepted with equanimity, and almost as a matter of course, the rough times which come to people in the great outside world.

"I do not believe," I said, in answer to Mrs. Lecks, "that that is the island to which the captain would have taken us; but, whatever it is, it is dry land, and we must get there as soon as we can."

"That's true," said Mrs. Aleshine, "for I'd like to have ground nearer to my feet than six miles, and if we don't find anythin' to eat and any place to sleep when we get there, it's no more than can be said of where we are now."

"You're too particular, Barb'ry Aleshine," said Mrs. Lecks, "about your comforts. If you find the ground too hard to sleep on when you get there, you can put on your life-preserver, and go to bed in the water."

"Very good," said Mrs. Aleshine; "and if these islands are made of coral, as I've heard they was, and if they're as full of small p'ints as some coral I've got at home, you'll be glad to take a berth by me, Mrs. Lecks."

I counseled my companions to follow me as rapidly as possible, and we all pushed vigorously forward. When we had approached near enough to the island to see what sort of place it really was, we perceived that it was a low-lying spot, apparently covered with verdure, and surrounded, as far as we could see as we rose on the swells, by a rocky reef, against which a tolerably high surf was running. I knew enough of the formation of these coral islands to suppose that within this reef was a lagoon of smooth water, into which there were openings through the rocky barrier. It was necessary to try to find one of these, for it would be difficult and perhaps dangerous to attempt to land through the surf.

Before us we could see a continuous line of white-capped breakers; and so I led my little party to the right, hoping that we would soon see signs of an opening in the reef.

We swam and paddled, however, for a long time, and still the surf rolled menacingly on the rocks before us. We were now as close to the island as we could approach with safety, and I determined to circumnavigate it, if necessary, before I would attempt, with these two women, to land upon that jagged reef. At last we perceived, at no great distance before us, a spot where there seemed to be no breakers; and when we reached it we found, to our unutterable delight, that here was smooth water flowing through a wide opening in the reef. The rocks were piled up quite high, and the reef, at this point at least, was a wide one; for as we neared the opening we found that it narrowed very soon and made a turn to the left, so that from the outside we could not see into the lagoon.

I swam into this smooth water, followed close by Mrs. Lecks and Mrs. Aleshine, who, however, soon became unable to use their oars, owing to the proximity of the rocks. Dropping these useful implements, they managed to paddle after me with their hands; and they were as much astonished as I was when, just after making the slight turn, we found stretched across the narrow passage a great iron bar about eight or ten inches above the water. A little farther on, and two or three feet above the water, another iron bar extended from one rocky wall to the other. Without uttering a word, I examined the lower bar, and found one end of it fastened by means of a huge padlock to a great staple driven into the rock. The lock was securely wrapped in what appeared to be tarred canvas. A staple through an eye-hole in the bar secured the other end of it to the rocks.

"These bars were put here," I exclaimed, "to keep out boats, whether at high or low water. You see they can only be thrown out of the way by taking off the padlocks."

"They won't keep us out," said Mrs. Lecks, "for we can duck under. I suppose whoever put 'em here didn't expect anybody to arrive on life-preservers."

FRANK R. STOCKTON.

PROTECTION.

WELL, what do you propose to do? You have heard the Prime-Minister declare that, if he could restore all the protection which you have had, that protection would not benefit agriculturists. Is that your belief? If so, why not proclaim it? and if it is not your conviction, you will have falsified your mission in this House, by following the right honorable baronet out into the lobby, and opposing inquiry into the condition of the very man who sent you here.

With mere politicians I have no right to expect to succeed in this motion. But I have no hesitation in telling you, that, if you give me a committee of this House, I will explode the delusion of agricultural protection! I will bring forward such a mass of evidence, and give you such a preponderance of talent and of authority, that when the Blue-Book is published and sent forth to the world, as we can now send it, by our vehicles of information, your system of protection shall not live in public opinion for two years afterward. Politicians do not want that. This cry of protection has been a very convenient handle for politicians. The cry of protection carried the counties at the last election, and politicians gained honors, emoluments, and place by it.

But is that old tattered flag of protection, tarnished and torn as it is already, to be kept hoisted still in the counties for the benefit of politicians; or will you come forward honestly and fairly to inquire into this question? I cannot believe that the gentry of England will be made mere drum-heads to be sounded upon by a Prime-Minister to give forth unmeaning and empty sounds, and to have no articulate voice of their own. No! You are the gentry of England who represent the counties. You are the aristocracy of England. Your fathers led our fathers; you may lead us if you will go the right way. But, although you have retained your influence with this country longer than any of her aristocracy, it has not been by opposing popular opinion, or by setting yourselves against the spirit of the age.

In other days, when the battle and the hunting-fields were the tests of manly vigor, your fathers were first and foremost there. The aristocracy of England were not like the noblesse of France, the mere minions of a court; nor were they like the hidalgos of Madrid, who dwindled into pigmies. You have been Englishmen. You have not shown a want of courage and firmness when any call has been made upon you.

This is a new era. It is the age of improvement, it is the age of social advancement, not the age for war or for feudal sports. You live in a mercantile age, when the whole wealth of the world is poured into your lap. You cannot have the advantages of commercial rents and feudal privileges; but you may be what you always have been, if you will identify yourselves with the spirit of the age. The English people look to the gentry and aristocracy of their country as their leaders. I, who am not one of you, have no hesitation in telling you that there is a deep-rooted, an hereditary prejudice, if I may so call it, in your favor in this country. But you never got it, and you will not keep it, by obstructing the spirit of the age. If you are indifferent to enlightened means of finding employment to your own peasantry; if you are found obstructing that advance which is calculated to knit nations more together in the bonds of peace by means of commercial intercourse; if you are found fighting against the discoveries which have almost given breath and life to material nature, and setting up yourselves as obstructives of that which destiny has decreed shall go on, — why, then, you will be the gentry of England no longer, and others will be found to take your place.

RICHARD COBDEN.

TO HIS WIFE.

DEAR PRUE: — I have yours of the 14th, and am infinitely obliged to you for the length of it. I do not know another whom I could commend for that circumstance; but where we entirely love, the continuance of anything they do to please us is a pleasure. As for your relations; once for all, pray take it for granted that my regard and conduct towards all and singular of them shall be as you direct.

I hope, by the grace of God, to continue what you wish me, every way an honest man. My wife and my children are the objects that have wholly taken up my heart; and as I am not invited or encouraged in anything which regards the public, I am easy under that neglect or envy of my past actions, and cheerfully contract that diffusive spirit within the interests of my own family. You are the head of us; and I stoop to a female reign, as being naturally made the slave of beauty. But, to prepare for our manner of living when we are again together, give me leave to say, while I am here at leisure, and come to lie at Chelsea, what I think may contribute to our better way of living. I very much approve Mrs. Evans and her husband, and, if you take my advice, I would have them have a being in our house, and Mrs. Clark the care and inspection of the nursery. I would have you entirely at leisure, to pass your time with me, in diversions, in books, in entertainments, and no manner of business intrude upon us but at stated times: for, though you are made to be the delight of my eyes, and food of all my senses and faculties, yet a turn of care and housewifery, and I know not what prepossession against conversation-pleasures, robs me of the witty and the handsome woman, to a degree not to be expressed. I will work my brains and fingers to procure us plenty of things, and demand nothing of you but to take delight in agreeable dresses, cheerful discourses, and gay sights, attended by me. This may be done by putting the kitchen and the nursery in the hands I propose; and I shall have nothing to do but to pass as much time at home as I possibly can in the best company in the world. . . .

Miss Moll grows a mighty beauty, and she shall be very prettily dressed, as likewise shall Betty and Eugene; and, if I throw away a little money in adorning my brats, I hope you will forgive me. They are, I thank God, all very well; and the charming form of their mother has tempered the likeness they bear to their rough sire, who is, with the greatest fondness, your most obliged and most obedient husband,

SIR RICHARD STEELE.

ON THE WAR OF 1812.

WE are told, by gentlemen in the opposition, that government has not done all that was incumbent on it to do, to avoid just cause of complaint on the part of Great Britain; that in particular the certificates of protection, authorized by the act of 1796, are fraudulently used.

Sir, government has done too much in granting those paper protections. I can never think of them without being shocked. They resemble the passes which the master grants to his negro slave: "Let the bearer, Mungo, pass and repass without molestation." What do they imply? That Great Britain has a right to seize all who are not provided with them. From their very nature, they must be liable to abuse on both sides. If Great Britain desires a mark, by which she can know her own subjects, let her give them an ear-mark. The colors that float from the masthead should be the credentials of our seamen. There is no safety to us, and the gentlemen have shown it, but in the rule that all who sail under the flag (not being enemies), are protected by the flag. It is impossible that this country should ever abandon the gallant tars who have won for us such splendid trophies. Let me suppose that the genius of Columbia should visit one of them in his oppressor's prison, and attempt to reconcile him to his forlorn and wretched condition. She would say to him, in the language of gentlemen on the other side: "Great Britain intends you no harm; she did not mean to impress you, but one of her own subjects; having taken you by mistake, I will remonstrate, and try to prevail upon her, by peaceable means, to release you; but I cannot, my son, fight for you." If he did not consider this mere mockery, the poor tar would address her judgment and say: "You owe me, my country, protection; I owe you, in return, obedience. I am no British subject; I am a native of old Massachusetts, where lived my aged father, my wife, my children. I have faithfully discharged my duty. Will you refuse to do yours?" Appealing to her passions, he would continue: "I lost this eye in fighting under Truxton, with the Insurgente; I got this scar before Tripoli; I broke this leg on board the Constitution, when the Guerriere struck." . . . I will not imagine the dreadful catastrophe to which he would be driven by an abandonment of him to his oppressor. It will not be, it cannot be, that his country will refuse him protection. . . .

An honorable peace is attainable only by an efficient war. My plan would be to call out the ample resources of the country, give them a judicious direction, prosecute the war with the utmost vigor, strike wherever we can reach the enemy, at sea or on land, and negotiate the terms of a peace at Quebec or at Halifax. We are told that England is a proud and lofty nation, which, disdaining to wait for danger, meets it half way. Haughty as she is we triumphed over her once, and, if we do not listen to the counsels of timidity and despair, we shall again prevail. In such a cause, with the aid of Providence, we must come out crowned with success; but, if we fail, let us fail like men, lash ourselves to our gallant tars, and expire together in one common struggle, fighting for free trade and seamen's rights.

HENRY CLAY.

A TALENT FOR MUSIC.

It may have been a quarter of an hour later on, that my attention was suddenly arrested by the sound of music issuing from the back room, where Mr. Finkelstein remained alone. I recognized the tune as the Carnival of Venice; and it brought my heart into my mouth, for that was one of the tunes that my grandmother had used to play upon her piano. But now the instrument was not a piano. Unless my ears totally deceived me, it was a hand-organ. This struck me as very odd; and I went to the door of the parlor, and looked in. There sat Mr. Finkelstein, a newspaper open before him, and a cigar between his fingers, reading and smoking; while on the floor in front of him, surely enough, stood a hand-organ; and, with his foot upon the crank of it, he was operating the instrument just as you would operate the wheel of a bicycle. Well, I couldn't help smiling, though I knew that it was unmannerly of me to do so. The scene was really too ludicrous for anything. Mr. Finkelstein appeared a little embarrassed when he spied me looking at him, and stopped his playing, and said rather sheepishly, with somewhat of the air of a naughty child surprised in mischief, "Vail, Kraikory, I suppose you tink I'm crazy, hey? Vail, I cain't help it; I'm so fond of music. But look at here, Kraikory; don't you say nodings to Solly about it, will you? Dere's a goot poy. Don't you mention it to him. He vouldn't naifer let me hear de laist of it."

I having pledged myself to secrecy, Mr. Finkelstein picked the hand-organ up, and locked it away out of sight in a closet. But after we had had our dinner, he brought it forth again, and, not without some manifest hesitation, addressed me thus: "Look at here, Kraikory; dere's a proverp which says dot man is a creature of haibits. Vail, Kraikory, I got a sort of a haibit to lie down and take a short naip every day aifter my meals. And say, Kraikory, you know how fond of music I am, don't you? I simply dote on it, Kraikory. I guess maybe I'm de fondest man of music in de United States of America. And — vail, look at here, Kraikory, as you ain't got nodings in particular to do, I tought maybe you vouldn't mind to sit here a few minutes, and — and shust turn dot craink a little while I go to sleep — hey?"

I assented willingly; so Mr. Finkelstein lay down upon his lounge, and I began to turn the crank, thereby grinding out the rollicking measures of Finnigan's Ball.

"My kracious, Kraikory, you do it splendid," the old gentleman exclaimed, by way of encouragement. "You got a graind tailent for music, Kraikory." Then

GREGORY SURPRISES MR. FINKELSTEIN AT THE HAND ORGAN.

I heard him chuckle softly to himself, and murmur, "I cain't help it, I aictually cain't. I must haif my shoke." Very soon he was snoring peacefully.

Mr. Finkelstein, when he first noticed me poring over my school-books in the shop, expressed the liveliest kind of satisfaction with my conduct.

"Dot's right, Kraikory," he cried. "Dot's maiknificent. Go ahead mit your education. Dere ain't nodings like it. A first-claiss education — vail, sir, it's de graindest advaintage a feller can haif in de baittle of life. Yes, sir, dot's a faict. You go ahead mit your education, and you study real hard, and you'll get to be —

why, you might get to be an alderman, no mistake about it. But look at here, Kraikory; tell me; where you got de books, hey? You bought 'em? You don't say so? Vail, what you pay for dem, hey, Kraikory? Two tollars! Two aictual tollars! My kracious! Vail, look at here, Kraikory; I like to make you a little present of dem books, so here's a two-tollar pill to reimburse you. Oh! dot's all right. Don't mention it. Put it in de baink. Do what you please mit it. I got anudder."

And every now and then during the summer he would inquire, "Vail, Kraikory, how you getting on mit your education? Vail, I suppose you must know pretty much aiferydings by dis time, hey? Vail, now I give you a sum. If I can buy fife barrels of aipples for six tollars and a quowter, how much will seventeen barrels of potatoes coast me, hey? . . . Ach, I was only shoking, was I? Vail, dot's a faict; I was only shoking; and you was pretty smart to find it out. But now, shoking aside, I tell you what you do. You keep right on mit your education, and you study real hard, and you'll get to be — why, you might get to be as big a man as Horace Greeley, aictually." Horace Greeley was a candidate for the presidency that year, and he had no more ardent partisan than my employer.

After the summer had passed, and September came, Mr. Finkelstein called me into the parlor one day, and began, "Now, look at here, Kraikory; I got somedings important to talk to you about. I been tinking about dot little maitter of your education a good deal lately; and I talked mit Solly about it, and got his advice; and at laist I made up my mind dot you oughter go to school. You got so much aimbition about you, dot if you get a first-claiss education while you're young, you might get to be vun of de biggest men in New York City aifter you're grown up. Vail, me and Solly, we talked it all ofer, and we made up our minds dot you better go to school right away.

"Vail, now I tell you what I do. I found out de public schools open for de season next Monday morning. Vail, next Monday morning I take you up to de public school in Fifty-first Street, and I get you aidmitted. And now I tell you what I do. If you study real hard, and get A-number-vun marks, and cratchuate all right when de time comes — vail, den I send you to college! Me and Solly, we talked it all ofer, and dot's what we made up our minds we oughter do. Dere ain't nodings like a good education, Kraikory; you can bet ten tousand tollars on dot. When I was your age I didn't haif no chaince at vun; and dot's why I'm so ecknorant. But now you got de chaince, Kraikory; and you go ahead and take advaintage of it. My kracious! When I see you cratchuate from college, I'll be so prout I von't know what to do."

HENRY HARLAND (*Sidney Luska*).

THE MILITIA BILL.

AGAINST whom are these charges brought? Against men, who in the war of the Revolution were in the councils of the nation, or fighting the battles of your country. And by whom are they made? By runaways chiefly from the British dominions, since the breaking out of the French troubles. It is insufferable. It cannot be borne. It must and ought, with severity, to be put down in this House; and out of it to meet the lie direct. We have no fellow-feeling for the suffering and oppressed Spaniards! Yet even them, we do not reprobate. Strange! that we should have no objection to any other people or government, civilized or savage, in the whole world! The great autocrat of all the Russias receives the homage of our high consideration. The Dey of Algiers and his divan of pirates are very civil, good sort of people, with whom we find no difficulty in maintaining the relations of peace and amity. "Turks, Jews, and infidels;" Melimelli, or the Little Turtle; barbarians and savages of every clime and color, are welcome to our arms. With chiefs of banditti, negro or mulatto, we can treat and trade. Name, however, but England, and all our antipathies are up in arms against her. Against whom? Against those whose blood runs in our veins; in common with whom, we claim Shakespeare, and Newton, and Chatham, for our countrymen; whose form of government is the freest on earth, our own only excepted; from whom every valuable principle of our own institutions has been borrowed — representation, jury trial, voting the supplies, writ of habeas corpus, our whole civil and criminal jurisprudence;—against our fellow Protestants, identified in blood, in language, in religion, with ourselves. In what school did the worthies of our land, the Washingtons, Henrys, Hancocks, Franklins, Rutledges of America, learn those principles of civil liberty which were so nobly asserted by their wisdom and valor? American resistance to British usurpation has not been more warmly cherished by these great men and their compatriots; not more by Washington, Hancock, and Henry, than by Chatham and his illustrious associates in the British Parliament. It ought to be remembered, too, that the heart of the English people was with us. It was a selfish and corrupt ministry, and their servile tools, to whom we were not more opposed than they were. I trust that none such may ever exist among us; for tools will never be wanting to subserve the purposes, however ruinous or wicked, of kings and ministers of state. I acknowledge the influence of a Shakespeare and a Milton upon my imagination, of a Locke upon my understanding, of a Sidney upon my political principles, of a Chatham upon qualities which, would to God I possessed in common with that illustrious man! of a Tillotson, a Sherlock, and a Porteus upon my religion. This is a British influence which I can never shake off.

JOHN RANDOLPH.

ON CONVERSATION.

... Let a man have read, thought, studied, as much as he may, rarely will he reach his possible advantages as a ready man, unless he has exercised his powers much in conversation — that was Lord Bacon's idea. Now, this wise and useful remark points in a direction not objective, but subjective — that is, it does not promise any absolute extension to truth itself, but only some greater facilities to the man who expounds or diffuses the truth. Nothing will be done for truth objectively that would not at any rate be done, but subjectively it will be done with more fluency, and at less cost of exertion to the doer. On the contrary, my own growing reveries on the latent powers of conversation (which, though a thing that then I hated, yet challenged at times unavoidably my attention) pointed to an absolute birth of new insight into the truth itself, as inseparable from the finer and more scientific exercise of the talking art. It would not be the brilliancy, the ease, or the adroitness of the expounder that would benefit, but the absolute interests of the thing expounded. A feeling dawned on me of a secret magic lurking in the peculiar life, velocities, and contagious ardor of conversation, quite separate from any which belonged to books; arming a man with new forces, and not merely with a new dexterity in wielding the old ones. I felt, and in this I could not be mistaken, as too certainly it was a fact of my own experience, that in the electric kindling of life between two minds, and far less from the kindling natural to conflict (though that also is something) than from the kindling through sympathy with the object discussed, in its momentary coruscation of shifting phases, there sometimes arise glimpses and shy revelations of affinity, suggestion, relation, analogy, that could not have been approached through any avenues of methodical study. Great organists find the same effect of inspiration, the same result of power creative and revealing, in the mere movement and velocity of their own voluntaries, like the heavenly wheels of Milton, throwing off fiery flakes and bickering flames; these *impromptu* torrents of music create rapturous *fioriture*, beyond all capacity in the artist to register, or afterward to imitate. The reader must be well aware that many philosophic instances exist where a change in the degree makes a change in the kind. Usually this is otherwise; the prevailing rule is, that the principle subsists unaffected by any possible variation in the amount or degree of the force. But a large class of exceptions must have met the reader, though from want of a pencil he has improperly omitted to write them down in his pocket-book — cases, namely, where upon passing beyond a certain point in the graduation, an alteration takes place suddenly in the kind of affect, a new direction is given to the power. Some illustration of this truth occurs in conversation, where a velocity in the movement of thought is made possible (and often natural), greater than ever can arise in methodical books; and where, secondly, approximations are more obvious and easily affected between things too remote for a steadier contemplation. One remarkable evidence of a specific power lying hid in conversation may be seen in such writings as have moved by impulses most nearly resembling those of conversation; for instance, in those of Edmund Burke. For one moment, reader, pause upon the spectacle of two contrasted intellects, Burke's and Johnson's: one an intellect essentially going forward, governed by the very necessity of growth — by the law of motion in advance; the latter, essentially an intellect retrogressive, retrospective, and throwing itself back on its own steps. This original difference was aided accidentally in Burke by the tendencies of political partisanship, which, both from moving amongst moving things and uncertainties, as compared with the more stationary aspects of moral philosophy, and also from its more fluctuating and fiery passions, must unavoidably reflect in greater life the tumultuary character of conversation. The result from these original differences of intellectual constitution, aided by these secondary differences of pursuit, is, that Dr. Johnson never, in any instance, grows a truth before your eyes, whilst in the act of delivering it, or moving toward it. All that he offers up to the end of the chapter he had when he began. But to Burke, such was the prodigious elasticity of his thinking, equally in his conversation and in his writings, the mere act of movement became the principle or cause of movement. Motion propagated motion, and life threw off life. The very violence of a projectile, as thrown by him, caused it to rebound in fresh forms, fresh angles, splintering, coruscating, which gave out thoughts as new (and that would at the beginning have been as startling) to himself as they are to his reader. In this power, which might be illustrated largely from the writings of Burke, is seen something allied to the powers of a prophetic seer, who is compelled oftentimes into seeing things as unexpected by himself as by others. Now, in conversation, considered as to its tendencies and capacities, there sleeps an intermitting spring of such sudden revelation, showing much of the same general character; a power putting on a character essentially differing from the character worn by the power of books.

Thomas De Quincey.

FOLLOWING THE LEADER.

TOURISTS ON THE CONTINENT.

. . . Three weeks of London were more than enough for me, and I feel as if I had had enough of it and pleasure. Then I remained a month with my parents; then I brought my girls on a little pleasuring tour. We spent ten days at Baden, when I set intrepidly to work again, and have been five days in Switzerland now; not bent on going up mountains, but on taking things easily. How beautiful it is! How pleasant! How great and affable, too, the landscape is! It's delightful to be in the midst of such scenes; the ideas get generous reflections from them. I don't mean to say my thoughts grow mountainous and enormous like the Alpine chain yonder; but in fine, it is good to be in the presence of this noble nature; it is keeping good company; keeping away mean thoughts. I see in the papers, now and again, accounts of fine parties in London. *Bon dieu!* Is it possible any one ever wanted to go to fine London parties; and are there now people sweating in Mayfair routs? The European continent swarms with your people. They are not all as polished as Chesterfield. I wish some of them spoke French a little better. I saw five of them at supper at Basle the other night with their knives down their throats. It was awful. My daughter saw it; and I was obliged to say: "My dear, your great-great-grandmother, one of the finest ladies of the old school I ever saw, always applied cold steel to her wittles. It's no crime to eat with a knife"; which is all very well; but I wish five of 'em at a time wouldn't. . . .

WILLIAM MAKEPEACE THACKERAY.

THE MIRACLE OF NATURE.

I WILL confess to you, though, that in those first heats of youth, this little England — or rather this little patch of moor in which I have struck roots as firm as the wild fir-trees do — looked at moments rather like a prison than a palace; that my foolish young heart would sigh, "O! that I had wings" — not as a dove, to fly home to its nest and croodle there — but as an eagle, to swoop away over land and sea, in a rampant and self-glorifying fashion, on which I now look back as altogether unwholesome and undesirable. But the thirst for adventure and excitement was strong in me, as perhaps it ought to be in all at twenty-one. Others went out to see the glorious new worlds of the West, the glorious old worlds of the East — why should not I? Others rambled over Alps and Apennines, Italian picture-galleries and palaces, filling their minds with fair memories — why should not I? Others discovered new wonders in botany and zoology — why should not I? Others too, like you, fulfilled to the utmost that strange lust after the burra shikar, which even now makes my pulse throb as often as I see the stags' heads in our friend

A——'s hall — why should not I? It is not learned in a day, the golden lesson of the old Collect, to "love the thing which is commanded, and desire that which is promised." Not in a day, but in fifteen years one can spell out a little of its worth; and when one finds one's self on the wrong side of forty, and the first gray hairs begin to show on the temples, and one can no longer jump as high as one's third button — scarcely, alas! to any button at all; and what with innumerable sprains, bruises, soakings, and chillings, one's lower limbs feel in a cold thaw, much like an old post-horse's, why, one makes a virtue of necessity; and if one still lusts after sights, takes the nearest, and looks for wonders, not in the Himalayas or Lake Ngami, but in the turf on the lawn and the brook in the park; and with good Alphonse Karr enjoys the macro-microcosm in one "*Tour autour de mon jardin.*"

A SIGHT TO MAKE ONE'S PULSES THROB.

For there it is, friend, the whole infinite miracle of nature in every tuft of grass, if we have only eyes to see it, and can disabuse our minds of that tyrannous phantom of size. Only recollect that great and small are but relative terms; that, in truth, nothing is great or small, save in proportion to the quantity of creative thought which has been exercised in making it; that the fly who basks upon one of the trilithons of Stonehenge, is in truth infinitely greater than all Stonehenge together, though he may measure the tenth of an inch, and the stone on which he sits five-and-twenty feet. You differ from me? Be it so. Even if you prove me wrong I will believe myself in the right: I cannot afford to do otherwise. If you rob me of my faith in "minute philosophy," you rob me of a continual source of content, surprise, delight.

CHARLES KINGSLEY.

MRS. POTIPHAR'S "CABINET SHOP."

The furnishing was certainly performed with great splendor and expense. My drawing-rooms strongly resembled the warehouse of an ideal cabinet-maker. Every whim of table — every caprice of chair and sofa, is satisfied in those rooms. There are curtains like rainbows, and carpets, as if the curtains had dripped all over the floor. There are heavy cabinets of carved walnut, such as belong in the heavy wainscotted rooms of old palaces, set against my last French pattern of wall-paper. There are lofty chairs, like the thrones of archbishops in Gothic cathedrals, standing by the side of the elaborately gilded frames of mirrors. Marble statues of Venus and Apollo support my mantels, upon which *ormolu* Louis Quatorze clocks ring the hours. In all possible places there are statues, statuettes, vases, plates, teacups, and liquor vases. The wood-work, when white, is elaborated in Moresco carving — when oak and walnut, it is heavily moulded. The contrasts are pretty, but rather sudden. In truth, my house is a huge curiosity-shop of valuable articles — clustered without taste, or feeling, or reason. They are there, because my house was large and I was able to buy them; and because, as Mrs. P. says, one must have a buhl and *ormolu*, and new forms of furniture, and do as well as one's neighbors, and show that one is rich, if he is so. They are there, in fact, because I couldn't help it. I didn't want them, but then I don't know what I did want. Somehow, I don't feel as if I had a home, merely because orders were given to the best upholsterers and fancy-men in town to send a sample of all their wares to my house. To pay a morning call at Mrs. Potiphar's is, in some ways, better than going shopping. You see more new and costly things in a shorter time. People say, "What a love of a chair!" "What a darling table!" "What a heavenly sofa!" and they all go and tease their husbands to get things precisely like them. When Kurz Pacha, the Sennaar minister, came to a dinner at my house, he said:

"Bless my soul! Mr. Potiphar, your house is just like your neighbor's."

I know it. I am perfectly aware that there is no more difference between my house and Crœsus's, than there is in two ten-dollar bills of the same bank. He might live in my house and I in his without any confusion. He has the same curtains, carpets, chairs, tables, Venuses, Apollos, busts, vases, etc. And he goes into his room and thinks it's all a devilish bore, just as I do. We have each got to refurnish every few years, and, therefore, have no possible opportunity for attaching ourselves to the objects about us. Unfortunately Kurz Pacha particularly detested precisely what Mrs. P. most liked, because it is the fashion to like them. I mean the Louis Quatorze and the Louis Quinze things.

"Taste, dear Mrs. Potiphar," said the Pacha, "was a thing not known in the days of those kings. Grace was entirely supplanted by grotesqueness, and now, instead of pure and beautiful Greek forms, we must collect these hideous things. If you are going backward to find models, why not go as far as the good ones? My dear madam, an *ormolu* Louis Quatorze clock would have given Pericles a fit.

Your drawing-rooms would have thrown Aspasia into hysterics. Things are not beautiful because they cost money; nor is any grouping handsome without harmony. Your house is like a woman dressed in Ninon de l'Enclos's bodice, with Queen Anne's hooped skirt, who limps in Chinese shoes, and wears an Elizabethan ruff around her neck, and a Druse's horn on her head. My dear madam, this is the kind of thing we go to see in museums. It is the old stock joke of the world."

By Jove! how mad Mrs. Potiphar was! She rose from the table, to the great dismay of Kurz Pacha, and I could only restrain her by reminding her that the Sennaar minister had but an imperfect idea of our language, and that in Sennaar people probably said what they thought when they conversed.

"You'd better go to Sennaar then, yourself, Mr. Potiphar," said my wife, as she smoothed her rumpled feathers.

"'Pon my word, madam, it's my own opinion," replied I.

Kurz Pacha, who is a philosopher (of the Sennaar school), asks me if people have no ideas of their own in building houses. I answer, none that I know of, except that of getting the house built. The fact is, it is as much as Paul Potiphar can do to make the money to erect his palatial residence, and then to keep it going. There are a great many fine statues in my house, but I know nothing about them; I don't see why we should have such heathen images in reputable houses. But Mrs. P. says:

"Pooh! have you no love for the fine arts?"

There it is! It doesn't do not to love the fine arts; so Polly is continually clattering up the halls and staircases with marble, and sending me heavy bills for the same.

When the house was ready, and my wife had purchased the furniture, she came and said to me:

"Now, my dear P., there is one thing we haven't thought of."

"What's that?"

"Pictures, you know, dear."

GUESTS OF THE POTIPHARS.

"What do you want pictures for?" growled I, and rather surlily, I am afraid.

"Why, to furnish the walls; what do you suppose we want pictures for?"

"I tell you, Polly," said I, "that pictures are the most extravagant kind of furniture. Pshaw! a man rubs and dabbles a little upon a canvas two feet square, and then coolly asks three hundred dollars for it."

"Dear me, Pot," she answered, "I don't want home-made pictures. What an idea! Do you think I'd have pictures on my walls that were painted in this country? — No, my dear husband, let us have some choice specimens of the old masters. A landscape by Rayfel, for instance; or one of Angel's fruit-pieces, or a cattle scene by Veryness, or a Madonna of Giddo's, or a boar-hunt of Hannibal Crackeye's."

What was the use of fighting against this sort of thing? I told her to have it her own way. Mrs. P. consulted Singe, the pastry-cook, who told her his cousin had just come out from Italy with a lot of the very finest pictures in the world, which he had bribed one of the Pope's guard to steal from the Vatican, and which he would sell at a bargain.

They hang on my walls, now. They represent nothing in particular; but in certain lights, if you look very closely, you can easily recognize something in them that looks like a lump of something brown. There is one very ugly woman with a convulsive child in her arms, to which Mrs. P. directly takes all her visitors, and asks them to admire the beautiful Shay douver of Giddo's. When I go out to dinner with people that talk of pictures and books, and that kind of thing, I don't like to seem behind; so I say, in a critical way, that Giddo was a good painter. None of them contradict me, and one day when somebody asked: "Which of his pictures do you prefer?" I answered straight, "His Shay douver," and no more questions were asked.

They hang all about the house now. The Giddo is in the dining-room. I asked the Sennaar minister if it wasn't odd to have a religious picture in the dining-room. He smiled, and said that it was perfectly proper if I liked it, and if the picture of such an ugly woman didn't take away my appetite.

"What difference does it make," said he, in the Sennaar manner; " it would be equally out of keeping with every other room in your house. My dear Potiphar, it is a perfectly unprincipled house, this of yours. If your mind were in the condition of your house, so ill-assorted, so confused, so over-loaded with things that don't belong together, you would never make another cent. You have order, propriety, harmony, in your dealings with the Symmes's Hole Bore Co., and they are the secrets of your success. Why not have the same elements in your house? Why pitch every century, country, and fashion, higgledy-piggledy into your parlors and dining-rooms? Have everything you can get, in heaven's name, but have everything in its place. If you are a plodding tradesman, knowing and caring nothing about pictures, or books, or statuary, or *objets de vertu*, don't have them. Suppose your neighbor chooses to put them in his house. If he has them merely because he had the money to pay for them, he is the butt of every picture and book he owns.

"When I meet Mr. Crœsus in Wall Street, I respect him as I do a king in his palace, or a scholar in his study. He is master of the occasion. He commands like Nelson at the Nile. I, who am merely a diplomatist, skulk and hurry along, and if Mr. Crœsus smiles, I inwardly thank him for his charity. Wall Street is Crœsus' sphere, and all his powers play there perfectly. But when I meet him in his house, surrounded by objects of art, by the triumphs of a skill which he does not understand, and for which he cares nothing,—of which, in fact, he seems afraid, because he knows any chance question about them would trip him up,—my feeling is very much changed. If I should ask him what *ormolu* is, I don't believe he could answer, though his splendid *ormolu* clock rang, indignant, from the mantel. But if I should say: 'Invest me this thousand dollars,' he would secure me eight per cent. It certainly isn't necessary to know what *ormolu* is, nor to have any other *objet de vertu* but your wife. Then why should you barricade yourself behind all these things that you really cannot enjoy, because you don't understand? If you could not read Italian, you would be a fool to buy Dante, merely because you knew he was a great poet. And, in the same way, if you know nothing about matters of art, it is equally foolish of you to buy statues and pictures, although you hear on all sides that, as Mrs. P. says, one must love art.

GEORGE WILLIAM CURTIS.

AN APPEAL FOR UNION.

NOT the reception of the treaty of peace negotiated at Ghent, nor any other event which has occurred during my progress in public life, ever gave such unbounded and universal satisfaction as the settlement of the Missouri Compromise. We may argue from like causes like effects. Then, indeed, there was great excitement. Then, indeed, all the legislatures of the North called out for the exclusion of Missouri, and all the legislatures of the South called out for her admission as a State. Then, as now, the country was agitated like the ocean in the midst of a turbulent storm. But now, more than then, has this agitation been increased. Now, more than then, are the dangers which exist, if the controversy remains unsettled, more aggravated and more to be dreaded. The idea of disunion was then scarcely a low whisper. Now, it has become a familiar language in certain portions of the country. The public mind and the public heart are becoming familiarized with that most dangerous and fatal of all events—the disunion of the States. People begin to contend that this is not so bad a thing as they had supposed. Like the progress in all human affairs, as we approach danger it disappears, it diminishes in our conception, and we no longer regard it with that awful apprehension of consequences that we did before we came into contact with it. Everywhere now there is a state of things, a degree of alarm and apprehension, and determination to fight, as they regard it, against the aggressions of the North. That did not so demonstrate itself at the period of the Missouri Compromise. It was followed, in consequence of the adoption of the measure which settled the difficulty of Missouri, by peace, harmony, and tranquillity. So, now, I infer, from the greater amount of agitation, from the greater amount of danger, that, if you adopt the measures under consideration, they, too, will be followed by the same amount of contentment, satisfaction, peace, and tranquillity, which ensued after the Missouri Compromise. . . .

The responsibility of this great measure passes from the hands of the committee, and from my hands. They know, and I know, that it is an awful and tremendous responsibility. I hope that you will meet it with a just conception and a true appreciation of its magnitude, and the magnitude of the consequences that may ensue from your decision one way or the other. The alternatives, I fear, which the measure presents, are concord and increased discord; a servile civil war, originating in its causes on the lower Rio Grande, and terminating possibly in its consequences on the upper Rio Grande in the Santa Fé country, or the restoration of harmony and fraternal kindness. I believe from the bottom of my soul that the measure is the reunion of this Union. I believe it is the dove of peace, which, taking its aërial flight from the dome of the Capitol, carries the glad tidings of assured peace and restored harmony to all the remotest extremities of this distracted land. I believe that it will be attended with all these beneficent effects. And now let us discard all resentment, all passions, all petty jealousies, all personal desires, all love of place, all hankerings after the gilded crumbs which fall from the table of power. Let us forget popular fears, from whatever quarter they may spring. Let us go to the limpid fountain of unadulterated patriotism, and, performing a solemn lustration, return divested of all selfish, sinister, and sordid impurities, and think alone of our God, our country, our consciences, and our glorious Union — that Union without which we shall be torn into hostile fragments, and sooner or later become the victims of military despotism, or foreign domination.

Mr. President, what is an individual man? An atom, almost invisible without a magnifying glass — a mere speck upon the surface of the immense universe; not a second in time, compared to immeasurable, never-beginning, and never-ending eternity; a drop of water in the great deep, which evaporates and is borne off by the winds; a grain of sand, which is soon gathered to the dust from which it sprung. Shall a being so small, so petty, so fleeting, so evanescent, oppose itself to the onward march of a great nation, which is to subsist for ages and ages to come; oppose itself to that long line of posterity which, issuing from our loins, will endure during the existence of the world? Forbid it, God. Let us look to our country and our cause, elevate ourselves to the dignity of pure and disinterested patriots, and save our country from all impending dangers. What if, in the march of this nation to greatness and power, we should be buried beneath the wheels that propel it onward! What are we — what is any man — worth who is not ready and willing to sacrifice himself for the benefit of his country when it is necessary? . . .

If this Union shall become separated, new unions, new confederacies will arise. And with respect to this, if there be any — I hope there is no one in the Senate — before whose imagination is flitting the idea of a great Southern Confederacy to take possession of the Balize and the mouth of the Mississippi, I say in my place, never! never! never! will we who occupy the broad waters of the Mississippi and its upper tributaries consent that any foreign flag shall float at the Balize or upon the turrets of the Crescent City — never! never! I call upon all the South.

Sir, we have had hard words, bitter words, bitter thoughts, unpleasant feelings toward each other in the progress of this great measure. Let us forget them. Let us sacrifice these feelings. Let us go to the altar of our country and swear, as the oath was taken of old, that we will stand by her; that we will support her; that we will uphold her Constitution; that we will preserve her Union; and that we will pass this great, comprehensive, and healing system of measures, which will hush all the jarring elements, and bring peace and tranquillity to our homes.

Henry Clay.

JUSTICE FOR THE SLAVE.

We do not play politics, anti-slavery is no half-jest with us; it is a terrible earnest, with life or death, worse than life or death, on the issue. It is no lawsuit, where it matters not to the good feeling of opposing counsel which way the verdict goes, and where advocates can shake hands after the decision as pleasantly as before. When we think of such a man as Henry Clay, his long life, his mighty influence cast always into the scale against the slave, of that irresistible fascination with which he moulded every one to his will; when we remember that, his conscience acknowledging the justice of our cause, and his heart open on every other side to the gentlest impulses, he could sacrifice so remorselessly his convictions and the welfare of millions to his low ambition; when we think how the slave trembled at the sound of his voice, and that, from a multitude of breaking hearts there went up nothing but gratitude to God when it pleased him to call that great sinner from this world, we cannot find it in our hearts, we could not shape our lips to ask any man to do him honor. No amount of eloquence, no sheen of official position, no loud grief of partisan friends, would ever lead us to ask monuments or walk in fine processions for pirates; and the sectarian zeal or selfish ambition which gives up, deliberately and in full knowledge of the facts, three million of human beings, to hopeless ignorance, daily robbery, systematic prostitution, and murder, which the law is neither able nor undertakes to prevent or avenge, is more monstrous, in our eyes, than the love of gold which takes a score of lives with merciful quickness on the high seas. Haynau on the Danube is no more hateful to us than Haynau on the Potomac. Why give mobs to one and monuments to the other?

Wendell Phillips.

RALEIGH'S LAST WORDS TO HIS WIFE.

You shall receive, my dear wife, my last words in these my last lines; my love I send you, that you may keep when I am dead, and my counsel, that you may remember it when I am no more. I would not with my will present you sorrows, dear Bess; let them go to the grave with me, and be buried in the dust. And seeing that it is not the will of God that I shall see you any more, bear my destruction patiently, and with an heart like yourself.

First, I send you all the thanks which my heart can conceive, or my words express, for your many travails and cares for me; which though they have not taken effect as you wished, yet my debt to you is not the less; but pay it I never shall in this world.

Secondly, I beseech you, for the love you bare me living, that you do not hide yourself many days, but by your travails seek to help the miserable fortunes and the right of your poor child. Your mourning cannot avail me that am but dust.

Thirdly, you shall understand, that my lands were conveyed *bona fide* to my child; the writings were drawn at midsummer was twelve months, as divers can witness: and I trust my blood will quench their malice who desired my slaughter, that they will not seek also to kill you and yours with extreme poverty. To what friend to direct you I know not, for all mine have left me in the true time of trial. Most sorry am I, that, being thus surprised by death, I can leave you no better estate; God hath prevented all my determinations,—that great God which worketh all in all; and if you can live free from want, care for no more, for the rest is but a vanity: love God, and begin betimes — in him you shall find true, everlasting, and endless comfort; when you have travailed and wearied yourself with all sorts of worldly cogitations, you shall sit down by sorrow in the end. Teach your son also to serve and fear God whilst he is young, that the fear of God may grow up in him; then will God be an husband to you, and a father to him — an husband and a father that can never be taken from you.

SIR WALTER RALEIGH.

Baylie oweth me a thousand pounds, and Aryan six hundred; in Jernesey also, I have much owing me. Dear wife, I beseech you, for my soul's sake, pay all poor men. When I am dead, no doubt you shall be much sought unto, for the world thinks I was very rich: have a care to the fair pretences of men, for no greater misery can befall you in this life, than to become a prey unto the world, and after to be despised. I speak (God knows) not to dissuade you from marriage, for it will be best for you, both in respect of God and the world. As for me, I am no more yours, nor you mine; death hath cut us asunder, and God hath divided me from the world, and you from me. Remember your poor child for his father's sake, who loved you in his happiest estate. I sued for my life, but God knows it was for you and yours that I desired it: for know it, my dear wife, your child is the child of a true man, who in his own respect despiseth death and his misshapen and ugly forms. I cannot write much; God knows how hardly I steal this time when all sleep; and it is also time for me to separate my thoughts from the world. Beg my dead body, which living was denied you, and either lay it in Sherbourne, or Exeter church by my father and mother. I can say no more; time and death call me away. The everlasting God, powerful, infinite, and inscrutable God Almighty, who is goodness itself, the true light and life, keep you and yours, and have mercy upon me, and forgive my persecutors and false accusers, and send us to meet in his glorious kingdom. My dear wife, farewell; bless my boy, pray for me, and let my true God hold you both in his arms.

Yours that was, but now not mine own,

WALTER RALEIGH.

DEATH, THE CONQUEROR.

IT is a mighty change that is made by the death of every person, and it is visible to us who are alive. Reckon but from the sprightfulness of youth and the fair cheeks and full eyes of childhood, from the vigorousness and strong flexure of the joints of five-and-twenty, to the hollowness and dead paleness, to the loathsomeness and horror of a three days' burial, and we shall perceive the distance to be very great and very strange. But so have I seen a rose newly springing from the clefts of its hood, and at first it was fair as the morning, and full with the dew of heaven, as the lamb's fleece; but when the ruder breath had forced open its virgin modesty, and dismantled its too youthful and unripe retirements, it began to put on darkness and to decline to softness and the symptoms of a sickly age; it bowed the head and broke its stalk, and at night, having lost some of its leaves, and all its beauty, it fell into the portion of weeds and out-worn faces. So does the fairest beauty change, and it will be as bad with you and me; and then what servants shall we have to wait upon us in the grave? What friends to visit us?

What officious people to cleanse away the moist and unwholesome clouds reflected upon our faces from the sides of the weeping vaults, which are the longest weepers for our funerals?

A man may read a sermon, the best and most passionate that ever man preached, if he shall but enter into the sepulchers of kings. In the same Escurial where the Spanish princes live in greatness and power, and decree war or peace, they have wisely placed a cemetery where their ashes and their glory shall sleep till time shall be no more: and where our kings have been crowned, their ancestors lie interred, and they must walk over their grandsire's head to take his crown. There is an acre sown with royal seed, the copy of the greatest change from rich to naked, from ceiled roofs to arched coffins, from living like gods to die like men. There is enough to cool the flames of lust, to abate the heights of pride, to appease the itch of covetous desires, to sully and dash out the dissembling colors of a lustful, artificial, and imaginary beauty. There the warlike and the peaceful, the fortunate and the miserable, the beloved and the despised princes, mingle their dust, and pay down their symbol of mortality, and tell all the world that when we die, our ashes shall be equal to kings, and our accounts easier, and our pains for our crimes shall be less. To my apprehension, it is a sad record which is left by Athenæus concerning Ninus the great Assyrian monarch, whose life and death is summed up in these words: "Ninus, the Assyrian had an ocean of gold, and other riches more than the sand in the Caspian Sea; he never saw the stars, and perhaps he never desired it; he never stirred up the holy fire among the Magi, nor touched his god with the sacred rod, according to the laws: he never offered sacrifice, nor worshipped the deity, nor administered justice, nor spake to the people; nor numbered them; but he was most valiant to eat and drink, and having mingled his wines, he threw the rest upon the stones. This man is dead, behold his sepulcher, and now hear where Ninus is. Sometime I was Ninus, and drew the breath of a living man, but now am nothing but clay. I have nothing but what I did eat, and what I served to myself in lust is all my portion: the wealth with which I was blessed, my enemies meeting together shall carry away, as the mad Thyades carry a raw goat. I am gone to hell: and when I went thither, I neither carried gold, nor horse, nor silver chariot. I, that wore a mitre, am now a little heap of dust."

JEREMY TAYLOR.

GREATNESS AND ABILITY.

In general, greatness is eminence of ability ; so there are as many different forms thereof as there are qualities wherein a man may be eminent. These various forms of greatness should be distinctly marked, that, when we say a man is great, we may know exactly what we mean.

In the rudest ages, when the body is man's only tool for work or war, eminent strength of body is the thing most coveted. Then, and so long as human affairs are controlled by brute force, the giant is thought to be the great man, — is had in honor for his eminent brute strength.

When men have a little outgrown that period of force, cunning is the quality most prized. The nimble brain outwits the heavy arm, and brings the circumvented giant to the ground. He who can overreach his antagonist, plotting more subtly, winning with more deceitful skill ; who can turn and double on his unseen track, "can smile and smile, and be a villain," — he is a great man.

Brute force is merely animal; cunning is the animalism of the intellect, — the mind's least intellectual element.

As men go on in their development, finding qualities more valuable than the strength of the lion or the subtlety of the fox, they come to value higher intellectual faculties, — great understanding, great imagination, great reason. Power to think is then the faculty men value most; ability to devise means for attaining ends desired ; the power to originate ideas, to express them in speech, to organize them into institutions; to organize things into a machine, men into an army or a state, or a gang of operatives ; to administer these various organizations. He who is eminent in this ability is thought the great man.

But there are qualities nobler than the mere intellect, — the moral, the affectional, the religious faculties, — the power of justice, of love, of holiness, of trust in God, and of obedience to his law, — the eternal right. These are the highest qualities of man : whoso is most eminent therein is the greatest of great men. He is as much above the merely intellectual great men, as they above the men of mere cunning or force.

Thus, then, we have four different kinds of greatness. Let me name them bodily greatness, crafty greatness, intellectual greatness, religious greatness. Men in different degrees of development will value the different kinds of greatness. Belial cannot yet honor Christ. How can the little girl appreciate Aristotle and Kant ? The child thinks as a child. You must have manhood in you to honor it in others, even to see it.

Yet how we love to honor men eminent in such modes of greatness as we can understand ! Indeed, we must do so. Soon as we really see a real great man, his magnetism draws us, will we or no. Do any of you remember when, for the first time in adult years, you stood beside the ocean, or some great mountain of New Hampshire, or Virginia, or Pennsylvania, or the mighty mounts that rise in Switzerland? Do you remember what emotions came upon you at the awful presence? But if you are confronted by a man of vast genius, of colossal history and achievements, immense personal power of wisdom, justice, philanthropy, religion, of mighty power of will and mighty act; if you feel him as you feel the mountain and the sea, what grander emotions spring up! It is like making the acquaintance of one of the elementary forces of the earth, — like associating with gravitation itself! The stiffest neck bends over; down go the democratic knees; human nature is loyal then! A New-England shipmaster, wrecked on an island in the Indian Sea, was seized by his conquerors, and made their chief. Their captive became their king. After years of rule, he managed to escape. When he once more visited his former realm, he found that the savages had carried him to heaven, and worshipped him as a god greater than their fancied deities : he had revolutionized divinity, and was himself enthroned as a god. Why so? In intellectual qualities, in religious qualities, he was superior to their idea of God, and so they worshipped him. Thus loyal is human nature to its great men.

Talk of Democracy! — we are all looking for a master ; a man manlier than we. We are always looking for a great man to solve the difficulty too hard for us, to break the rock which lies in our way, — to represent the possibility of human nature as an ideal, and then to realize that ideal in his life. Little boys in the country, working against time, with stints to do, long for the passing-by of some tall brother, who in a few minutes shall achieve what the smaller boy took hours to do. And we are all of us but little boys, looking for some great brother to come and help us end our tasks.

But it is not quite so easy to recognize the greatest kind of greatness. A Nootka-Sound Indian would not see much in Leibnitz, Newton, Socrates, or Dante; and if a great man were to come as much before us as we are before the Nootka-Sounders, what should we say of him? Why, the worst names we could devise, Blasphemer, Hypocrite, Infidel, Atheist. Perhaps we should dig up the old cross, and make a new martyr of the man posterity will worship as a deity. It is the men who are up that see the rising sun, not the sluggards. It takes greatness to see greatness, and know it at the first; I mean to see greatness of the highest kind. Bulk anybody can see ; bulk of body or mind. The loftiest form of greatness is never popular in its time. Men cannot understand or receive it. Guinea negroes would think a juggler a greater man than Franklin. What would be thought of Martin Luther at Rome, of Washington at St. Petersburg, of Fenelon among the Sacs and Foxes? Herod and Pilate were popular in their day, — men of property and standing. They got nominations and honor enough. Jesus of Nazareth got no nomination, got a cross between two thieves, was crowned with thorns, and, when he died, eleven Galileans gathered together to lament their Lord. Any man can measure a walking-stick, — so many hands long, and so many nails beside ; but it takes a mountain intellect to measure the Andes and the Altai.

THEODORE PARKER.

ANNIE AND LAWRENCE.

"A solemn solitude like this would, to my thinking, be much more likely to lower your spirits. I don't like solitude myself, and therefore, I suppose it is that I thought an impressible nature, like yours, would find something sad in the loneliness of these silent woods."

Annie turned and fixed on him her large blue eyes. "But I'm not alone," she said.

As Lawrence looked into her eyes he saw that they were as clear as the purest crystal, and that he could look through them straight into her soul, and there he saw that this woman loved him. The vision was as sudden as if it had been a night-scene lighted up by a flash of lightning, but it was as clear and plain as if it had been that same scene under the noonday sun. . . .

Never before had he looked into the eyes of a woman who loved him; and, leaning toward this one, he put his arm around her and drew her toward him. "And never shall you be alone," he said.

She looked up at him with tears starting to her eyes, and then she put her head against his breast. She was too happy to say anything, and she did not try.

FRANK R. STOCKTON.

THE ETHICS OF LAUGHTER.

ANATOMIKALLY konsidered, laffing iz the sensation ov pheeling good all over, and showing it principally in one spot.

Morally konsidered, it iz the next best thing tew the 10 commandments.

Theoretikally konsidered, it kan out-argy all the logik in existence.

Pyroteknikally konsidered, it iz the fire-works of the soul. . . .

But i don't intend this essa for laffing in the lump, but for laffing on the half-shell.

Laffing iz just az natral tew cum tew the surface az a rat iz tew cum out ov hiz hole when he wants tew.

Yu kant keep it back by swallowing enny more than yu kan the heekups.

If a man kan't laff there iz sum mistake made in putting him together, and if he won't laff he wants az mutch keeping away from, az a bear-trap when it iz sot.

I have seen people who laffed altogether too mutch for their own good or for ennyboddy else's; they laft like a barrell ov nu sider with the tap pulled out, a perfekt stream.

This iz a grate waste ov natral juice.

THE ETHICS OF LAUGHTER.

I have seen other people who didn't laff enuff tew giv themselfs vent; they waz like a barrell ov nu sider too, that waz bunged up tite, apt tew start a hoop and leak all away on the sly.

Thare ain't neither ov theze 2 ways right, and they never ought tew be pattented. . . .

Genuine laffing iz the vent of the soul, the nostrils of the heart, and iz just az necessary for health and happiness az spring water iz for a trout.

Thare iz one kind ov a laff that i always did rekommend ; it looks out ov the eye fust with a merry twinkle, then it kreeps down on its hands and kneze and plays

around the mouth like a pretty moth around the blaze ov a kandle, then it steals over into the dimples ov the cheeks and rides around into thoze little whirlpools for a while, then it lites up the whole face like the mello bloom on a damask roze, then it swims oph on the air with a peal az klear and az happy az a dinner-bell, then it goes bak agin on golden tiptoze like an angel out for an airing, and laze down on its little bed ov violets in the heart where it cum from.

Thare iz another laff that noboddy kan withstand ; it iz just az honest and noizy az a distrikt skool let out tew play, it shakes a man up from hiz toze tew hiz temples, it dubbles and twists him like a whiskee phit, it lifts him oph from hiz cheer, like feathers, and lets him bak agin like melted led, it goes all thru him like a pikpocket, and finally leaves him az weak and az krazy az tho he had bin soaking all day in a Rushing bath and forgot tew be took out.

This kind ov a laff belongs tew jolly good phellows who are az healthy az quakers, and who are az eazy tew pleaze az a gall who iz going tew be married to-morrow.

In konclushion i say laff every good chance yu kan git, but don't laff unless yu feal like it, for there ain't nothing in this world more harty than a good honest laff, nor nothing more hollow than a hartless one.

When yu do laff open your mouth wide enuff for the noize tew git out without squealing, thro yure hed bak az tho yu waz going tew be shaved, hold on tew yure false hair with both hands and then laff till yure soul gets thoroly rested.

But i shall tell yu more about theze things at sum fewter time.

HENRY W. SHAW (Josh Billings).

ROGER WILLIAMS.

WHILE the State was thus connecting by the closest bonds the energy of its faith with its form of government, there appeared in its midst one of those clear minds which sometimes bless the world by their power of receiving moral truth in its purest light, and of reducing the just conclusions of their principles to a happy and consistent practice. In February of the first year of the colony, but a few months after the arrival of Winthrop, and before either Cotton or Hooker had embarked for New England, there arrived at Nantasket, after a stormy passage of sixty-six days, "a young minister, godly and zealous, having precious" gifts. It was Roger Williams. He was then but a little more than thirty years of age; but his mind had already matured a doctrine which secures him an immortality of fame, as its application has given religious peace to the American world. He was a Puritan, and a fugitive from English persecution; but his wrongs had not clouded his accurate understanding; in the capacious recesses of his mind he had revolved the nature of intolerance, and he, and he alone, had arrived at the great principle which is its sole effectual remedy. He announced his discovery under the simple proposition of the sanctity of conscience. The civil magistrate should restrain crime, but never control opinion; should punish guilt, but never violate the freedom of the soul. The doctrine contained within itself an entire reformation of theological jurisprudence; it would blot from the statute-book the felony of nonconformity; would quench the fires that persecution had so long kept burning; would repeal every law compelling attendance on public worship; would abolish tithes and all forced contributions to the maintenance of religion; would give an equal protection to every form of religious faith; and never suffer the authority of the civil government to be enlisted against the mosque of the Mussulman or the altar of the fire-worshipper, against the Jewish synagogue or the Roman cathedral.

THE OLD MILL.

It is wonderful with what distinctness Roger Williams deduced these inferences from his great principle; the consistency with which, like Pascal and Edwards, — those bold and profound reasoners on other subjects, — he accepted every fair inference from his doctrines; and the circumspection with which he repelled every unjust imputation. In the unwavering assertion of his views he never changed his position; the sanctity of conscience was the great tenet which, with all its consequences, he defended, as he first trod the shores of New England; and in his extreme old age it was the last pulsation of his heart. But it placed the young emigrant in direct opposition to the whole system on which Massachusetts was founded; and, gentle and forgiving as was his temper, prompt as he was to concede every thing which honesty permitted, he always asserted his belief with temperate firmness and unbending benevolence.

GEORGE BANCROFT.

THE DEATH OF COLONEL NEWCOME.

BUT our Colonel, we all were obliged to acknowledge, was no more our friend of old days. He knew us again, and was good to every one round him, as his wont was; especially when Boy came, his old eyes lighted up with simple happiness, and, with eager trembling hands, he would seek under his bed-clothes, or the pockets of his dressing-gown, for toys or cakes, which he had caused to be purchased for his grandson. There was a little laughing, red-cheeked, white-headed gown-boy of the school, to whom the old man had taken a great fancy. One of the symptoms of his returning consciousness and recovery, as we hoped, was his calling for this child, who pleased our friend by his archness and merry ways; and who, to the old gentleman's unfailing delight, used to call him, "Codd Colonel." "Tell little F——, that Codd Colonel wants to see him;" and the little gown-boy was brought to him; and the Colonel would listen to him for hours; and hear all about his lessons and his play; and prattle, almost as childishly, about Dr. Raine, and his own early school-days. The boys of the school, it must be said, had heard the noble old gentleman's touching history, and had all got to know and love him. They came every day to hear news of him; sent him in books and papers to amuse him; and some benevolent young souls, — God's blessing on all honest boys, say I, — painted theatrical characters, and sent them in to Codd Colonel's grandson. The little fellow was made free of gown-boys, and once came thence to his grandfather in a little gown, which delighted the old man hugely. Boy said he would like to be a little gown-boy; and I make no doubt, when he is old enough, his father will get him that post, and put him under the tuition of my friend Dr. Senior.

So, weeks passed away, during which our dear old friend still remained with us. His mind was gone at intervals, but would rally feebly; and with his consciousness returned his love, his simplicity, his sweetness. He would talk French with Madame de Florac, at which time, his memory appeared to awaken with surprising vividness, his cheek flushed, and he was a youth again, — a youth all love and hope, — a stricken old man, with a beard as white as snow covering the noble careworn face. At such times he called her by her Christian name of Léonore; he addressed courtly old words of regard and kindness to the aged lady; anon he wandered in his talk, and spoke to her as if they still were young. Now, as in those early days, his heart was pure; no anger remained in it; no guile tainted it; only peace and good-will dwelt in it.

Rosey's death had seemed to shock him for a while when the unconscious little boy spoke of it. Before that circumstance, Clive had even forbore to wear mourning, lest the news should agitate his father. The Colonel remained silent and was very much disturbed all that day, but he never appeared to comprehend the fact quite; and, once or twice afterwards, asked, Why she did not come to see him? She was prevented, he supposed — she was prevented, he said, with a look of terror: he never once otherwise alluded to that unlucky tyrant of his household, who had made his last years so unhappy.

The circumstances of Clive's legacy he never understood: but more than once spoke of Barnes to Ethel, and sent his compliments to him, and said he should like to shake him by the hand. Barnes Newcome never once offered to touch that honored hand, though his sister bore her uncle's message to him. They came often from Bryanston Square: Mrs. Hobson even offered to sit with the Colonel, and read to him, and brought him books for his improvement. But her presence disturbed him; he cared not for her books; the two nurses whom he loved faithfully watched him; and my wife and I were admitted to him sometimes, both of whom he honored with regard and recognition. As for F. B., in order to be near his Colonel, did not that good fellow take up his lodging in Cistercian Lane, at the "Red Cow"? He is one whose errors, let us hope, shall be pardoned, *quia multum amavit.* I am sure he felt ten times more joy at hearing of Clive's legacy, than if thousands had been bequeathed to himself. . May good health and good fortune speed him!

The days went on, and our hopes, raised sometimes, began to flicker and fail. One evening the Colonel left his chair for his bed in pretty good spirits, but passed a disturbed night, and the next morning was too weak to rise. Then he remained in his bed, and his friends visited him there. One afternoon he asked for his little gown-boy, and the child was brought to him, and sat by the bed with a very awe-stricken face: and then gathered courage, and tried to amuse him by telling him how it was a half-holiday, and they were having a cricket-match with the St. Peter's boys in the green, and Gray Friars was in and winning. The Colonel quite understood about it; he would like to see the game; he had played many a game on that green when he was a boy. He grew excited; Clive dismissed his father's little friend, and put a sovereign into his hand; and away he ran to say that Codd Colonel had come into a fortune, and to buy tarts, and to see the match out. I, *curre,* little white-haired gown-boy! Heaven speed you, little friend.

After the child had gone, Thomas Newcome began to wander more and more. He talked louder; he gave the word of command, spoke Hindustanee as if to his men. Then he spoke words in French rapidly, seizing a hand that was near him, and crying, "*Toujours, toujours!*" But it was Ethel's hand which he took. Ethel and Clive and the nurse were in the room with him; the nurse came to us, who were sitting in the adjoining apartment; Madame de Florac was there, with my wife and Bayham

At the look in the woman's countenance Madame de Florac started up. "He is very bad, he wanders a great deal," the nurse whispered. The French lady fell instantly on her knees, and remained rigid in prayer.

Sometime afterwards Ethel came in with a scared face to our pale group. "He is calling for you again, dear lady," she said, going up to Madame de Florac, who was still kneeling; "and just now he said he wanted Pendennis to take care of his boy. He will not know you." She hid her tears as she spoke.

She went into the room where Clive was at the bed's foot; the old man within it talked on rapidly for awhile: then again he would sigh and be still: once more I heard him say hurriedly, "Take care of him when I'm in India;" and then with a heartrending voice he called out, "Léonore, Léonore!" She was kneeling by his side now. The patient's voice sank into faint murmurs; only a moan now and then announced that he was not asleep.

At the usual evening hour the chapel bell began to toll, and Thomas Newcome's hands outside the bed feebly beat time. And just as the last bell struck, a peculiar sweet smile shone over his face, and he lifted up his head a little, and quickly said, "*Adsum!*" and fell back. It was the word we used at school, when names were called; and lo, he, whose heart was as that of a little child, had answered to his name, and stood in the presence of The Master.

WILLIAM MAKEPEACE THACKERAY.

TO GROSVENOR C. BEDFORD.

LET not Gifford suppose me a troublesome man to deal with, pertinacious about trifles, or standing upon punctilios of authorship. No, Grosvenor, I am a quiet, patient, easy-going hack of the mule breed; regular as clock-work in my pace, surefooted, bearing the burden which is laid on me, and only obstinate in choosing my own path. If Gifford could see me by this fireside, where, like Nicodemus, one candle suffices me in a large room, he would see a man in a coat "still more threadbare than his own" when he wrote his "Imitation," working hard and getting little — a bare maintenance, and hardly that; writing poems and history for posterity, with his whole heart and soul; one daily progressive in learning, not so learned as he is poor, and not so poor as proud, not so proud as happy. Grosvenor, there is not a lighter-hearted nor a happier man on the face of this wide world.

Your godson thinks that I have nothing to do but to play with him, and anybody who saw what reason he has for his opinion would be disposed to agree with him. I wish you could see my beautiful boy!

ROBERT SOUTHEY.

GEORGE MACDONALD.

MAKING A FRIEND.

WHEN, having followed the road, I stood at last on the bridge, and, looking up and down the river through the misty air, saw two long rows of these pollards diminishing till they vanished in both directions, the sight of them took from me all power of enjoying the water beneath me, the green fields around me, or even the old-world beauty of the little bridge upon which I stood, although all sorts of bridges have been from very infancy a delight to me. For I am one of those who never get rid of their infantile predilections, and to have once enjoyed making a mud bridge, was to enjoy all bridges forever.

I saw a man in a white smock-frock coming along the road beyond, but I turned my back to the road, leaned my arms on the parapet of the bridge, and stood gazing where I saw no visions, namely, at those very poplars. I heard the man's footsteps coming up the crown of the arch, but I would not turn to greet him. I was in a selfish humor if ever I was; for surely if ever one man ought to greet another, it was upon such a comfortless afternoon. The footsteps stopped behind me, and I heard a voice:

"I beg yer pardon, sir; but be you the new vicar?"

I turned instantly and answered, "I am. Do you want me?"

"I wanted to see yer face, sir, that wur all, if ye'll not take it amiss."

Before me stood a tall old man with his hat in his hand, clothed as I have said, in a white smock-frock. He smoothed his short gray hair with his curved palm down over his forehead as he stood. His face was of a red brown, from much exposure to the weather. There was a certain look of roughness, without hardness, in it, which spoke of endurance rather than resistance, although he could evidently set his face as a flint. His features were large and a little coarse, but the smile that parted his lips when he spoke, shone in his gray eyes as well, and lighted up a countenance in which a man might trust.

"I wanted to see yer face, sir, if ye'll not take it amiss."

"Certainly not," I answered, pleased with the man's address, as he stood square before me, looking as modest as fearless. "The sight of a man's face is what everybody has a right to; but, for all that, I should like to know why you want to see my face."

"Why, sir, you be the new vicar. You kindly told me so when I axed you."

"Well, then, you'll see my face on Sunday in church — that is, if you happen to be there."

For, although some might think it the more dignified way, I could not take it as a matter of course that he would be at church. A man might have better reasons for staying away from church than I had for going, even though I was the parson, and it was my business. Some clergymen separate between themselves and their office to a degree which I cannot understand. To assert the dignities of my office seems to me very like exalting myself; and when I have had a twinge of conscience about it, as has happened more than once, I have then found comfort in these two texts: "The Son of Man came not to be ministered unto but to minister;" and "It is enough that the servant should be as his master." Neither have I ever been able to see the very great difference between right and wrong in a clergyman, and right and wrong in another man. All that I can pretend to have yet discovered comes to this: that what is right in another man is right in a clergyman; and what is wrong in another man is much worse in a clergyman. Here, however, is one more proof of approaching age. I do not mean the opinion, but the digression.

"Well, then," I said, "you'll see my face in church on Sunday, if you happen to be there."

"Yes, sir; but you see, sir, on the bridge here, the parson is the parson like, and I'm Old Rogers; and I looks in his face, and he looks in mine, and I says to myself, 'This is my parson.' But o' Sundays he's nobody's parson; he's got his work to do, and it mun be done, and there's an end on't."

That there was a real idea in the old man's mind was considerably clearer than the logic by which he tried to bring it out.

"Did you know parson that's gone, sir?" he went on.

"No," I answered.

"O, sir! he wur a good parson. Many's the time he come and sit at my son's bedside — him that's dead and gone, sir — for a long hour, on a Saturday night, too. And then when I see him up in the desk the next mornin', I'd say to myself, 'Old Rogers, that's the same man as sat by your son's bedside last night. Think o' that, Old Rogers!' But, somehow, I never did feel right sure o' that same. He didn't seem to have the same cut, somehow; and he didn't talk a bit the same. And when he spoke to me after sermon, in the churchyard, I was always of a mind to go into the church again and look up to the pulpit to see if he wur really out ov it; for this warn't the same man, you see. But you'll know all about it better than I can tell you, sir. Only I always liked parson better out o' the pulpit, and that's how I come to want to make you look at me, sir, instead o' the water down there, afore I see you in the church to-morrow mornin'."

The old man laughed a kindly laugh; but he had set me thinking, and I did not know what to say to him all at once. So after a short pause, he resumed —

"You'll be thinking me a queer kind of a man, sir, to speak to my betters before my betters speaks to me. But mayhap you don't know what a parson is to us poor folk that has ne'er a friend more larned than theirselves but the parson. And besides, sir, I'm an old salt, — an old man-o'-war's man, — and I've been all round the world, sir; and I ha' been in all sorts o' company, pirates and all, sir; and I ain't a bit frightened of a parson. No; I love a parson, sir. And I'll tell you why, sir. He's got a good telescope, and he gits to the masthead, and he looks out. And he sings out, 'Land ahead!' or 'Breakers ahead!' and gives directions accordin'. Only I can't always make out what he says. But when he shuts up his spyglass, and comes down the riggin', and talks to us like one man to another, then I don't know what I should do without the parson. Good-evenin' to you, sir, and welcome to Marshmallows."

The pollards did not look half so dreary. The river began to glimmer a little; and the old bridge had become an interesting old bridge. The country altogether was rather nice than otherwise. I had found a friend already! — that is, a man to whom I might possibly be of some use; and that was the most precious friend I could think of in my present situation and mood. I had learned something from him too; and I resolved to try all I could to be the same man in the pulpit that I was out of it. Some may be inclined to say that I had better have formed the resolution to be the same man out of the pulpit that I was in it. But the one will go quite right with the other. Out of the pulpit I would be the same man I was in it — seeing and feeling the realities of the unseen; and in the pulpit I would be the same man I was out of it — taking facts as they are, and dealing with things as they show themselves in the world. GEORGE MACDONALD.

TO LADY HOLLAND.

I HEAR you laugh at me for being happy in the country, and upon this I have a few words to say. In the first place, whether one lives or dies, I hold, and have always held, to be of infinitely less moment than is generally supposed; but if life is to be, then it is common sense to amuse yourself with the best you can find where you happen to be placed. I am not leading precisely the life I should choose, but that which (all things considered, as well as I could consider them) appeared to me to be the most eligible. I am resolved, therefore, to like it, and to reconcile myself to it; which is more manly than to feign myself above it, and to send up complaints by the post, of being thrown away, and being desolate, and such like trash. I am prepared, therefore, either way. If the chances of life ever enable me to emerge, I will show you that I have not been wholly occupied by small and sordid pursuits. If (as the greater probability is) I am come to the end of my career, I give myself quietly up to horticulture, etc. In short, if it be my lot to crawl, I will crawl contentedly; if to fly, I will fly with alacrity; but, as long as I can possibly avoid it, I will never be unhappy. If, with a pleasant wife, three children, a good house and farm, many books, and many friends, who wish me well, I cannot be happy, I am a very silly, foolish fellow, and what becomes of me is of very little consequence. I have at least this chance of doing well in Yorkshire, that I am heartily tired of London.

SYDNEY SMITH.

TO BERNARD BARTON.

DEAR B. B.:— Do you know what it is to succumb under an insurmountable day-mare, — "a whoreson lethargy," Falstaff calls it, — an indisposition to do anything, or to be anything; a total deadness and distaste a suspension of vitality; an indifference to locality; a numb, soporifical good-for-nothingness; an ossification all over; an oyster-like insensibility to the passing events; a mind-stupor; a brawny defiance to the needles of a thrusting-in conscience? Did you ever have a very bad cold, with a total irresolution to submit to water-gruel processes? This has been for many weeks my lot, and my excuse; my fingers drag heavily over this paper, and, to my thinking, it is three-and-twenty furlongs from here to the end of this demi-sheet. I have not a thing to say; nothing is of more importance than another; I am flatter than a denial or a pancake; emptier than Judge——'s wig when the head is in it; duller than a country stage when the actors are off it; a cipher, an O! I acknowledge life at all only by an occasional convulsional cough, and a permanent phlegmatic pain in the chest. I am weary of the world; life is weary of me. My day is gone into twilight, and I don't think it worth the expense of candles. My wick hath a thief in it, but I can't muster courage to snuff it. I inhale suffocation; I can't distinguish veal from mutton; nothing interests me. 'Tis twelve o'clock and Thurtell is just now coming out upon the New Drop, Jack Ketch, alertly tucking up his greasy sleeves to do the last office of mortality, yet cannot I elicit a groan or a moral reflection. If you told me the world will be at an end to-morrow, I should just say: "Will it?" I have not volition enough left to dot my i's, much less to comb my eyebrows; my eyes are set in my head; my brains are gone out to see a poor relation in Moorfields, and they did not say when they'd come back again; my skull is a Grub Street attic to let, — not so much as a joint-stool or a cracked jorden left in it; my hand writes, not I, from habit, as chickens run about a little when their heads are off. O for a vigorous fit of gout, colic, toothache, — an ear-wig in my auditory, a fly in my visual organs. Pain is life, — the sharper the more evidence of life; but this apathy, this death! Did you ever have an obstinate cold, — a six or seven weeks' unintermitting chill and suspension of hope, fear, conscience, and everything? Yet do I try all I can to cure it; I try wine and spirits, and smoking, and snuff in unsparing quantities; but they all only seem to make me worse instead of better. I sleep in a damp room, but it does me no good; I come home late o' nights, and do not find any visible amendment! Who shall deliver me from the body of this death? It is just fifteen minutes after twelve; Thurtell is by this time a good way on his journey, baiting at Scorpion, perhaps! Ketch is bargaining for his cast-coat and waist-coat; the Jew demurs at first at three half-crowns, but on consideration that he may get somewhat by showing them in the town, finally closes.

CHARLES LAMB.

EVERY MAN GREAT.

EVERY man, in every condition, is great. It is only our own diseased sight which makes him little. A man is great as a man, be he where or what he may. The grandeur of his nature turns to insignificance all outward distinctions. His powers of intellect, of conscience, of love, of knowing God, of perceiving the beautiful, of acting on his own mind, on outward nature, and on his fellow-creatures, — these are glorious prerogatives. Through the vulgar error of undervaluing what is common, we are apt, indeed, to pass these by as of little worth. But, as in the outward creation, so in the soul, the common is the most precious. Science and art may invent splendid modes of illuminating the apartments of the opulent; but these are all poor and worthless, compared with the common light which the sun sends into all our windows, which he pours freely, impartially, over hill and valley, which kindles daily the eastern and western sky: and so the common lights of reason, and conscience, and love, are of more worth and dignity than the rare endowments which give celebrity to a few. Let us not disparage that nature which is common to all men; for no thought can measure its grandeur. It is the image of God, the image even of his infinity, for no limits can be set to its unfolding. He who possesses the divine powers of the soul is a great being, be his place what it may. You may clothe him with rags, may immure him in a dungeon, may chain him to slavish tasks. But he is still great. You may shut him out of your houses; but God opens to him heavenly mansions. He makes no show, indeed, in the streets of a splendid city; but a clear thought, a pure affection, a resolute act of a virtuous will, have a dignity of quite another kind, and far higher than accumulations of brick, and granite, and plaster, and stucco, however cunningly put together.

The truly great are to be found everywhere; nor is it easy to say in what condition they spring up most plentifully. Real greatness has nothing to do with a man's sphere. It does not lie in the magnitude of his outward agency, in the extent of the effects which he produces. The greatest men may do comparatively little abroad. Perhaps the greatest in our city at this moment are buried in obscurity. Grandeur of character lies wholly in force of soul, — that is, in the force of thought, moral principle, and love; and this may be found in the humblest condition of life. A man brought up to an obscure trade, and hemmed in by the wants of a growing family, may, in his narrow sphere, perceive more clearly, discriminate more keenly, weigh evidence more wisely, seize on the right means more decisively, and have more presence of mind in difficulty, than another who has accumulated vast stores of knowledge by laborious study; and he has more of intellectual greatness. Many a man, who has gone but a few miles from home, understands human nature better, detects motives and weighs character more sagaciously, than another who has traveled over the known world, and made a name by his reports of different countries. It is force of thought which measures intellectual, and so it is force of principle which measures moral, greatness, — that highest of human endowments, that brightest manifestation of the Divinity. The greatest man is he who chooses the Right with invincible resolution, who resists the sorest temptations from within and without, who bears the heaviest burdens cheerfully, who is calmest in storms and most fearless under menace and frowns, whose reliance on truth, on virtue, on God, is most unfaltering. I believe this greatness to be most common among the multitude, whose names are never heard. Among common people will be found more of hardship borne manfully, more of unvarnished truth, more of religious trust, more of that generosity which gives what the giver needs himself, and more of a wise estimate of life and death, than among the more prosperous. In these remarks you will see why I feel and express a deep interest in the obscure, — in the mass of men. The distinctions of society vanish before the light of these truths. I attach myself to the multitude, not because they are voters and have political power, but because they are men, and have within their reach the most glorious prizes of humanity.

WILLIAM ELLERY CHANNING.

THE ALHAMBRA BY MOONLIGHT.

The moon, which then was invisible, has gradually gained upon the nights, and now rolls in full splendor above the towers, pouring a flood of tempered light into every court and hall. The garden beneath my window is gently lighted up, the orange and citron trees are tipped with silver, the fountain sparkles in the moonbeams, and even the blush of the rose is faintly visible.

I have sat for hours at my window inhaling the sweetness of the garden, and musing on the checkered features of those whose history is dimly shadowed out in the elegant memorials around. Sometimes I have issued forth at midnight when every thing was quiet, and have wandered over the whole building. Who can do justice to a moonlight night in such a climate and in such a place? The temperature of an Andalusian midnight, in summer, is perfectly ethereal. We seem lifted up into a purer atmosphere; there is a serenity of soul, a buoyancy of spirits, an elasticity of frame, that render mere existence enjoyment. The effect of moonlight, too, on the Alhambra has something like enchantment. Every rent and chasm of time, every mouldering tint and weather-stain, disappears, the marble resumes its original whiteness, the long colonnades brighten in the moonbeams, the halls are illuminated with a softened radiance, until the whole edifice reminds one of the enchanted palace of an Arabian tale.

At such time I have ascended to the little pavilion, called the Queen's Toilette, to enjoy its varied and extensive prospect. To the right, the snowy summits of the Sierra Nevada would gleam like silver clouds against the darker firmament, and all the outlines of the mountain would be softened, yet delicately defined. My delight, however, would be to lean over the parapet of the tocador, and gaze down upon Granada, spread out like a map below me, all buried in deep repose, and its white palaces and convents sleeping as it were in the moonshine.

Sometimes I would hear the faint sounds of castanets from some party of dancers lingering in the Alameda; at other times I have heard the dubious tones of a guitar, and the notes of a single voice rising from some solitary street, and have pictured to myself some youthful cavalier serenading his lady's window,—a gallant custom of former days, but now sadly on the decline, except in the remote towns and villages of Spain.

Such are the scenes that have detained me for many an hour loitering about the courts and balconies of the castle, enjoying that mixture of reverie and sensation which steal away existence in a Southern climate,—and it has been almost morning before I have retired to my bed, and been lulled to sleep by the falling waters of the fountain of Lindaraxa.

Washington Irving.

NIPPED IN THE BUD.

Mr. Thomas Watts had already conceived a passion that was ardent, and pointed, and ambitious to a degree which Susan characterized as "perfectly redickerlous."

But who was the young lady who had thus concentrated upon herself all the first fresh worship of that young but manly heart? Was it Miss Jones, or Miss Sharp? Was it Miss Holland or Miss Hutchins? Not one of these. Mr. Thomas Watts had with one tremendous bound leaped clear over the heads of these secondary characters, and cast himself at the very foot of the throne. To be plain, Mr. Watts fondly, entirely, madly, loved Miss Julia Louisa Wilkins, the mistress and head of the Dukesborough Female Institution.

Probably this surprising reach might be attributed to the ambitious nature of his father, from whom he had inherited this and some other qualities. Doubtless, however, the recollection of having been kept long in frocks had engendered a desire to convince the world that they had sadly mistaken their man. Whatever was the motive power, such was the fact. Now, notwithstanding this state of his own feelings, he had never made a declaration in so many words to Miss Wilkins. But he did not doubt for a moment that she thoroughly understood his looks, and sighs, and devoted services. For the habit which all of us have of enveloping beloved objects in our hearts, and making them, so to speak, understand and reciprocate our feelings, had come to Mr. Watts even to a greater degree, perhaps, than if he had been older. He was as little inclined and as little able to doubt Miss Wilkins as to doubt himself. Facts seemed to bear him out. She had not only smiled upon him time and time again, and patted him sweetly on the back of his head, and praised his roach to the very skies; but once, when he had carried her a great armful of good, fat pine-knots, she was so overcome as to place her hand under his chin, look him fully in the face, and declare if he wasn't a man, there wasn't one in this wide, wide world.

Such was the course of his true love when its smoothness suffered that interruption which so strangely obtrudes itself among the fondest affairs of the heart. Miss Susan had threatened so often without fulfilment to give information to their mother, that he had begun to presume there was little or no danger from that quarter. Besides, Mr. Watts had now grown so old and manlike that he was getting to be without apprehension from any quarter. He reflected that within a few weeks more he would be fourteen years old, when legal rights would accrue. Determining not to choose any "gardzeen," it would follow that he must become his own. Yet he did not intend to act with unnecessary notoriety. His plans were, to consummate his union on the very day he should be fourteen; but to do so clandestinely, and then run away, not stopping until he should get with his bride plump into Vermont. For even the bravest find it necessary sometimes to retreat.

Of the practicability of this plan he had no doubt, because he knew that Miss

MISS JULIA LOUISA WILKINS.

Wilkins had five hundred dollars in hard cash — a whole stocking full. This sum seemed to him immensely adequate for their support in becoming style for an indefinitely long period of time.

As the day of his majority approached, he grew more and more reserved in his intercourse with his family. This was scarcely to be avoided now, when he was already beginning to consider himself as not one of them. If his conscience ever upbraided him as he looked upon his toiling mother and his helpless brothers and sisters, and knew that he alone was to rise into luxury, while they were to be left in their lowly estate, he reflected that it was a selfish world at best, and that every man must take care of himself. But one day, after a season of unusual reserve, and when he had behaved to Miss Susan in a way which she considered outrageously supercilious, the latter availed herself of his going into the village, fulfilled her threat, and gave her mother full information of the state of his feelings. That resolute woman was in the act of ironing a new homespun frock she had just made for Susan.

She laid down her iron, sat down in a chair, and looked up at Susan.

"Susan, don't be foolin' 'long o' me."

"Ma, I tell you it's the truth."

"Susan, do you want me to believe that Tom's a fool? I know'd the child didn't have no great deal of sense; but I didn't think he was a clean-gone fool." . . .

"Yes, we lives and larns. But, bless me, it won't do to tarry here. Susan, have that frock ironed all right, stiff and starch, by the time I git back. I sha'n't be gone long."

The lady arose, and, without putting on her bonnet, walked rapidly down the streets.

"What are you looking for, Mrs. Watts?" inquired an acquaintance whom she met on her way.

"I'm a-looking for a person of the name of Mr. Watts," she answered, and rushed madly on. The acquaintance hurried home; but told other acquaintances, on the way, that the Widow Watts have lost her mind, and gone ravin' distracted. Soon afterwards, as Mr. Watts was slowly returning, his mind full of great thoughts, and his head somewhat bowed, he suddenly became conscious that his hat was removed, and his roach rudely seized. Immediately afterwards he found himself carried along the street, his head foremost, and his legs and feet performing the smallest possible part in the act of locomotion. The villagers looked on with wonder. The conclusion was universal. Yes, the Widow Watts have lost her mind.

When she had reached her cabin with her charge, a space was cleared in the middle, by removing the stools and the children. Then Mr. Watts was ordered to remove such portions of his attire as might oppose any hindrance whatever to the application of a leather strap to those parts of his person which his mother might select.

"Oh, mother, mother!" began Mr. Watts.

"No motherin' o' me, sir. Down with 'em." And down they came; and down came the strap, rapidly, violently.

"Oh, mammy, mammy!"

"Ah, now! that sounds a little like old times, when you used to be a boy," she exclaimed in glee, as the sounds were repeated amid the unslackened descent of the strap. Mrs. Watts seemed disposed to carry on a lively conversation during this flagellation. She joked her son pleasantly about Miss Wilkins, inquired when it was to be and who was to be invited? Oh, no! she forgot it was not to be a big wedding, but a private one. But how long were they going to be gone before they would make a visit? But Mr. Watts not only could not see the joke, but was not able to join in the conversation at all, except to continue to scream louder and louder, "Oh, mammy, mammy!" Mrs. Watts, finding him not disposed to be talkative, except in mere ejaculatory remarks, appealed to little Jack, and Mary Jane, and Polly Ann, and to all, down even to the baby. She asked them, Did they know that Buddy Tommy were a man grown, and were going to git married and have a wife, and then go away off yonder to the Vermontes? Little Jack, and Polly Ann, and baby, and all, evidently did not precisely understand; for they all cried and laughed tumultuously.

How long this exercise, varied as it was by most animated conversation, might have continued if the mother had not become exhausted, there is no calculating. Things were fast approaching that condition when the son declared that his mother would kill him if she didn't stop.

"That," she answered between breaths, "is — what — I — aims — to do — if — I can't git it — all — all — every — spang — passel — outen you."

Tom declared that it was all gone.

"Is you — a man — or — is you — a boy?"

"Boy! boy! mammy!" cried Tom. "Let me up, mammy — and — I'll be a boy — as long — as I live."

She let him up.

"Susan, whar's that frock? Ah, there it is. Lookee here. Here's your clo'es, my man. Mary Jane, put away them pantaloonses."

Tom was making ready to resume the frock. But Susan remonstrated. It wouldn't look right, now; and she would go Tom's security that he wouldn't be a man any more.

He was cured. From being an ardent lover, he grew to become a hearty hater of the principal of Dukesborough Female Institution, the more implacable upon his hearing that she had laughed immoderately at his whipping. Before many months she removed from the village; and when, two years afterwards, a rumor (whether true or not we never knew) came that she was dead, Tom was accused of being gratified by the news. Nor did he deny it.

"Well, fellers," said he, "I know it weren't right; but I couldn't keep from being glad, if it had a-kilt me."

RICHARD MALCOLM JOHNSTON.

TO ROBERT AINSLIE.

There is one thing for which I set great store by you as a friend, and it is this — that I have not a friend upon earth, besides yourself, to whom I can talk nonsense without forfeiting some degree of his esteem. Now to one like me, who never cares for speaking any thing else but nonsense, such a friend as you is an invaluable treasure. I was never a rogue, but have been a fool all my life; and, in spite of all my endeavors, I see now plainly that I shall never be wise. Now it rejoices my heart to have met with such a fellow as you, who, though you are not just such a hopeless fool as I, yet I trust you will never listen so much to the temptations of the devil as to grow so very wise that you will in the least disrespect an honest fellow because he is a fool. In short, I have set you down as the staff of my old age, when the whole list of my friends will, after a decent share of pity, have forgot me.

> Though in the morn comes sturt and strife,
> Yet joy may come at noon;
> And I hope to live a merry, merry life,
> When a' thir days are done.

Write me soon, were it but a few lines just to tell me how that good, sagacious man, your father, is — that kind, dainty body, your mother — that strapping chiel, your brother Douglas — and my friend Rachel, who is as far before Rachel of old, as she was before her blear-eyed sister Leah.

Robert Burns.

THE ADVENT OF PEACE.

The times that tried men's souls are over, and the greatest and completest revolution the world ever knew, gloriously and happily accomplished.

But to pass from the extremes of danger to safety, from the tumult of war to the tranquillity of peace, — though sweet in contemplation, requires a gradual composure of the senses to receive it. Even calmness has the power of stunning, when it opens too instantly upon us. The long and raging hurricane that should cease in a moment would leave us in a state rather of wonder than enjoyment; and some moments of recollection must pass before we could be capable of tasting the felicity of repose. There are but few instances in which the mind is fitted for sudden transitions; it takes in its pleasures by reflection and comparison, and those must have time to act before the relish for new scenes is complete.

In the present case, the mighty magnitude of the object, the various uncertainties of fate it has undergone, the numerous and complicated dangers we have suffered or escaped, the eminence we now stand on, and the vast prospect before us, must all conspire to impress us with contemplation.

To see it in our power to make a world happy, to teach mankind the art of being so, to exhibit on the theater of the universe a character hitherto unknown, and to have, as it were, a new creation entrusted to our hands, are honors that command reflection, and can neither be too highly estimated, nor too gratefully received.

In this pause then of reflection, while the storm is ceasing, and the long agitated mind vibrating to a rest, let us look back on the scenes we have passed, and learn from experience what is yet to be done.

Never, I say, had a country so many openings to happiness as this. Her setting out in life, like the rising of a fair morning, was unclouded and promising. Her cause was good. Her principles just and liberal. Her temper serene and firm. Her conduct regulated by the wisest steps, and everything about her wore the mark of honor. It is not every country (perhaps there is not another in the world) that can boast so fair an origin. Even the first settlement of America corresponds with the character of the revolution. Rome, once the proud mistress of the universe, was originally a band of ruffians. Plunder and rapine made her rich, and her oppression of millions made her great. But America need never be ashamed to tell her birth, nor relate the stages by which she rose to empire.

The remembrance then of what is past, if it operates rightly, must inspire her with the most laudable of all ambitions, that of adding to the fair fame she began with. The world has seen her great in adversity; struggling, without a thought of yielding, beneath accumulated difficulties, bravely, nay, proudly encountering distress, and rising in resolution as the storm increased. All this is justly due to her, for her fortitude has merited the character. Let then the world see that she can bear prosperity; and that her honest virtue in time of peace is equal to the bravest virtue in time of war.

She is now descending to the scenes of quiet and domestic life,—not under the cypress shade of disappointment, but to enjoy, in her own land, and under her own vine, the sweet of her labors, and the reward of her toil. In this situation may she never forget that a fair national reputation is of as much importance as independence, that it possesses a charm that wins upon the world, and makes even enemies civil, that it gives a dignity which is often superior to power, and commands reverence where pomp and splendor fail.

THOMAS PAINE.

HOMER'S INVENTIVE POWER.

ON whatever side we contemplate Homer, what principally strikes us is his invention. It is that which forms the character of each part of his work; and accordingly we find it to have made his fable more extensive and copious than any other, his manners more lively and strongly marked, his speeches more affecting and transporting, his sentiments more warm and sublime, his images and descriptions more full and animated, his expression more raised and daring, and his numbers more rapid and various. I hope, in what has been said of Virgil, with regard to any of these heads, I have no way derogated from his character. Nothing is more absurd or endless, than the common method of comparing eminent writers by an opposition of particular passages in them, and forming a judgment from thence of their merit upon the whole. We ought to have a certain knowledge of the principal character and distinguished excellence of each: it is in that we are to consider him, and in proportion to his degree in that we are to admire him. No author or man ever excelled all the world in more than one faculty: and as Homer has done this in invention, Virgil has in judgment. Not that we are to think Homer wanted judgment, because Virgil had it in a more eminent degree; or that Virgil wanted invention, because Homer possessed a larger share of it: each of these great authors had more of both than perhaps any man besides, and are only said to have less in comparison with one another. Homer was the greater genius; Virgil, the better artist. In one we most admire the man; in the other, the work. Homer hurries and transports us with a commanding impetuosity; Virgil leads us with an attractive majesty: Homer scatters with a generous profusion; Virgil bestows with a careful magnificence: Homer, like the Nile, pours out his riches with a boundless overflow; Virgil, like a river in its banks, with a gentle and constant stream. When we behold their battles, methinks the two poets resemble the heroes they celebrate: Homer, boundless and irresistible as Achilles, bears all before him, and shines more and more as the tumult increases; Virgil, calmly daring like Æneas, appears undisturbed in the midst of the action; disposes all about him, and conquers with tranquillity. And when we look upon their machines, Homer seems like his own Jupiter in his terrors, shaking Olympus, scattering the lightnings, and firing the heavens; Virgil, like the same power in his benevolence, counselling with the gods, laying plans for empires, and regularly ordering his whole creation.

HOMER.
(*Bust in British Museum, London.*)

ALEXANDER POPE.

TO H. S. WILLIAMS.

. . . I could not help wondering whether Cornhill will ever change for me, as Oxford has changed for you. I have some pleasant associations connected with it now — will these alter their character some day?

Perhaps they may — though I have faith to the contrary, because, I think, I do not exaggerate my partialities; I think I take faults along with excellences — blemishes together with beauties. And, besides, in the matter of friendship, I have observed that disappointment here arises chiefly, not from liking our friends too well, or thinking of them too highly, but rather from an over-estimate of their liking for and opinion of us; and that if we guard ourselves with sufficient scrupulousness of care from error in this direction, and can be content and even happy to give more affection than we receive — can make just comparison of circumstances, and be severely accurate in drawing inferences thence, and never let self-love blind our eyes, — I think we may manage to get through life with consistency and constancy, unembittered by that misanthropy which springs from revulsions of feeling. All this sounds a little metaphysical, but it is good sense if you consider it. The moral of it is, that if we would build on a sure foundation in friendship, we must love our friends for their sakes rather than for our own; we must look at their truth to themselves, full as much as their truth to us. In the latter case, every wound to self-love would be a cause of coldness; in the former, only some painful change in the friend's character and disposition — some fearful breach in his allegiance to his better self — could alienate the heart. . . .

CHARLOTTE BRONTË.

MR. CASAUBON'S ROMANCE.

MY DEAR MISS BROOKE: — I have your guardian's permission to address you on a subject than which I have none more at heart. I am not, I trust, mistaken in the recognition of some deeper correspondence than that of date, in the fact that a consciousness of need in my own life had arisen contemporaneously with the possibility of my becoming acquainted with you. For in the first hour of meeting you, I had an impression of your eminent and perhaps exclusive fitness to supply that need (connected, I may say, with such activity of the affections as even the preoccupations of a work too special to be abdicated could not uninterruptedly dissimulate); and each succeeding opportunity for observation has given the impression an added depth by convincing me more emphatically of that fitness which I had preconceived, and thus evoking more decisively those affections to which I have but now referred. Our conversations have, I think, made sufficiently clear to you the tenor of my life and purposes, — a tenor unsuited, I am aware, to the commoner order of minds. But I have discerned in you an elevation of thought and a capability of devotedness, which I had hitherto not conceived to be compatible either with the early bloom of youth or with those graces of sex that may be said at once to win and to confer distinction when combined, as they notably are in you, with the mental qualities above indicated. It was, I confess, beyond my hope to meet with this rare combination of elements both solid and attractive, adapted to supply aid in graver labors and to cast a charm over vacant hours; and but for the event of my introduction to you (which, let me again say, I trust not to be superficially coincident with foreshadowing needs, but providentially related thereto as stages toward the completion of a life's plan), I should presumably have gone on to the last without any attempt to lighten my solitariness by a matrimonial union.

Such, my dear Miss Brooke, is the accurate statement of my feelings; and I rely on your kind indulgence in venturing now to ask you how far your own are of a nature to confirm my happy presentiment. To be accepted by you as your husband and the earthly guardian of your welfare, I should regard as the highest of providential gifts. In return, I can at least offer you an affection hitherto unwasted, and the faithful consecration of a life which, however short in the sequel, has no backward pages whereon, if you choose to turn them, you will find records such as might justly cause you either bitterness or shame. I wait the expression of your sentiments with an anxiety which it would be the part of wisdom (were it possible) to divert by a more arduous labor than usual. But in this order of experience I am still young, and in looking forward to an unfavorable possibility, I cannot but feel that resignation to solitude will be more difficult after the temporary illumination of hope. In any case, I shall remain,

Yours, with sincere devotion,
EDWARD CASAUBON.

MY DEAR MR. CASAUBON : — I am very grateful to you for loving me, and thinking me worthy to be your wife. I can look forward to no better happiness than that which would be one with yours. If I said more, it would only be the same thing written out at greater length, for I cannot now dwell on any other thought than that I may be through life

Yours devotedly, DOROTHEA BROOKE.
GEORGE ELIOT.

THE BROAD OPEN COUNTRY.

THE WHITE ROSE ROAD.

DRIVING through the long woodland way, shaded and chilly when you are out of the sun; across the Great Works River and its pretty elm-grown intervale; across the short bridges of brown brooks; delayed now and then by the sight of ripe strawberries in sunny spots by the roadside, one comes to a higher open country, where farm joins farm, and the cleared fields lie all along the highway, while the woods are pushed back a good distance on either hand. The wooded hills, bleak here and there with granite ledges, rise beyond. The houses are beside the road, with green door-yards and large barns, almost empty now, and with wide doors standing open, as if they were already expecting the hay crop to be brought in. The tall green grass is waving in the fields as the wind goes over, and there is a fragrance of whiteweed and ripe strawberries and clover blowing through the sunshiny barns, with their lean sides and their festoons of brown, dusty cobwebs; dull, comfortable creatures they appear to imaginative eyes, waiting hungrily for their yearly meal. The eave-swallows are teasing their sleepy shapes, like the birds which flit about great beasts; gay, movable, irreverent, almost derisive, those barn swallows fly to and fro in the still, clear air. The noise of our wheels brings fewer faces to the windows than usual, and we lose the pleasure of seeing some of our friends who are apt to be looking out, and to whom we like to say good-day. Some funeral must be taking place, or perhaps the women may have gone out into the fields. It is hoeing-time and strawberry-time, and already we have seen some of the younger women at work among the corn and potatoes. One sight will be charming to remember. On a green hillside sloping to the west, near one of the houses, a thin little girl was working away lustily with a big hoe on a patch of land perhaps fifty feet by twenty. There were all sorts of things growing there, as if a child's fancy had made the choice, — straight rows of turnips and carrots and beets, a little of everything, one might say; but the only touch of color was from a long border of useful sage in full bloom of dull blue, on the upper side. I am sure this was called Katy's or Becky's piece by the elder members of the family. One can imagine how the young creature had planned it in the spring, and persuaded the men to plow and harrow it, and since then had stoutly done all the work herself, and meant to send the harvest of the piece to market, and pocket her honest gains, as they came in, for some great end. She was as thin as a grasshopper, this busy little gardener, and hardly turned to give us a glance, as we drove slowly up the hill close by. The sun will brown and dry her like a spear of grass on that hot slope, but a spark of fine spirit is in the small body, and I wish her a famous crop. I hate to say that the piece looked backward, all except the sage, and that it was a heavy bit of land for the clumsy hoe to pick at. The only puzzle is, what she proposes to do with so long a row of sage. Yet there may be a large family with a downfall of measles yet ahead, and she does not mean to be caught without sage-tea.

Along this road every one of the old farmhouses has at least one tall bush of white roses by the door, — a most lovely sight, with buds and blossoms, and unvexed green leaves. I wish that I knew the history of them, and whence the first bush was brought. Perhaps from England itself, like a red rose that I know in Kittery, and the new shoots from the root were given to one neighbor after another all through the district. The bushes are slender, but they grow tall without climbing against the wall, and sway to and fro in the wind with a grace of youth and an inexpressible charm of beauty. How many lovers must have picked them on Sunday evenings, in all the bygone years, and carried them along the roads or by the pasture footpaths, hiding them clumsily under their Sunday coats if they caught sight of any one coming. Here, too, where the sea wind nips many a young life before its prime, how often the white roses have been put into paler hands, and withered there!

We met now and then a man or woman, who stopped to give us hospitable greeting; but there was no staying for visits, lest the daylight might fail us. It was delightful to find this old-established neighborhood so thriving and populous, for a few days before I had driven over three miles of road, and passed only one house that was tenanted, and six cellars or crumbling chimneys where good farm-houses had been, the lilacs blooming in solitude, and the fields, cleared with so much difficulty a century or two ago, all going back to the original woodland from which they were won. What would the old farmers say to see the fate of their worthy bequest to the younger generation? They would wag their heads sorrowfully, with sad foreboding.

* * * * * *

The very fields looked busy with their early summer growth, the horses began to think of the clack of the oat-bin cover, and we were hurried along between the silvery willows and the rustling alders, taking time to gather a handful of stray-away conserve roses by the roadside; and where the highway made a long bend eastward among the farms, two of us left the carriage, and followed a footpath along the green river bank and through the pastures, coming out to the road again only a minute later than the horses. I believe that it is an old Indian trail followed from the salmon falls farther down the river, where the up-country Indians came to dry the

plentiful fish for their winter supplies. I have traced the greater part of this deep-worn footpath, which goes straight as an arrow across the country, the first day's trail being from the falls (where Mason's settlers came in 1627, and built their Great Works of a saw-mill with a gang of saws, and presently a grist mill beside) to Emery's Bridge. I should like to follow the old footpath still farther. I found part of it by accident a long time ago. Once, as you came close to the river, you were sure to find fishermen scattered along, — sometimes I myself have been discovered; but it is not much use to go fishing any more. If some public-spirited person would kindly be the Frank Buckland of New England, and try to have the laws enforced that protect the inland fisheries, he would do his country great service. Years ago, there were so many salmon that, as an enthusiastic old friend once assured me, "You could walk across on them below the falls;" but now they are unknown, simply because certain substances which would enrich the farms are thrown from factories and tanneries into our clear New England streams. Good river fish are growing very scarce. The smelts, and bass, and shad have all left this upper branch of the Piscataqua, as the salmon left it long ago, and the supply of one necessary sort of good cheap food is lost to a growing community, for the lack of a little thought and care in the factory companies and saw-mills, and the building in some cases of fish-ways over the dams. I think that the need of preaching against this bad economy is very great. The sight of a proud lad with a string of undersized trout will scatter half the idlers in town into the pastures next day, but everybody patiently accepts the depopulation of a fine clear river, where the tide comes fresh from the sea to be tainted by the spoiled stream, which started from its mountain sources as pure as heart could wish. Man has done his best to ruin the world he lives in, one is tempted to say at impulsive first thought; but after all, as I mounted the last hill before reaching the village, the houses took on a new look of comfort and pleasantness; the fields that I knew so well were a fresher green than before, the sun was down, and the provocations of the day seemed very slight compared to the satisfaction. I believed that with a little more time we should grow wiser about our fish and other things beside.

It will be good to remember the white rose road and its quietness in many a busy town day to come. As I think of these slight sketches, I wonder if they will have to others a tinge of sadness; but I have seldom spent an afternoon so full of pleasure and fresh and delighted consciousness of the possibilities of rural life.

SARAH ORNE JEWETT.

MISS MALONEY ON THE CHINESE QUESTION.

Och! don't be talkin'. Is it howld on, ye say? An' didn't I howld on till the heart of me was clane broke entirely, and me wastin' that thin ye could clutch me wid yer two hands. To think o' me toilin' like a nager for the six year I've been in Ameriky — bad luck to the day I iver left the owld counthry! — to be bate by the likes o' them! (faix, an' I'll sit down when I'm ready, so I will, Ann Ryan; an' ye'd better be listenin' than drawin' yer remarks). An' is it meself, with five good charácters from respectable places, would be herdin' wid the haythens? The saints forgive me, but I'd be buried alive sooner'n put up wid it a day longer. Sure, an' I was the granehorn not to be lavin' at once-t when the missus kim into me kitchen wid her perlaver about the new waiter-man which was brought out from Californy. "He'll be here the night," says she. "And, Kitty, it's meself looks to you to be kind and patient wid him; for he's a furriner," says she, a kind o' lookin' off. "Sure, an' it's little I'll hinder nor interfare wid him, nor any other, mum," says I, a kind o' stiff; for I minded me how these French waiters, wid their paper collars and brass rings on their fingers, isn't company for no gurril brought up dacent and honest. Och! sorra a bit I knew what was comin' till the missus walked into me kitchen, smilin', and says, kind o' schared, "Here's Fing Wing, Kitty; an' ye'll have too much sinse to mind his bein' a little strange." Wid that she shoots the doore; and I, misthrustin' if I was tidied up sufficient for me fine buy wid his paper collar, looks up, and — Howly fathers! may I niver brathe another breath, but there stud a rale haythen Chineser, a-grinnin' like he'd just come off a tay-box. If ye'll belave me, the crayther was that yeller it 'ud sicken ye to see him; and sorra stich was on him but a black night-gown over his trowsers, and the front of his head shaved claner nor a copper-biler, and a black tail a-hangin' down from it behind, wid his two feet stook into the haythenestest shoes ye ever set eyes on. Och! but I was up stairs afore ye could turn about, a-givin' the missus warnin', an' only stopt wid her by her raisin' me wages two dollars, an' playdin' wid me how it was a Christian's duty to bear wid haythens, an' taitch 'em all in our power — the saints save us! Well, the ways and trials I had wid that Chineser, Ann Ryan, I couldn't be tellin'. Not a blissid thing cud I do, but he'd be lookin' on wid his eyes cocked up'ard like two poomp-handles; an' he widdout a speck or smitch o' whiskers on him, an' his finger-nails full a yard long. But it's dyin' ye'd be to see the missus a-larnin' him, an' he a-grinnin', an' waggin' his pig-tail (which was pieced out long wid some black stoof, the haythen chate!) and gettin' into her ways wonderful quick, I don't deny, imitatin' that sharp, ye'd be shurprised, an' ketchin' an' copyin' things the best of us will do a-hurried wid work, yet don't want comin' to the knowledge o' the family — bad luck to him!

THE WAITER-MAN.

Is it ate wid him? Arrah, an' would I be sittin' wid a haythen, an' he a-atin' wid drum-sticks? — yes, an' atin' dogs an' cats unknownst to me, I warrant ye, which it is the custom of them Chinesers, till the thought made me that sick I could die. An' didn't the crayther proffer to help me a wake ago come Toosday, an' me foldin' down me clane clothes for the ironin', an' fill his haythen mouth wid water, an' afore I could hinder, squirrit it through his teeth stret over the best linen tablecloth, and fold it up tight, as innercent now as a baby, the derrity baste! But the worrest of all was the copyin' he'd be doin' till ye'd be dishtracted. It's yerself knows the tinder feet that's on me since ever I've bin in this countdry. Well, owin' to that, I fell into a way o' slippin' me shoes off when I'd be settin' down to pale the praties, or the likes o' that; an' do ye mind, that haythen would do the same thing after me whiniver the missus set him to parin' apples or tomaterses. The saints in heaven couldn't ha' made him belave he cud kape the shoes on him when he'd be palin' any thing.

Did I lave for that? Faix, an' I didn't. Didn't he get me into throuble wid my missus, the haythen! Ye're aware yerself how the boondles comin' in from the grocery often contains more 'n 'll go into any thing dacently. So, for that matter, I'd now and then take out a sup o' sugar, or flour, or tay, an' wrap it in paper, and put it in me bit of a box tucked under the ironin'-blanket, the how it cuddent be bodderin' any one. Well, what shud it be, but this blessed Saturday morn, the missus was a-spakin' pleasant an' respec'ful wid me in me kitchen, when the grocer buy comes in, and stands fornenst her wid his boondles; an' she motions like to

Fing Wing (which I never would call him by that name nor any other but just haythen)— she motions to him, she does, for to take the boondles, an' emty out the sugar and what not where they belongs. If ye'll belave me, Ann Ryan, what did that blatherin' Chineser do but take out a sup o' sugar, an' a han'ful o' tay, an' a bit o' chaze, right afore the missus, wrap 'em into bits o' paper, an' I spacheless wid shurprise, an' he the next minute up wid the ironin'-blanket, an' pullin' out me box wid a show o' bein' sly to put them in. Och! the Lord forgive me, but I clutched it, an' the missus sayin', "O, Kitty!" in a way that ud cruddle your blood. "He's a haythen nager," says I. "I've found yer out," says she. "I'll arrist him," says I. "It's yerself ought to be arristid," says she. "Yer won't," says I. "I will," says she. And so it went, till she give me such sass as I cuddent take from no lady, an' I give her warnin' an' left that instant, an' she a-pointin' to the doore.

Mary Mapes Dodge.

TO MRS. JANE LAWDER.

. . . It is probable you may one of these days see me turned into a perfect hunks, and as dark and intricate as a mouse-hole. I have already given my land-lady orders for an entire reform in the state of my finances. I declaim against hot suppers, drink less sugar in my tea, and check my grate with brick-bats. Instead of hanging my room with pictures, I intend to adorn it with maxims of frugality. Those will make pretty furniture enough, and won't be a bit too expensive; for I shall draw them all out with my own hands, and my landlady's daughter shall frame them with the parings of my black waistcoat. Each maxim is to be inscribed on a sheet of clean paper, and wrote with my best pen, of which the following will serve as a specimen: "Look sharp;" "Mind the main chance;" "Money is money now;" "If you have a thousand pounds, you can put your hands by your sides and say you are worth a thousand pounds every day of the year;" "Take a farthing from a hundred, and it will be a hundred no longer." Thus, which way soever I turn my eyes they are sure to meet one of those friendly monitors; and as we are told of an actor who hung his room round with looking-glass to correct the defects of his person, my apartment shall be furnished in a peculiar manner, to correct the errors of my mind.

Faith! Madam, I heartily wish to be rich, if it were only for this reason — to say without a blush how much I esteem you; but alas! I have many a fatigue to encounter before that happy time comes, when your poor old simple friend may again give a loose to the luxuriance of his nature, sitting by Kilmore fireside, recount the various adventures of a hard-fought life, laugh over the follies of the day, join his flute to your harpsicord, and forget that ever he starved in those streets where Butler and Otway starved before him. . . . Oliver Goldsmith.

THE DEATH OF LITTLE NELL.

She was dead. No sleep so beautiful and calm, so free from trace of pain, so fair to look upon! She seemed a creature fresh from the hand of God, and waiting for the breath of life; not one who had lived and suffered death.

She was dead. Dear, gentle, patient, noble Nell, was dead. Her little bird — a poor slight thing the pressure of a finger would have crushed — was stirring nimbly in its cage; and the strong heart of its child-mistress was mute and motionless forever.

Where were the traces of her early cares, her sufferings and fatigues? All gone. Sorrow was dead indeed in her, but peace and perfect happiness were born; imaged in her tranquil beauty and profound repose.

And still her former self lay there, unaltered in this change. Yes. The old fireside had smiled upon that sweet face; it had passed, like a dream, through haunts of misery and care — there had been the same mild lovely look. So shall we know the angels, in their majesty, after death.

She was dead, and passed all help or need of it.

It is not in this world that Heaven's justice ends. Think what earth is, compared to the world to which her young spirit has winged its early flight; and say, if one deliberate wish expressed in solemn terms above this bed would call her back to life, which of us would utter it?

In that calm time, when outward things and inward thoughts teem with assurances of immortality, and worldly hopes and fears are humbled in the dust before them — then, with tranquil and submissive hearts they turned away, and left the child with God.

Oh! it is hard to take to heart the lesson that such deaths will teach, but let no man reject it, for it is one that all must learn, and is a mighty universal truth.

When Death strikes down the innocent and young, for every fragile form from which he lets the panting spirit free, a hundred virtues rise, in shapes of mercy, charity and love, to walk the world, and bless it. Of every tear that sorrowing mortals shed on such green graves, some good is born, some gentler nature comes. In the Destroyer's steps there spring up bright creations that defy his power, and his dark path becomes a way of light to heaven.

DICKENS.

ON AMERICAN INSTITUTIONS.

Mr. Chairman :— Viewed from the standpoint of a foreigner, our government may be said to be the feeblest on earth. From our standpoint, and with our experience it is the mightiest. But why would a foreigner call it the feeblest? He can point out a half dozen ways in which it can be destroyed without violence. Of course, all governments may be overturned by the sword ; but there are several ways in which our government may be annihilated without the firing of a gun.

For example, if the people of the United States should say, we will elect no Representatives to the House of Representatives. Of course, this is a violent supposition ; but suppose that they do not, is there any remedy? Does our Constitution provide any remedy whatever ? In two years there would be no House of Representatives ; of course no support of the government, and no government. Suppose, again, the States should say, through their Legislatures, we will elect no Senators. Such abstention alone would absolutely destroy this government ; and our system provides no process of compulsion to prevent it.

Again, suppose the two Houses were assembled in their usual order, and a majority of one in this body or in the Senate should firmly band themselves

JAMES ABRAM GARFIELD.

together and say, we will vote to adjourn the moment the hour of meeting arrives, and continue so to vote at every session during our two years of existence ; the government would perish, and there is no provision of the Constitution to prevent it. Or again, if a majority of one of either body should declare that they would vote down, and did vote down, every bill to support the government by appropriations, can you find in the whole range of our judicial or our executive authority any remedy whatever ? A Senator or a member of this House is free, and may vote "no" on every proposition. Nothing but his oath and his honor restrains him. Not so with executive and judicial officers. They have no power to destroy this government.

Let them travel an inch beyond the line of the law, and they fall within the power of impeachment. But, against the people who create Representatives; against the Legislatures who create Senators; against Senators and Representatives in these Halls, there is no power of impeachment; there is no remedy, if, by abstention or by adverse votes, they refuse to support the government.

JAMES ABRAM GARFIELD.

ON THE WAR.

GENTLEMEN :— I am not insensible to the patriotic motives which prompted you to do me the honor to invite me to address you, on this occasion, upon the momentous issues now presented in the condition of the country. With a heart filled with sadness and grief I comply with your request.

For the first time since the adoption of this Federal Constitution, a wide-spread conspiracy exists to destroy the best government the sun of heaven ever shed its rays upon. Hostile armies are now marching upon the Federal capital, with a view of planting a revolutionary flag upon its dome. . . . The boast has gone forth by the secretary of war of this revolutionary government that on the first day of May the revolutionary flag shall float from the walls of the Capitol at Washington, and that on the fourth day of July the revolutionary army shall hold possession of the Hall of Independence. The simple question presented to us is whether we will wait for the enemy to carry out this boast of making war on our soil, or whether we will rush as one man to the defence of this government, and its capital, to defend it from the hands of all assailants who have threatened it. Already the piratical flag has been unfurled against the commerce of the United States. Letters of marque have been issued, appealing to the pirates of the world to assemble under that revolutionary flag, and commit depredations on the commerce carried on under the stars and stripes. Hostile batteries have been planted upon its fortresses; custom-houses have been established; and we are required now to pay tribute and taxes without having a voice in making the laws imposing them, or having a share in the distribution of them after they have been collected. The question is whether this war of aggression shall proceed, and we remain with folded arms inactive spectators, or whether we shall meet the aggressors at the threshold and turn back the tide. . . .

STEPHEN ARNOLD DOUGLAS.

FOR FREEDOM OF TRADE.

WITH the opportunity of unrestricted exchange of these products, how limitless the horizon of our possibilities! Let American adventurousness and genius be free upon the high seas, to go wherever they please and bring back whatever they please, and the oceans will swarm with American sails, and the land will laugh with the plenty within its borders. The trade of Tyre and Sidon, the far extending commerce of the Venetian republic, the wealth-producing traffic of the Netherlands, will be as dreams in contrast with the stupendous reality which American enterprise will develop in our own generation. Through the humanizing influence of the trade thus encouraged, I see nations become the friends of nations, and the causes of war disappear. I see the influence of the great republic in the amelioration of the condition of the poor and the oppressed in every land, and in the moderation of the arbitrariness of power. Upon the wings of free trade will be carried the seeds of free government, to be scattered everywhere to grow and ripen into harvests of free peoples in every nation under the sun.

FRANK H. HURD.

JOHN KEATS TO WILLIAM REYNOLDS.

BY this post I write to Rice, who will tell you why we have left Shanklin, and how we like the place. I have indeed scarcely anything else to say, leading so monotonous a life, unless I was to give you a history of sensations and day nightmares. You would not find me at all unhappy in it, as all my thoughts and feelings, which are of the selfish nature, home speculations, every day continue to make me more iron. I am convinced more and more, every day, that fine writing is, next to fine doing, the top thing in the world; the "Paradise Lost" becomes a greater wonder. The more I know what my diligence may in time probably effect, the more does my heart distend with pride and obstinacy. I feel it in my power to become a popular writer. I feel it in my power to refuse the poisonous suffrage of a public. My own being, which I know to be, becomes of more consequence to me than the crowds of shadows in the shape of men and women that inhabit a kingdom. The soul is a world of itself, and has enough to do in its own home. Those whom I know already and who have grown as it were a part of myself, I could not do without; but for the rest of mankind, they are as much a dream to me as Milton's "Hierarchies." I think if I had a free and healthy and lasting organization of heart, and lungs as strong as an ox, so as to be able to bear unhurt the shock of extreme thought and sensation without weariness, I could pass my life very nearly alone, though it should last eighty years. But I feel my body too weak to support me to this height; I am obliged continually to check myself, and be nothing.

It would be vain for me to endeavor after a more reasonable manner of writing to you. I have nothing to speak of but myself, and what can I say but what I feel? If you should have any reason to regret this state of excitement in me, I will turn the tide of your feelings in the right channel, by mentioning that it is the only state for the best sort of poetry — that is all I care for, all I live for. Forgive me for not filling up the whole sheet; letters become so irksome to me, that the next time I leave London I shall petition them all to be spared me. To give me credit for constancy, and at the same time waive letter-writing, will be the highest indulgence I can think of.

JOHN KEATS.

MRS. CARLYLE TO HER HUSBAND.

. . . I could swear you never heard of Madame —— de ——. But she has heard of you; and if you were in the habit of thanking God "for the blessing made to fly over your head," you might offer a modest thanksgiving for the honor that stunning lady did you in galloping madly all around Hyde Park in chase of your "brown wide-awake" the last day you rode there; no mortal could predict what the result would be if she came up with you. To seize your bridle and look at you till she was satisfied was a trifle to what she was supposed capable of. She only took to galloping after you when more legitimate means had failed.

She circulates everywhere, this madcap "Frenchwoman." She met "the Rev. John" (Barlow), and said, when he was offering delicate attentions, "There is just one thing I wish you would do for me — to take me to see Mr. Carlyle." "Tell me to ask the Archbishop of Canterbury to dance a polka with you," said Barlow, aghast, "and I would dare it, though I have not the honor of his acquaintance; but take anybody to Mr. Carlyle — impossible!"

"That silly old Barlow won't take me to Carlyle," said the lady to George Cook; "you must do it then." "Gracious Heavens!" said George Cook; "ask me to take you up to the Queen, and introduce you to her, and I would do it, and 'take the six months' imprisonment,' or whatever punishment was awarded me; but take anybody to Mr. Carlyle — impossible!"

Soon after this, George Cook met her riding in the Park, and said: "I passed Mr. Carlyle a little way on, in his brown wide-awake." The lady lashed her horse and set off in pursuit, leaving her party out of sight, and went all round the Park at full gallop, looking out for the wide-awake. . . . .

MRS. THOMAS CARLYLE.

SWEETNESS AND LIGHT.

The pursuit of perfection, then, is the pursuit of sweetness and light. He who works for sweetness and light, works to make reason and the will of God prevail. He who works for machinery, he who works for hatred, works only for confusion. Culture looks beyond machinery, culture hates hatred; culture has one great passion, the passion for sweetness and light. It has one even yet greater!—the passion for making them prevail. It is not satisfied till we all come to a perfect man; it knows that the sweetness and light of the few must be imperfect until the raw and unkindled masses of humanity are touched with sweetness and light. If I have not shrunk from saying that we must work for sweetness and light, so neither have I shrunk from saying that we must have a broad basis, must have sweetness and light for as many as possible. Again and again I have insisted how those are the happy moments of humanity, how those are the marking epochs of a people's life, how those are the flowering times for literature and art and all the creative power of genius, when there is a national glow of life and thought, when the whole of society is in the fullest measure permeated by thought, sensible to beauty, intelligent and alive. Only it must be real thought and real beauty; real sweetness and real light. Plenty of people will try to give the masses, as they call them, an intellectual food prepared and adapted in the way they think proper for the actual condition of the masses. The ordinary popular literature is an example of this way of working on the masses. Plenty of people will try to indoctrinate the masses with the set of ideas and judgments constituting the creed of their own profession or party. Our religious and political organizations give an example of this way of working on the masses. I condemn neither way; but culture works differently. It does not try to teach down to the level of inferior classes; it does not try to win them for this or that sect of its own, with ready-made judgments and watch-words. It seeks to do away with classes; to make the best that has been taught and known in the world current everywhere; to make all men live in an atmosphere of sweetness and light, where they may use ideas, as it uses them itself, freely,— nourished, and not bound by them.

This is the social idea; and the men of culture are the true apostles of equality. The great men of culture are those who have had a passion for diffusing, for making prevail, for carrying from one end of society to the other, the best knowledge, the best ideas of their time; who have labored to divest knowledge of all that was harsh, uncouth, difficult, abstract, professional, exclusive; to humanize it, to make it efficient outside the clique of the cultivated and learned, yet still remaining the best knowledge and thought of the time, and a true source, therefore, of sweetness and light. Such a man was Abelard in the Middle Ages, in spite of all his imperfections; and thence the boundless emotion and enthusiasm which Abelard excited. Such were Lessing and Herder in Germany, at the end of the last century; and their services to Germany were in this way inestimably precious. Generations will pass, and literary monuments will accumulate, and works far more perfect than the works of Lessing and Herder will be produced in Germany; and yet the names of these two men will fill a German with a reverence and enthusiasm such as the names of the most gifted masters will hardly awaken. And why? Because they humanized knowledge; because they broadened the basis of life and intelligence; because they worked powerfully to diffuse sweetness and light, to make reason and the will of God prevail. With St. Augustine they said: "Let us not leave Thee alone to make in the secret of thy knowledge, as thou didst before the creation of the firmament, the division of light from darkness; let the children of thy spirit, placed in their firmament, make their light shine upon the earth, mark the division of night and day, and announce the revolution of the times; for the old order is passed, and the new arises; the night is spent, the day is come forth; and thou shalt crown the year with thy blessing, when thou shalt send forth laborers into thy harvest sown by other hands than theirs; when thou shalt send forth new laborers to new seed-times, whereof the harvest shall be not yet."

MATTHEW ARNOLD.

FASHIONABLE LIFE AT KINKAIRD HOUSE.

. . . I see something of fashionable people here, and truly to my plebeian conception there is not a more futile class of persons on the face of the earth. If I were doomed to exist as a man of fashion, I do honestly believe I should swallow ratsbane, or apply to hemp or steel before three months were over. From day to day and year to year the problem is, not how to use time, but how to waste it least painfully. They have their dinners and their routs. They move heaven and earth to get every thing arranged and enacted properly; and when the whole is done, what is it? Had the parties all wrapped themselves in warm blankets and kept their beds, much peace had been among several hundreds of His Majesty's subjects, and the same result, the uneasy destruction of half a dozen hours, had been quite as well attained. No wonder poor women take to opium and scandal. The wonder is rather that these queens of the land do not some morning, struck by the hopelessness of their condition, make a general finish by simultaneous consent, and exhibit to coroners and juries the spectacle of the whole world of *ton* suspended by their garters, and freed at last from *ennui* in the most cheap and complete of all possible modes. There is something in the life of a sturdy peasant toiling from sun to sun for a plump wife and six eating children; but as for the Lady Jerseys and the Lord Petershams, peace be with them.

THOMAS CARLYLE.

PETITION OF THUGS.

WE, the most religious fraternity of Thugs, having heard it reported throughout the whole extent of India, that toleration is granted by the wisdom of the British Parliament to every diversity of creed, do most humbly submit our grievances to the patient consideration of your Honorable House. We claim a much higher antiquity than the earliest of devotional institutions known in Britain. We are the first-born of Cain. We profit by the holy book he left behind him, covering with fig-leaf what we consider to be unessential or liable to misinterpretation. Our humanity teaches us to confine no dissidents in unhealthy prisons, or to separate no husband from his wife, no father from his children, but merely to offer up man's life-blood to Him who gave man life. Our forefather, Cain, did not cast his brother Abel into a dark cavern infested by bats and serpents, but slew him as manfully and dexterously, and instantaneously, as could have been done by the best swordsman in the service of Hyder Ali.

It is reported to us, that there are religions by which it is declared lawful and right to disobey the prince they have sworn to obey, and even to select out of the rabble a leader of singing boys in flowing stoles, sable and white, purple and scarlet; and to place him in opposition to the rightful ruler of the land. . . .

We lay our cause with confidence at the bar of your Honorable House, claiming and deserving no more than has already been granted by it, to the three or four last religions which have consecutively been dominant in Great Britain. We hear that these religions are rolling over one another at this instant, and exercising a prodigious volubility of limb and tongue; the elderly and decrepit thrown on its back, cursing and swearing, but holding down the younger by the throat. We take no delight, no interest, in these prolusions; and we demand only simple protection, in meet reward for undivided allegiance.

No prayers do we offer up to God that it may please his Divine Majesty to assist us in sweeping our enemies from his earth; no thanksgiving for having bestrewn it with limbs and carcasses to satiate the hyena and the vulture. We invite our fellow-men to die as becomes them in His service. We lead Death by the hand in quiet and silence to his own door, and we depart in peace. Therefore we, conscious of our innocence and purity, venture to remind our generous protectors that the few we sacrifice are sacrificed to our God alone, and neither to gratify pride nor vengeance; that if we slay a few hundreds in the space of a year, our gracious protectors slay occasionally as many thousands between the rising and setting sun. We do not, indeed, with the same fervor and magnificence as our gracious protectors, sing hymns, beat drums, blow trumpets, and swing bells from lofty towers in jubilee; but we wash our hands, lay aside our daggers, bend our knees, and pray.

Confidently, then, do we approach our gracious protectors, and entreat the same favor, the same liberty of worship, as our fellow-subjects.

WALTER SAVAGE LANDOR.

THE BATTLE OF TLASCALA.

As a battle was now inevitable, Cortéz resolved to march out and meet the enemy in the field. This would have a show of confidence, that might serve the double purpose of intimidating the Tlascalans, and inspiriting his own men, whose enthusiasm might lose somewhat of its heat, if compelled to await the assault of their antagonists, inactive in their own intrenchments. The sun rose bright on the following morning, the 5th of September, 1519, an eventful day in the history of the Spanish Conquest. The general reviewed his army, and gave them, preparatory to marching, a few words of encouragement and advice.

The infantry he instructed to rely on the point rather than the edge of their swords, and to endeavor to thrust their opponents through the body. The horsemen were to charge at half speed, with their lances aimed at the eyes of the Indians. The artillery, the arquebusiers, and crossbow-men, were to support one another, some loading while others discharged their pieces, that there should be an unintermitted firing kept up through the action. Above all, they were to maintain their ranks close and unbroken, as on this depended their preservation.

They had not advanced a quarter of a league, when they came in sight of the Tlascalan army. Its dense array stretched far and wide over a vast plain or meadow ground, about six miles square. Its appearance justified the report which had been given of its numbers.

Nothing could be more picturesque than the aspect of these Indian battalions with the naked bodies of the common soldiers gaudily painted, the fantastic helmets of the chiefs glittering with gold and precious stones, and the glowing panoplies of feather-work which decorated their persons. Innumerable spears and darts tipped with points of transparent itztli, or fiery copper, sparkled bright in the morning sun, like the phosphoric gleams playing on the surface of a troubled sea, while the rear of the mighty host was dark with the shadows of banners, on which were emblazoned the armorial bearings of the great Tlascalan and Otomir chieftains. Among these, the white heron on the rock, the cognizance of the house of Xicotencatl, was conspicuous, and, still more, the golden eagle with outspread wings, in the fashion of a Roman signum, richly ornamented with emeralds and silver-work, the great standard of the republic of Tlascala.

The common file wore no covering except a girdle around the loins. Their bodies were painted with the appropriate colors of the chieftain whose banner they followed. The feather-mail of the higher class of warriors exhibited, also, a similar selection of colors for the like object, in the same manner as the color of the tartan indicates the peculiar clan of the Highlander. The caciques and principal warriors were clothed in a quilted cotton tunic, two inches thick, which, fitting close to the body, protected also the thighs and the shoulders. Over this the wealthier Indians wore cuirasses of thin gold plate, or silver. Their legs were defended by leathern boots or sandals, trimmed with gold. But the most brilliant part of their costume was a rich mantle of the plumage of feather-work, embroidered with curious art, and furnishing some resemblance to the gorgeous surcoat worn by the European knight, over his armor in the Middle Ages. This graceful and picturesque dress was surmounted by a fantastic head-piece made of wood or leather, representing the head of some wild animal, and frequently displaying a formidable array of teeth. With this covering the warrior's head was enveloped, producing a most grotesque and hideous effect. From the crown floated a splendid panache of the richly variegated plumage of the tropics, indicating, by its form and colors, the rank and family of the wearer. To complete their defensive armor, they carried shields or targets, made sometimes of wood covered with leather, but more usually of a light frame of reeds quilted with cotton, which were preferred, as rougher and less liable to fracture than the former. They had other bucklers, in which the cotton was covered with an elastic substance, enabling them to be shut up in a more compact form, like a fan or umbrella. These shields were decorated with showy ornaments, according to the taste or wealth of the wearer, and fringed with a beautiful pendant of feather-work.

Their weapons were slings, bows and arrows, javelins, and darts. They were accomplished archers, and would discharge two or even three arrows at a time. But they most excelled in throwing the javelin. One species of this, with a thong attached to it, which remained in the slinger's hand, that he might recall the weapon, was especially dreaded by the Spaniards. These various weapons were pointed with bone, or the mineral itztli (obsidian), the hard, vitreous substance, already noticed, as capable of taking an edge like a razor, though easily blunted. Their spears and arrows were also frequently headed with copper. Instead of a sword, they bore a two-handed staff, about three feet and a half long, in which, at regular distances, were inserted transversely, sharp blades of itztli, — a formidable weapon, which, an eye-witness assures us, he had seen fell a horse at a blow.

Such was the costume of the Tlascalan warrior, and indeed of that great family of nations generally, who occupied the plateau of Anahuac. Some parts of it, as the targets and the cotton-mail, or escaupil, as it was called in Castilian, were so excellent, that they were subsequently adopted by the Spaniards, as equally effectual in the way of protection, and superior, on the score of lightness and convenience, to their own. They were of sufficient strength to turn an arrow, or the stroke of a javelin, although impotent as a defence against fire-arms. But what armor is not? Yet it is probably no exaggeration to say, that, in convenience, gracefulness, and strength, the arms of the Indian warrior were not very inferior to those of the polished nations of antiquity. As soon as the Castilians came in sight, the Tlascalans set up their yell of defiance, rising high above the wild barbaric minstrelsy of shell, atabal, and trumpet, with which they proclaimed their triumphant anticipation of victory over the paltry forces of the invaders.

When the latter had come within bowshot, the Indians hurled a tempest of missiles, that darkened the sun for a moment as with a passing cloud, strewing the earth around with heaps of stones and arrows. Slowly and steadily the little band

THE BATTLE OF TLASCALA.

of Spaniards held on its way amidst this arrowy shower, until it had reached what appeared the proper distance for delivering its fire with full effect.

Cortéz then halted, and, hastily forming his troops, opened a general well-directed fire along the whole line. Every shot bore its errand of death; and the ranks of the Indians were mowed down faster than their comrades in the rear could carry off their bodies, according to custom, from the field. The balls in their passage through the crowded files, bearing splinters of the broken harness, and mangled limbs of the warriors, scattered havoc and desolation in their path. The mob of barbarians stood petrified with dismay, till, at length, galled to desperation by their intolerable sufferings, they poured forth simultaneously their hideous war-shriek, and rushed impetuously on the Christians.

On they came like an avalanche, a mountain torrent, shaking the solid earth, and sweeping away every obstacle in its path. The little army of Spaniards opposed a bold front to the overwhelming mass. But no strength could withstand it. They faltered, gave way, were borne along before it, and their ranks were broken and thrown into disorder. It was in vain the general called on them to close again and rally. His voice was drowned by the din of fight, and the fierce cries of the assailants. For a moment, it seemed that all was lost. The tide of battle had turned against them, and the fate of the Christians was sealed.

But every man had that within his bosom which spoke louder than the voice of the general. Despair gave unnatural energy to his arm. The naked body of the Indian afforded no resistance to the sharp Toledo steel; and with their good swords, the Spanish infantry at length succeeded in staying the human torrent. The heavy guns from a distance thundered on the flank of the assailants, which, shaken by the iron tempest, was thrown into disorder. Their very numbers increased the confusion, as they were precipitated on the masses in front. The horse at the same moment, charging gallantly under Cortéz, followed up the advantage, and at length compelled the tumultuous throng to fall back with greater precipitation and disorder than that with which they had advanced.

More than once in the course of the action a similar assault was attempted by the Tlascalans, but each time with less spirit, and greater loss. They were too deficient in military science to profit by their vast superiority in numbers. They were distributed into companies, it is true, each serving under its own chieftain and banner. But they were not arranged by rank and file, and moved in a confused mass, promiscuously heaped together. They knew not how to concentrate numbers on a given point, or even how to sustain an assault, by employing successive detachments to support and relieve one another. A very small part only of their array could be brought into contact with an enemy inferior to them in amount of forces. The remainder of the army, inactive and worse than useless, in the rear, served only to press tumultuously on the advance, and embarrass its movements by mere weight of numbers, while, on the least alarm, they were seized with a panic and threw the whole body into inextricable confusion. It was, in short, the combat of the ancient Greeks and Persians over again.

Still the great numerical superiority of the Indians might have enabled them, at a severe cost of their own lives, indeed, to wear out, in time, the constancy of the Spaniards, disabled by wounds and incessant fatigue. But, fortunately for the latter, dissensions arose among their enemies. A Tlascalan chieftain, commanding one of the great divisions, had taken umbrage at the haughty demeanor of Xicotencatl, who had charged him with misconduct or cowardice in the late action. The injured cacique challenged his rival to single combat. This did not take place. But, burning with resentment, he chose the present occasion to indulge it, by drawing off his forces, amounting to ten thousand men, from the field. He also persuaded another of the commanders to follow his example.

Thus reduced to about half his original strength, and that greatly crippled by the losses of the day, Xicotencatl could no longer maintain his ground against the Spaniards. After disputing the field with admirable courage for four hours, he retreated and resigned it to the enemy. The Spaniards were too much jaded, and too many were disabled by wounds, to allow them to pursue; and Cortez, satisfied with the decisive victory he had gained, returned in triumph to his position on the hill of Tzompach.

The number of killed in his own ranks had been very small, notwithstanding the severe loss inflicted on the enemy. These few he was careful to bury where they could not be discovered, anxious to conceal not only the amount of the slain, but the fact that the white were mortal. But very many of the men were wounded, and all the horses. The trouble of the Spaniards was much enhanced by the want of many articles important to them in their present exigency. They had neither oil nor salt, which, as before noticed, was not to be obtained in Tlascala. Their clothing, accommodated to a softer climate, was ill adapted to the rude air of the mountains; and bows and arrows, as Bernal Diaz sarcastically remarks, formed an indifferent protection against the inclemency of the weather.

Still, they had much to cheer them in the events of the day; and they might draw from them a reasonable ground for confidence in their own resources, such as no other experience could have supplied. Not that the results could authorize anything like contempt for their Indian foe. Singly and with the same weapons, he might have stood his ground against the Spaniard. But the success of the day established the superiority of science and discipline over mere physical courage and numbers. It was fighting over again, as we have said, the old battle of the European and the Asiatic. But the handful of Greeks who routed the hosts of Xerxes and Darius, it must be remembered, had not so obvious an advantage on the score of weapons as was enjoyed by the Spaniards in these wars. The use of fire-arms gave an ascendency which cannot easily be estimated; one so great, that a contest between nations equally civilized, which should be similar in all other respects to that between the Spaniards and the Tlascalans, would probably be attended with a similar issue. To all this must be added the effect produced by the cavalry. The nations of Anahuac had no large domesticated animals, and were unacquainted with any beast of burden. Their imaginations were bewildered when they beheld the strange apparition of the horse and his rider moving in unison and obedient to one impulse, as if possessed of a common nature; and when they saw the terrible animal, with his "neck clothed in thunder," bearing down their squadrons and trampling them in the dust, no wonder they should have regarded him with the mysterious terror felt for a supernatural being. A very little reflection on the manifold grounds of superiority, both moral and physical, possessed by the Spaniards in this contest, will surely explain the issue, without any disparagement to the courage or capacity of their opponents.

Cortéz, thinking the occasion favorable, followed up the important blow he had struck by a new mission to the capital, bearing a message of similar import with that recently sent to the camp. But the senate was not yet sufficiently humbled. The late defeat caused, indeed, general consternation. Maxixcatzin, one of the four great lords who presided over the republic, reiterated with greater force the arguments before urged by him for embracing the proffered alliance of the strangers. The armies of the state had been beaten too often to allow any reasonable hope of successful resistance; and he enlarged on the generosity shown by the politic Conqueror to his prisoners, — so unusual in Anahuac, — as an additional motive for an alliance with men who knew how to be friends as well as foes.

But in these views he was overruled by the war party, whose animosity was sharpened, rather than subdued, by the late discomfiture. Their hostile feelings were further exasperated by the younger Xicotencatl, who burned for an opportunity to retrieve his disgrace, and to wipe away the stain which had fallen for the first time on the arms of the republic.

In their perplexity, they called in the assistance of the priests, whose authority was frequently invoked in the deliberations of the American chiefs. The latter inquired, with some simplicity, of these interpreters of fate, whether the strangers were supernatural beings, or men of flesh and blood like themselves. The priests, after some consultation, are said to have made the strange answer, that the Spaniards, though not gods, were children of the Sun; that they derived their strength from that luminary, and, when his beams were withdrawn, their powers would also fail. They recommended a night attack, therefore, as one which afforded the best chance of success. This apparently childish response may have had in it more of cunning than credulity. It was not improbably suggested by Xicotencatl himself, or by the caciques in his interest, to reconcile the people to a measure which was contrary to military usages, — indeed, it may be said, to the public law of Anahuac. Whether the fruit of artifice or superstition, it prevailed; and the Tlascalan general was empowered, at the head of a detachment of ten thousand warriors, to try the effect of an assault by night on the Christian camp.

The affair was conducted with such secrecy, that it did not reach the ears of the Spaniards. But their general was not one who allowed himself, sleeping or waking, to be surprised on his post. Fortunately, the night appointed was illumed by the full beams of an autumnal moon; and one of the videttes perceived by its light, at a considerable distance, a large body of Indians moving towards the Christian lines. He was not slow in giving the alarm to the garrison.

The Spaniards slept, as has been said, with their arms by their side; while their horses, picketed near them, stood ready saddled, with the bridle hanging at the bow. In five minutes, the whole camp was under arms; when they beheld the dusky columns of the Indians cautiously advancing over the plain, their heads just peering above the tall maize with which the land was partially covered, Cortéz determined not to abide the assault in his intrenchments, but to sally out and pounce on the enemy when he had reached the bottom of the hill.

Slowly and stealthily the Indians advanced, while the Christian camp, hushed in profound silence, seemed to them buried in slumber. But no sooner had they reached the slope of the rising ground, than they were astounded by the deep battle cry of the Spaniards, followed by the instantaneous apparition of the whole army, as they sallied forth from the works, and poured down the sides of the hill. Brandishing aloft their weapons, they seemed to the troubled fancies of the Tlascalans, like so many specters or demons hurrying to and fro in mid-air, while the uncertain light magnified their numbers, and expanded the horse and his rider into gigantic and unearthly dimensions.

Scarcely waiting the shock of their enemy, the panic-struck barbarians let off a feeble volley of arrows, and, offering no other resistance, fled rapidly and tumultuously across the plain. The horse easily overtook the fugitives, riding them down and cutting them to pieces, without mercy, until Cortéz, weary with slaughter, called off his men, leaving the field loaded with the bloody trophies of victory.

WILLIAM HICKLING PRESCOTT.

AUNT MARIA AND THE AUTOPHONE.

"You see, stranger, we're the musicalest family in the whole county. When I married ma, she says, 'Abner' (that's me) 'Abner,' says she, 'I kin do without a rag carpet in the kitchen, but I cann't live without a melodjun in the parlor.'

"So we had a melodjun in the parlor, and the children came naturally by their love for music. Why, bless your soul! I may say they took to it with their first breaths, and kept it up always after. The girls had the melodjun, and the boys had every thing from a willow whistle to a fiddle, and when Martha and Stella was draggin' a duet out of the melodjun in the parlor, and Jehiel and Jonathan scrapin' out the 'Arkansaw Traveller' in the kitchen on a fiddle and banjo, it was a musical abode.

"Everything went along all right until Aunt Maria came. Lordy! how that woman did hate music! Nobody had any peace in the house, and what's the worst, a sort of bad luck came over the harmless instruments themselves. Jonathan's fiddle strings was always getting broke before he'd half tuned up, and the pesky melodjun took to leaking so that both gals together, one on the pedals and the other on the keys, could hardly pump 'Old Hundred' out of her Sundays. Some did suspect Maria, but," said the old man, looking cautiously around, "I don't think she was altogether to blame; howsomever," with a significant wink, "she got the credit of it.

"When John Henry — he's the youngest — came, Maria's heart seemed to kind of soften. His first drum lasted a week, and I noticed she never had anything to say agin his vocal accomplishments. Well, when John Henry was four years old, the old woman began to look around and see what instrument he'd be likely to take to. Aunt Maria said it was a burning shame to make that innocent child a stumblin'-block in the way of Christians, but I said I guessed John Henry could stand it — if we could.

"The next day ma went down to the village to sell her butter and eggs, and when she came home at night she had a small bundle, which she put away in the parlor until after supper. I know'd what it was — leastways, not exactly, but I guessed by the way the old woman slung the dishes on the table that night that we should hear some news soon. When the dishes was washed up, 'Ma,' says I, 'didn't I see you bring in a bundle jest now?' 'You did, Abner,' says she, and she smiled from one ear to the other. 'Abner,' says she, 'I've found an instrument at last for John Henry.' Aunt Maria fetched a kind of cross between a sigh and a groan, but nobody paid any attention to her. 'Well, ma,' says I, 'let's have it.' So out she brought the bundle, and there was sort of an accordjun on two legs, and a lot of bits of white paper as full of holes as the old woman's colander. We all got around the table while ma showed us how it worked. 'You see,' says she, 'you jest poke in the paper — here, John Henry, this is your'n, and you shall have the first try; there — you shove the paper in there, and work your hand so, and it plays all the music on the paper.' 'Ma,' says I, 'do you mean to say, as a member in good and regular standin', that that 'ere instrument plays them holes?' But John Henry had grabbed the instrument, and jest as sure as I set here, stranger, that four-year-old child squeezed out 'Old Hundred' jest as solemn and a derned sight faster than ma's melodjun. But you oughter to see Aunt Maria; she straightened up and glared at that innocent child as if she wished he had lived in Palestine about the year one, and bolted out of the room without a word.

"Well, stranger, it was a sight to see John Henry on the kitchen floor with that 'ere thing between his little knees, and playing the 'Sweet By and By' in a way to make tears come to everybody's eyes, exceptin' always Aunt Maria's. For a month our house was the most popularest house at the Corners, and John Henry gave a free concert every night for an hour before he went to bed. The strangest thing," said the old man, in a mysterious tone, "was that that 'ere instrument kept in playin' order all the time, whether it was because John Henry took it to bed with him every night, or whether it was from the superior build of the consarn, I can't say. Perhaps" — with a wink — "Aunt Maria didn't understand its innerd construction as well as she did a fiddle or a melodjun.

"Well, as I say, the instrument kept in playin' order all winter; the music, 'specially the pop'lar tunes, was a little the worse for wear, but that's all. 'I want to be an angel' and one or two others got tored in two about the middle of March, and John Henry asked Aunt Maria to mend them one day, and, bless you! she loved that darlin' child too much to refuse him anything, so she pasted the tunes together as well as she could, and next day John Henry took his instrument to Sunday-school. You see, he'd taken it a number of times, and the teacher thought it kind of 'livened up the exercises. But this day, jest as John Henry was slowly and surely grindin' out 'I want to be an angel,' and had got to the middle of the tune (where it was tored, you see), all at onst out he came with 'Whoa, Emma!' and the innocent child was too much surprised to stop until the teacher suspended the musical exercises for that day. John Henry didn't git no prize that year, but I hold that Aunt Maria was morally responsible. You see, she had so little music in her — leastwise we thought so then — that she couldn't even be trusted to paste two tunes together.

"Howsomever, as spring came on, we thought we kind of noticed a change in Maria. It wasn't that she was gittin' musical — that was, perhaps, too much to expect on this arth, as I said to ma — but she was growin' mellow somehow. I think it was all owin' to John Henry's tender influence. You ask how I knew she was gittin' mellow, stranger? Well, you see, John Henry's instrument still kept in workin' order. She and John Henry would disappear by the hour, and what they did no one knew. Ma said one day she thought she had heard John Henry playin' on his instrument in Maria's room, leastwise she had heard a noise there, but it didn't sound like any instrument in that house. 'Perhaps,' said I, 'it was Maria singin'.' But the more I thought it over, the more mysterious the thing seemed, and I made up my mind I'd git to the bottom of it. So one day, when ma and the girls had gone to town, and the boys was hoein' potatoes, I jest slipped into the house and listened awhile. By and by I thought I heard a sound in the direction of Maria's room, and so I took off my boots and crawled softly up the stairs; but, lordy! I might jest as well have kept them on, for when I got up near the door I heard the most dreadful noises you ever dreamed of. If I had had any hair, it would have stood up and run off my head. I first thought that Maria was torturin' that innocent child, and was goin' to bust in the door, but I thought I'd first take a peep through the keyhole. What do you think I saw, stranger? John Henry was in his favorite attitude in the middle of the floor, workin' the instrument with one hand and feedin' the music in with the other, and Aunt Maria sat in her rockin'-chair, rockin' slowly to and fro, and keepin' time with her hands. Her glasses was pushed up on her forrard, and tears of joy was runnin' down her cheeks, and John Henry kept playin' faster and faster; but what music! No tune that I had ever hearn — and we had all sorts in that house at one time or another — came from that instrument. I thought something was wrong, and in I rushed. Aunt Maria cried 'Oh!' and fell back in her chair, lookin' dreadful sheepish; but John Henry! Stranger, what do you think that lamb did? Why, he jest winked at his pa, and when I asked him what that infernal row meant, he said, kind of under his breath, 'Why, you see, pa, one day I got one of them tunes in hindside foremost, and Aunt Maria was so pleased that I've gone on that way ever since, hindside foremost or upside down.'

"I said to ma that night when she got home: 'You see, ma, you was wrong about Maria; she's got as much music in her as the rest of the family, but she's obliged to take hers in a peculiar way. She can't take it straight, but jest give it to her hindside foremost or upside down, and she enjoys it as much as any-one.'"

Just then a whistle blew, and my friend's train came along. He got into the car with a dazed expression on his face, as if an idea was trying to crystallize into words. As the train was moving away he came rushing out on the rear platform, and putting up his hands in the form of a speaking-trumpet, he shouted, "Try your Browning hindside foremost," and as the train swept around a curve I heard faintly on the clear cold air, "or upside down."

THOMAS FREDERICK CRANE.

JOEL AT WORK.

"It's no use trying to sleep," declared Joel, in the middle of the night, and kicking the bedclothes for the dozenth time into a roll at the foot, "as long as I can see Mamsie's eyes. I'll just get up and tackle that Latin grammar now. Whew! haven't I got to work, though. Might as well begin at it," and he jumped out of bed.

Stepping softly over to the door that led into David's little room, he closed it carefully, and with a sigh, lighted the gas. Then he went over to the table where his school books ought to have been. But instead, the space was piled with a great variety of things — one or two balls, a tennis racket, and a confusion of fishing tackle, while in front, the last thing that had occupied him that day, lay a book of artificial flies.

Joel set his teeth together hard, and looked at them. "Suppose I sha'n't get much of this sort of thing this summer," he muttered. "Here goes!" and without trusting himself to take another look, he swept them all off down to the floor and into a corner.

"There," he said, standing up straight, "lie there, will you?" But they loomed up in a suggestive heap, and his fingers trembled to just touch them once.

"I must cover up the things, or else I know I'll be at them," he said, and hurrying over to the bed, he dragged off the coverlid. "Now," and he threw it over the fascinating mass, "I've got to study. Dear me, where are my books?"

For the next five minutes Joel had enough to do to collect his working instruments, and when at last he unearthed them from the corner of his closet where he had thrown them under a pile of boots, he was tired enough to sit down.

"I don't know which to go at first," he groaned, whirling the leaves of the upper book. "It ought to be Latin — but then it ought to be algebra just as much, and as for history — well there — here goes, I'll take them as they come."

With a very red face Joel plunged into the first one under his hand. It proved to be the Latin grammar, and with a grimace, he found the page, and resting his elbows on the table, he seized each side of his stubby head with his hand. "I'll hang on to my hair," he said, and plunged into his task.

And now there was no sound in the room but his hard breathing, and the noise he made turning the leaves, for he very soon found he was obliged to go back many lessons to understand how to approach the one before him; and with cheeks growing every instant more scarlet with shame and confusion, the drops of perspiration ran down his forehead and fell on his book.

"Whew!" he exclaimed, "it's horribly hot," and pushing back his book, he tiptoed over to the other window and softly raised it. The cool air blew into his face, and leaning far out into the dark night, he drew in deep breaths.

"I've skinned through and saved my neck a thousand times," he reflected, "and now I've got to dig like sixty to make up. There's Dave now, sleeping in there like a cat; he don't have anything to do, but to run ahead of the class like lightning — just because he" —

"Loves it," something seemed to sting the words into him. Joel drew in his head and turned abruptly away from the window.

"Pshaw! well, here goes," he exclaimed again, throwing himself into his chair. "She said 'I'd work myself to skin and bone, but I'd go through creditably.'" Joel bared his brown arm and regarded it critically. "I wonder how 'twould look all skin and bone," and he gave a short laugh.

"But this isn't studying." He pulled down his sleeve, and his head went over the book again.

MARGARET SIDNEY.

GRADLE.

. . . Our servant knows a few words of English, too; her name is *Gradle*, the short for Margaret. Jane wanted a fowl to boil for me. Now she has a theory that the more she makes her English un-English, the more it must be like German. Jane begins by showing Gradle a word in the dictionary.

Gradle. — "Ja! yees — hühn — henne — ja! yees."

Jane (a little through her nose). — "Hmn — hum — hem — yes — yaw, ken you geet a fowl — fool — foal, to boil — bile — bole for dinner?"

Gradle. — "Hot wasser?"

Jane. — "Yaw, in pit — pat — pot — hmn — hum — eh!"

Gradle (a little off the scent again). — "Ja, nein — wasser, pot — hot — nein."

Jane. — "Yes — no — good to eeat — chicken — cheeken — checking — choking — bird — bard — beard — lays eggs — eeggs — hune, heine — hin— make cheekin broth — soup — poultry — peltry — paltry!"

Gradle (quite at fault). — " Pfeltrighchtch ! — nein."

Jane (in despair). — "What shall I do! and Hood won't help me, he only laughs. This comes of leaving England!" (She casts her eyes across the street at the governor's poultry-yard, and a bright thought strikes her.) "Here, Gradle — come here — comb hair — hmn — hum — look there —dare — you see things walking — hmn, hum — wacking about — things with feathers — fathers — feethers."

Gradle (hitting it off again). — " Feethers — faders — ah hah! fedders — ja, ja, yees, sie bringen — fedders, ja, ja!"

Jane echoes " Fedders — yes — yaw, yaw!"

Exit Gradle, and after three quarters of an hour, returns triumphantly with two bundles of stationer's quills!!! . . .

THOMAS HOOD.

KIN BEYOND SEA.

BUT if there be those in this country who think that American democracy means public levity and intemperance, or a lack of skill and sagacity in politics, or the absence of self-command and self-denial, let them bear in mind a few of the most salient and recent facts of history which may profitably be recommended to their reflections. We emancipated a million of negroes by peaceful legislation; America liberated four or five millions by a bloody civil war: yet the industry and exports of the Southern States are maintained, while those of our negro colonies have dwindled; the South enjoys all its franchises, but we have, *proh pudor!* found no better method of providing for peace and order in Jamaica, the chief of our islands, than by the hard and vulgar, even where needful, expedient of abolishing entirely its representative institutions.

The Civil War compelled the States, both North and South, to train and embody a million and a half of men, and to present to view the greatest, instead of the smallest, armed forces in the world. Here there was supposed to arise a double danger. First, that on a sudden cessation of the war, military life and habits could not be shaken off, and, having become rudely and widely predominant, would bias the country toward an aggressive policy, or, still worse, would find vent in predatory or revolutionary operations. Secondly, that a military caste would grow up with its habits of exclusiveness and command, and would influence the tone of politics in a direction adverse to republican freedom. But both apprehensions

WILLIAM E. GLADSTONE.

proved to be wholly imaginary. The innumerable soldiery was at once dissolved. Cincinnatus, no longer an unique example, became the commonplace of every day, the type and mould of a nation. The whole enormous mass quietly resumed the habits of social life. The generals of yesterday were the editors, the secretaries, and the solicitors of to-day. The just jealousy of the State gave life to the now forgotten maxim of Judge Blackstone, who denounced as perilous the erection of a separate profession of arms in a free country. The standing army, expanded by the heat of civil contest to gigantic dimensions, settled down again into the framework of a miniature with the returning temperature of civil life, and became a power wellnigh invisible, from its minuteness, amidst the powers which sway the movements of a society exceeding forty millions.

More remarkable still was the financial sequel to the great conflict. The internal taxation for Federal purposes, which before its commencement had been unknown, was raised, in obedience to an exigency of life and death, so as to exceed every present and every past example. It pursued and worried all the transactions of life. The interest of the American debt grew to be the highest in the world, and the capital touched five hundred and sixty millions sterling. Here was provided for the faith and patience of the people a touchstone of extreme severity. In England, at the close of the great French war, the propertied classes, who were supreme in Parliament, at once rebelled against the Tory Government, and refused to prolong the income tax even for a single year. We talked big, both then and now, about the payment of our national debt; but sixty-three years have since elapsed, all of them except two called years of peace, and we have reduced the huge total by about one ninth; that is to say, by little over one hundred millions, or scarcely more than one million and a half a year. This is the conduct of a State elaborately digested into orders and degrees, famed for wisdom and forethought, and consolidated by a long experience. But America continued long to bear, on her unaccustomed and still smarting shoulders, the burden of the war taxation. In twelve years she has reduced her debt by one hundred and fifty-eight millions sterling, or at the rate of thirteen millions for every year. In each twelve months she has done what we did in eight years; her self-command, self-denial, and wise forethought for the future have been, to say the least, eightfold ours. These are facts which redound greatly to her honor; and the historian will record with surprise that an enfranchised nation tolerated burdens which in this country a selected class, possessed of the representation, did not dare to face, and that the most unmitigated democracy known to the annals of the world resolutely reduced at its own cost prospective liabilities of the State, which the aristocratic, and plutocratic, and monarchical government of the United Kingdom has been contented ignobly to hand over to posterity. And such facts should be told out. It is our fashion so to tell them, against as well as for ourselves; and the record of them may some day be among the means of stirring us up to a policy more worthy of the name and fame of England.

It is true, indeed, that we lie under some heavy and, I fear, increasing disadvantages, which amount almost to disabilities. Not, however, any disadvantage respecting power, as power is commonly understood. But, while America has a nearly homogeneous country, and an admirable division of political labor between the States individually and the Federal Government, we are, in public affairs, an overcharged and overweighted people.

We have undertaken the cares of empire upon a scale, and with a diversity, unexampled in history; and, as it has not yet pleased Providence to endow us with brain-force and animal strength in an equally abnormal proportion, the consequence is that we perform the work of government, as to many among its more important departments, in a very superficial and slovenly manner. The affairs of the three associated kingdoms, with their great diversities of law, interest, and circumstance, make the government of them, even if they stood alone, a business more voluminous, so to speak, than that of any other thirty-three millions of civilized men. To lighten the cares of the central legislature by judicious devolution, it is probable that much might be done; but nothing is done, or even attempted to be done. The greater colonies have happily attained to a virtual self-government; yet the aggregate mass of business connected with our colonial possessions continues to be very large. The Indian Empire is of itself a charge so vast, and demanding so much thought and care, that if it were the sole transmarine appendage to the Crown, it would amply tax the best ordinary stock of human energies. Notoriously it obtains from the Parliament only a small fraction of the attention it deserves. Questions affecting individuals, again, or small interests, or classes, excite here a greater interest, and occupy a larger share of time, than, perhaps, in any other community. In no country, I may add, are the interests of persons or classes so favored when they compete with those of the public; and in none are they more exacting, or more wakeful to turn this advantage to the best account. With the vast extension of our enterprise and our trade, comes a breadth of liability not less large, to consider every thing that is critical in the affairs of foreign states; and the real responsibilities thus existing for us, are unnaturally inflated for us by fast-growing tendencies toward exaggeration of our concern in these matters, and even toward setting up fictitious interests in cases where none can discern them except ourselves, and such continental friends as practice upon our credulity and our fears for purposes of their own. Last of all, it is not to be denied that in what I have been saying, I do not represent the public sentiment. The nation is not at all conscious of being overdone. The people see that their House of Commons is the hardest-working legislative assembly in the world: and, this being so, they assume it is all right. Nothing pays better, in point of popularity, than those gratuitous additions to obligations already beyond human strength, which look like accessions or assertion of power; such as the annexation of new territory, or the silly transaction known as the purchase of shares in the Suez Canal.

All my life long I have seen this excess of work as compared with the power to do it; but this evil has increased with the surfeit of wealth, and there is no sign that the increase is near its end. The people of this country are a very strong people; but there is no strength that can permanently endure, without provoking inconvenient consequences, this kind of political debauch. It may be hoped, but it cannot be predicted, that the mischief will be encountered and subdued at the point where it will have become sensibly troublesome, but will not have grown to be quite irremediable.

The main and central point of interest, however, in the institutions of a country is the manner in which it draws together and compounds the public forces in the balanced action of the State. It seems plain that the formal arrangements for this purpose in America are very different from ours. It may even be a question whether they are not, in certain respects, less popular; whether our institutions do not give more rapid effect, than those of the Union, to any formed opinion, and resolved intention, of the nation.

In the formation of the Federal Government we seem to perceive three stages of distinct advancement. First, the formation of the Confederation, under the pressure of the War of Independence. Secondly, the Constitution, which placed the Federal Government in defined and direct relation with the people inhabiting the several States. Thirdly, the struggle with the South, which for the first time, and definitely, decided that to the Union, through its Federal organization, and not to the State governments, were reserved all the questions not decided and disposed of by the express provisions of the Constitution itself. The great *arcanum imperii*, which with us belongs to the three branches of the Legislature, and which is expressed by the current phrase, "omnipotence of Parliament," thus became the acknowledged property of the three branches of the Federal Legislature; and the old and respectable doctrine of State independence is now no more than an archæological relic, a piece of historical antiquarianism. Yet the actual attributions of the State authorities cover by far the largest part of the province of government; and by this division of labor and authority, the problem of fixing for the nation a political center of gravity is divested of a large part of its difficulty and danger, in some proportions to the limitations of the working precinct.

Within that precinct, the initiation as well as the final sanction in the great business of finance is made over to the popular branch of the Legislature, and a most interesting question arises upon the comparative merits of this arrangement, and of our method, which theoretically throws upon the Crown the responsibility of initiating public charge, and under which, until a recent period, our practice was in actual and even close correspondence with this theory.

We next come to a difference still more marked. The Federal Executive is born anew of the nation at the end of each four years, and dies at the end. But, during the course of those years, it is independent, in the person both of the President and of his Ministers, alike of the people, of their representatives, and of that remarkable body, the most remarkable of all the inventions of modern politics, the Senate of the United States. In this important matter, whatever be the relative excellencies and defects of the British and American systems, it is most certain that nothing would induce the people of this country, or even the Tory portion of them, to exchange our own for theirs. It may, indeed, not be obvious to the foreign eye what is the exact difference of the two. Both the representative chambers hold the power of the purse. But in America its conditions are such that it does not operate in any way on behalf of the Chamber or of the nation, as against the Executive. In England, on the contrary, its efficiency has been such that it has worked out for itself channels of effective operation, such as to dispense with its direct use, and avoid the inconveniences which might be attendant upon that use. A vote of the House of Commons, declaring a withdrawal of its confidence, has always sufficed for the purpose of displacing a Ministry; nay, persistent obstruction of its measures, and even lighter causes, have conveyed the hint, which has been obediently taken. But the people, how is it with them? Do not the people in England part with their power, and make it over to the House of Commons, as completely as the American people part with it to the President? They give it over for four years: we for a period which on the average is somewhat more: they, to resume it at a fixed time; we, on an unfixed contingency, and at a time which will finally be determined, not according to the popular will, but according to the views of which a Ministry may entertain of its duty or convenience.

WILLIAM EWART GLADSTONE.

A TRUE CALEDONIAN.

I HAVE been trying all my life to like Scotchmen, and am obliged to desist from the experiment in despair. They cannot like me — and, in truth, I never knew one of that nation who attempted to do it. There is something more plain and ingenuous in their mode of proceeding. We know one another at first sight. There is an order of imperfect intellects (under which mine must be content to rank) which in its constitution is essentially anti-Caledonian. The owners of the sort of faculties I allude to, have minds rather suggestive than comprehensive. They have no pretenses to much clearness or precision in their ideas, or in their manner of expressing them. Their intellectual wardrobe (to confess fairly) has few whole pieces in it. They are content with fragments and scattered pieces of truth. She presents no full front to them — a feature or side-face at the most. Hints and glimpses, germs and crude essays at a system, is the utmost they pretend to. They beat up a little game peradventure — and leave it to knottier heads, more robust constitutions, to run it down. The light that lights them is not steady and polar, but mutable and shifting: waxing, and again waning. Their conversation is accordingly. They will throw out a random word in or out of season, and be content to let it pass for what it is worth. They cannot speak always as if they were upon their oath — but must be understood, speaking or writing, with some abatement. They seldom wait to mature a proposition, but e'en bring it to market in the green ear. They delight to impart their defective discoveries as they arise, without waiting for their full development. They are no systematizers, and would but err more by attempting it. Their minds, as I said before, are suggestive merely. The brain of a true Caledonian (if I am not mistaken) is constituted upon quite a different plan. His Minerva is born in panoply. You are never admitted to see his ideas in their growth — if, indeed, they do grow, and are not rather put together upon principles of clock-work. You never catch his mind in an undress. He never hints or suggests any thing, but unlades his stock of ideas in perfect order and completeness. He brings his total wealth into company, and gravely unpacks it. His riches are always about him. He never stoops to catch a glittering something in your presence to share it with you, before he quite knows whether it be true touch or not. You cannot cry halves to anything that he finds. He does not find, but bring. You never witness his first apprehension of a thing. His understanding is always at its meridian — you never see the first dawn, the early streaks. He has no falterings of self-suspicion. Surmises, guesses, misgivings, half-intuitions, semi-consciousnesses, partial illuminations, dim instincts, embryo conceptions, have no place in his brain or vocabulary. The twilight of dubiety never falls upon him. Is he orthodox — he has no doubts. Is he an infidel — he has none either. Between the affirmative and the negative there is no borderland with him. You cannot hover with him upon the confines of truth, or wander in the maze of a probable argument. He always keeps the path. You cannot make excursions with him — for he sets you right. His taste never fluctuates. His morality never abates. He cannot compromise, or understand middle actions. There can be but a right and a wrong. His conversation is as a book. His affirmations have the sanctity of an oath. You must speak upon the square with him. He stops a metaphor like a suspected person in an enemy's country. "A healthy book?" said one of his countrymen to me, who had ventured to give that appellation to "John Buncle," — "did I catch rightly what you said? I have heard of a man in health, and of a healthy state of body, but I do not see how that epithet can be properly applied to a book." Above all, you must beware of indirect expressions before a Caledonian. Clap an extinguisher upon your irony if you are unhappily blessed with a vein of it. Remember you are upon your oath. I have a print of a graceful female after Leonardo da Vinci, which I was showing off to Mr. ———. After he had examined it minutely, I ventured to ask him how he liked my beauty (a foolish name it goes by among my friends) — when he very gravely assured me, that " he had considerable respect for my character and talents" (so he was pleased to say), "but had not given himself much thought about the degree of my personal pretensions." The misconception staggered me, but did not seem much to disconcert him. Persons of this nation are particularly

A TYPICAL CALEDONIAN.

fond of affirming a truth — which nobody doubts. They do not so properly affirm, as annunciate it. They do indeed appear to have such a love for truth (as if, like virtue, it were valuable for itself) that all truth becomes equally valuable, whether the proposition that contains it be new or old, disputed, or such as is impossible to become a subject of disputation. I was present not long since at a party of North Britons, where a son of Burns was expected; and happened to drop a silly expression (in my South British way), that I wished it were the father instead of the son — when four of them started up at once to inform me, that "that was impossible, because he was dead."

CHARLES LAMB.

SIGHT AND INSIGHT.

THERE may be a meadow farm among the mountains. The heir to it gets a cabbage and a corn crop from it, suspecting no other latent fertility and produce. A man of science buys it, gets no less cabbages and hay, but reaps a geology-crop as well.

An artist buys it, and lo! a harvest of beauty and delight, budding even when the grain is garnered, dropping sweet into his eyes even from arctic dawns and blazing snows. A man of deepest insight lives on it, and the laws of his farm open to him the prudence and prodigality of Providence. In the way the grain grows, the enemies it has, the friendships of all good forces to its advance, in the chemistry of his farming, in the peace that sleeps on the hills, in the gathering and retreat of storms, in the soft approach of spring, and the melancholy death, — he reads lessons that become inmost wisdom. He has a faculty that is the sickle of more subtle crop-sheaves of spiritual truth. . . .

Just as there are spelling-classes for the youngest scholars in our schools, in which the separate letters are the chief things they see, where the great problem is to combine them into words, and where the mental organs are not capable of configuring words into propositions, — so very few of us on the planet ever get able to handle the letters of nature easily, ever get beyond the power of spelling them into single words. Some are able to read off the aspects of creation into science. They can put the stars together into paragraphs that state laws and harmonies and grandeurs. Some go farther, and rhyme the mighty vocabulary of science into beauty; but few get such command of the language that they see and rejoice in the highest, glorious truth which the volume holds. . . .

Insight, therefore, opens the intellectual world of law and harmony beneath the world of physical shows; within that, the world of beauty; within that again, the realm of spiritual language. In the human world it shows, deep behind deep, law working in society, controlling politics and shaping the destiny of nations; while, in the individual sphere, it unveils man as the epitome of the universe, clad continually in the electric vesture of his character.

Every man, as every animal, has sight ; but just according to the scale of his insight is the world he lives in a deep one, an awful one, a mystic and glorious world. We see what is, only as we see into what appears.

Out of three roots grows the great tree of nature, — truth, beauty, good. The man of science follows up its mighty stem, measures it, and sees its branches in the silver-leaved boughs of the firmament. The poet delights in the symmetry of its strength, the grace of its arches, the flush of its fruit. Only to the man with finer eye than both is the secret glory of it unveiled ; for his vision discerns how it is fed and in what air it thrives. To him it is only an expansion of the burning bush on Horeb, seen by the solemn prophet, glowing continually with the presence of Infinite Law and Love, yet standing forever unconsumed.

THOMAS STARR KING.

AN APOLOGY FOR ENGLISH.

IF any man would blame me either for taking such a matter in hand, or else for writing it in the English tongue, this answer I may make him, that when the best of the realm think it honest for them to use, I, one of the meanest sort, ought not to suppose it vile for me to write : and though to have written it in another tongue had been both more profitable for my study, and also more honest for my name, yet I can think my labour well bestowed, if with a little hinderance of my profit and name may come any furtherance to the pleasure or commodity of the gentlemen and yeomen of England, for whose sake I took this matter in hand. And as for the Latin or Greek tongue, every thing is so excellently done in them, that none can do better ; in the English tongue, contrary, every thing in a manner so meanly, both for the matter and handling, that no man can do worse. For therein the least learned, for the most part, have been always most ready to write. And they which had least hope in Latin have been most bold in English : when surely every man that is most ready to talk is not most able to write. He that will write well in any tongue, must follow this counsel of Aristotle, to speak as the common people do, to think as wise men do : as so should every man understand him, and the judgment of wise men allow him. Many English writers have not done so, but, using strange words, as Latin, French, and Italian, do make all things dark and hard. Once I communed with a man which reasoned the English tongue to be enriched and increased thereby, saying, Who will not praise that feast where a man shall drink at a dinner both wine, ale, and beer ? Truly, (quoth I) they be all good, every one taken by himself alone, but if you put *malvesye* and sack, red wine and white, ale and beer, and all in one pot, you shall make a drink not easy to be known, nor yet wholesome for the body.

ROGER ASCHAM.

ARRAYED FOR DEFENSE.

THE JUSTICE OF RIENZI THE TRIBUNE.

ALL that night the conspirators remained within that room, the doors locked and guarded ; the banquet unremoved, and its splendor strangely contrasting the mood of the guests.

The utter prostration and despair of these dastard criminals — so unlike the knightly nobles of France and England, has been painted by the historian in odious and withering colors. The old Colonna alone sustained his impetuous and imperious character. He strode to and fro the room like a lion in his cage, uttering loud threats of resentment and defiance ; and beating at the door with his clenched hands, demanding egress, and proclaiming the vengeance of the Pontiff.

The dawn came, slow and gray, upon that agonized assembly ; and just as the last star faded from the melancholy horizon, and by the wan and comfortless heaven, they regarded each other's faces, almost spectral with anxiety and fear, the great bell of the Capitol sounded the notes in which they well recognized the chime of death ! It was then that the door opened, and a drear and gloomy procession of cordeliers, one to each baron, entered the apartment ! At that spectacle, we are told, the terror of the conspirators was so great, that it froze up the very power of speech. The greater part at length, deeming all hope over, resigned themselves to their ghostly confessors. But when the friar appointed to Stephen approached that passionate old man, he waved his hand impatiently, and said, — " Tease me not ! tease me not ! "

"Nay, son, prepare for the awful hour."

"Son, indeed ! " quoth the baron. "I am old enough to be thy grandsire ; and for the rest, tell him who sent thee, that I neither am prepared for death, nor will prepare ! I have made up my mind to live these twenty years, and longer too ; if I catch not my death with the cold of this accursed night."

Just at that moment a cry that almost seemed to rend the Capitol asunder was heard, as, with one voice, the multitude below yelled forth, —

" Death to the conspirators ! — death ! death ! "

While this the scene in that hall, the Tribune issued from his chamber, in which he had been closeted with his wife and sister. The noble spirit of the one, the tears and grief of the other (who saw at one fell stroke perish the house of her betrothed), had not worked without effect upon a temper, stern and just indeed, but naturally averse from blood ; and a heart capable of the loftiest species of revenge.

He entered the council, still sitting, with a calm brow, and even a cheerful eye.

"Pandulfo di Guido," he said, turning to that citizen, " you are right ; you spoke as a wise man and a patriot, when you said that to cut off with one blow, however merited, the noblest heads of Rome, would endanger the state, sully our purple with an indelible stain, and unite the nobility of Italy against us."

"Such, Tribune, was my argument, though the council have decided otherwise."

"Hearken to the shouts of the populace, you cannot appease their honest warmth," said the demagogue Baroncelli.

Many of the council murmured applause.

"Friends," said the Tribune, with a solemn and earnest aspect, "let not posterity say that liberty loves blood ; let us for once adopt the example and imitate the mercy of our great Redeemer ! We have triumphed — let us forbear ; we are saved — let us forgive ! "

The speech of the Tribune was supported by Pandulfo, and others of the more mild and moderate policy ; and after a short but animated discussion, the influence of Rienzi prevailed, and the sentence of death was revoked, but by a small majority.

"And now," said Rienzi, "let us be more than just; let us be generous. Speak — and boldly. Do any of ye think that I have been over-hard, over-haughty with these stubborn spirits? — I read your answer in your brows! — I have! Do any of ye think this error of mine may have stirred them to their dark revenge? Do any of you deem that they partake, as we do, of human nature, — that they are softened by generosity, — that they can be tamed and disarmed by such vengeance as is dictated to noble foes by Christian laws?"

"I think," said Pandulfo, after a pause, "that it will not be in human nature if the men you pardon, thus offending and thus convicted, again attempt your life!"

"Methinks," said Rienzi, "we must do even more than pardon. The first great Cæsar, when he did not crush a foe, strove to convert him to a friend" —

"And perished by the attempt," said Baroncelli, abruptly.

Rienzi started and changed color.

"If you would save these wretched prisoners, better not wait till the fury of the mob become ungovernable," whispered Pandulfo.

The Tribune aroused himself from his reverie.

"Pandulfo," said he, in the same tone, "my heart misgives me — the brood of serpents are in my hand — I do not strangle them — they may sting me to death, in return for my mercy — it is their instinct! No matter: it shall not be said that the Roman Tribune bought with so many lives his own safety: nor shall it be written upon my grave-stone, 'Here lies the coward, who did not dare forgive.' What, lo! there, officers, unclose the doors! My masters, let us acquaint the prisoners with their sentence."

With that, Rienzi seated himself on the chair of state, at the head of the table, and the sun, now risen, cast its rays over the blood-red walls, in which the barons, marshalled in order into the chamber, thought to read their fate.

"My lords," said the Tribune, "ye have offended the laws of God and man! but God teaches man the quality of mercy. Learn at last, that I bear a charmed life. Nor is he whom, for high purposes, Heaven hath raised from the cottage to the popular throne, without invisible aid and spiritual protection. If hereditary monarchs are deemed sacred, how much more one in whose power the divine hand hath writ his witness! Yes, over him who lives but for his country, whose greatness is his country's gift, whose life is his country's liberty, watch the souls of the just, and the unsleeping eyes of the sworded seraphim! Taught by your late failure and your present peril, bid your anger against me cease; respect the laws, revere the freedom of your city, and think that no state presents a nobler spectacle than men born as ye are — a patrician and illustrious order — using your power to protect your city, your wealth to nurture its arts, your chivalry to protect its laws! Take back your swords — and the first man who strikes against the liberties of Rome, let him be your victim; even though that victim be the Tribune. Your cause has been tried — your sentence is pronounced. Renew your oath to forbear all hostility, private or public, against the government and the magistrates of Rome, and ye are pardoned — ye are free!"

Amazed, bewildered, the barons mechanically bent the knee: the friars who had received their confessions, administered the appointed oath; and while, with white lips, they muttered the solemn words, they heard below the roar of the multitude for their blood.

The ceremony ended, the Tribune passed into the banquet-hall, which conducted to a balcony, whence he was accustomed to address the people; and never, perhaps, was his wonderful mastery over the passions of an audience ("*ad persuadendum efficax dictator, quoque dulcis ac lepidus*") more greatly needed or more eminently shown, than on that day; for the fury of the people was at its height, and it was long ere he succeeded in turning it aside. Before he concluded, however, every wave of the wild sea lay hushed. The orator lived to stand on the same spot, to plead for a life nobler than those he now saved, — and to plead unheard and in vain!

As soon as the Tribune saw the favorable moment had arrived, the barons were admitted into the balcony: — in the presence of the breathless thousands, they solemnly pledged themselves to protect the Good Estate. And thus the morning which seemed to dawn upon their execution, witnessed their reconciliation with the people.

The crowd dispersed, the majority soothed and pleased — the more sagacious, vexed and dissatisfied.

"He has but increased the smoke and the flame which he was not able to extinguish," growled Cecco del Vecchio; and the smith's appropriate saying passed into a proverb and a prophecy.

Lord Lytton.

AN ENCOUNTER WITH THE IROQUOIS.

"'Twould be neglecting a warning that is given for our good, to lie hid any longer," said Hawk-eye, "when such sounds are raised in the forest! These gentle ones may keep close, but the Mohicans and I will watch upon the rock, where I suppose a major of the 60th would wish to keep us company."

"Is then our danger so pressing?" asked Cora.

"He who makes strange sounds, and gives them out for man's information, alone knows our danger. I should think myself wicked, unto rebellion against his will, was I to burrow with such warnings in the air! Even the weak soul who passes his days in singing, is stirred by the cry, and, as he says, is 'ready to go forth to the battle.' If 'twere only a battle, it would be a thing understood by us all, and easily managed; but I have heard that when such shrieks are atween heaven and 'arth, it betokens another sort of warfare!"

"If all our reasons for fear, my friend, are confined to such as proceed from supernatural causes, we have but little occasion to be alarmed," continued the undisturbed

Cora; "are you certain that our enemies have not invented some new and ingenious method to strike us with terror that their conquest may become more easy?"

"Lady," returned the scout, solemnly, "I have listened to all the sounds of the woods for thirty years, as a man will listen whose life and death depend on the quickness of his ears. There is no whine of the panther; no whistle of the cat-bird; nor any invention of the devilish Mingoes, that can cheat me! I have heard the forest moan like mortal men in their affliction; often, and again, have I listened to the wind playing its music in the branches of the girdled trees; and I have heard the lightning cracking in the air, like the snapping of blazing brush, as it spitted forth sparks and forked flames; but never have I thought that I heard more than the pleasure of Him who sported with the things of his hand. But neither the Mohicans, nor I, who am a white man without a cross, can explain the cry just heard. We, therefore, believe it is a sign given for our good."

"It is extraordinary!" said Heyward, taking his pistols from the place where he had laid them on entering; "be it a sign of peace or a signal of war, it must be looked to. Lead the way, my friend; I follow."

READY FOR WAR.

On issuing from their place of confinement, the whole party instantly experienced a grateful renovation of spirits, by exchanging the pent air of the hiding-place for the cool and invigorating atmosphere, which played around the whirlpools and pitches of the cataract. A heavy evening breeze swept along the surface of the river, and seemed to drive the roar of the falls into the recesses of their own caverns, whence it issued heavily and constant, like thunder rumbling beyond the distant hills. The moon had risen, and its light was already glancing here and there on the waters above them; but the extremity of the rock where they stood still lay in shadow. With the exception of the sounds produced by the rushing waters, and an occasional breathing of the air, as it murmured past them in fitful currents, the scene was still as night and solitude could make it. In vain were the eyes of each individual bent along the opposite shore, in quest of some signs of life that might explain the nature of the interruption they had heard. Their anxious and eager looks were baffled by the deceptive light, or rested only on naked rocks, and straight and immovable trees.

"Here is nothing to be seen but the gloom and quiet of a lovely evening," whispered Duncan; "how much should we prize such a scene, and all this breathing solitude, at any other moment, Cora! Fancy yourselves in security, and what now, perhaps, increases your terror, may be made conducive to enjoyment" —

"Listen!" interrupted Alice.

The caution was unnecessary. Once more the same sound arose, as if from the bed of the river, and having broken out of the narrow bounds of the cliffs, was heard undulating through the forest, in distant and dying cadences.

"Can any here give a name to such a cry?" demanded Hawk-eye, when the last echo was lost in the woods; "if so, let him speak; for myself, I judge it not to belong to 'arth!"

"Here, then, is one who can undeceive you," said Duncan; "I know the sound full well, for often have I heard it on the field of battle, and in situations which are frequent in a soldier's life. 'Tis the horrid shriek that a horse will give in his agony; oftener drawn from him in pain, though sometimes in terror. My charger is either a prey to the beasts of the forest, or he sees his danger without the power to avoid it. The sound might deceive me in the cavern, but in the open air, I know it too well to be wrong."

The scout and his companions listened to this simple explanation with the interest of men who imbibe new ideas, at the same time that they get rid of old ones which had proved disagreeable inmates.

The two latter uttered their usual and expressive exclamation, "hugh!" as the truth first glanced upon their minds; while the former, after a short musing pause, took upon himself to reply.

"I cannot deny your words," he said, "for I am little skilled in horses, though born where they abound. The wolves must be hovering above their heads on the bank, and the timorsome creatures are calling on man for help, in the best manner they are able. Uncas"—he spoke in Delaware—"Uncas, drop down in the canoe, and whirl a brand among the pack; or fear may do what the wolves can't get at to perform, and leave us without horses in the morning, when we shall have so much need to journey swiftly!"

The young native had already descended to the water, to comply, when a long howl was raised on the edge of the river, and was borne swiftly off into the depths of the forest, as though the beasts, of their own accord, were abandoning their prey in sudden terror. Uncas, with instinctive quickness, receded, and the three foresters held another of their low, earnest conferences.

"We have been like hunters who have lost the points of the heavens, and from whom the sun has been hid for days," said Hawk-eye, turning away from his companions; "now we begin again to know the signs of our course, and the paths are cleared from briers! Seat yourselves in the shade which the moon throws from yonder beech—'tis thicker than that of the pines—and let us wait for that which the Lord may choose to send next. Let all your conversation be in whispers; though it would be better, and perhaps, in the end, wiser, if each one held discourse with his own thoughts for a time."

The manner of the scout was seriously impressive, though no longer distinguished by any signs of unmanly apprehension. It was evident that his momentary weakness had vanished with the explanation of a mystery which his own experience had not served to fathom; and though he now felt all the realities of their actual condition, that he was prepared to meet them with the energy of his hardy nature.

This feeling seemed also common to the natives, who placed themselves in positions which commanded a full view of both shores, while their own persons were effectually concealed from observation. In such circumstances common prudence dictated that Heyward and his companions should imitate a caution that proceeded from so intelligent a source. The young man drew a pile of the sassafras from the cave, and placing it in the chasm which separated the two caverns, it was occupied by the sisters; who were thus protected by the rocks from any missiles, while their anxiety was relieved by the assurance that no danger could approach without a warning.

Heyward himself was posted at hand, so near that he might communicate with his companions without raising his voice to a dangerous elevation; while David, in imitation of the woodsmen, bestowed his person in such a manner among the fissures of the rocks, that his ungainly limbs were no longer offensive to the eye. In this manner hours passed by, without further interruption. The moon reached the zenith, and shed its mild light perpendicularly on the lovely sight of the sisters slumbering peacefully in each other's arms.

Duncan cast the wide shawl of Cora before a spectacle he so much loved to contemplate, and then suffered his own head to seek a pillow on the rock. David began to utter sounds that would have shocked his delicate organs in more wakeful moments; in short, all but Hawk-eye and the Mohicans lost every idea of consciousness, in uncontrollable drowsiness. But the watchfulness of these vigilant protectors neither tired nor slumbered. Immovable as that rock of which each appeared to form a part, they lay, with their eyes roving, without intermission, along the dark margin of trees that bounded the adjacent shores of the narrow stream. Not a sound escaped them; the most subtle examination could not have told they breathed. It was evident that this excess of caution proceeded from an experience that no subtlety on the part of their enemies could deceive. It was, however, continued without any apparent consequences, until the moon had set, and a pale streak above the tree-tops, at the bend of the river a little below, announced the approach of day.

Then, for the first time, Hawk-eye was seen to stir. He crawled along the rock, and shook Duncan from his heavy slumbers. "Now is the time to journey," he whispered; "awake the gentle ones, and be ready to get into the canoe when I bring it to the landing-place."

"Have you had a quiet night?" said Heyward; "for myself, I believe sleep has got the better of my vigilance."

"All is yet still as midnight. Be silent, but be quick."

By this time Duncan was thoroughly awake, and he immediately lifted the shawl from the sleeping females. The motion caused Cora to raise her hand as if to repulse him, while Alice murmured, in her soft, gentle voice, "No, no, dear father, we were not deserted; Duncan was with us!"

"Yes, sweet innocence," whispered the youth; "Duncan is here, and while life continues or danger remains, he will never quit thee. Cora! Alice! awake! The hour has come to move!"

A loud shriek from the younger of the sisters, and the form of the other standing upright before him, in bewildered horror, was the unexpected answer he received. While the words were still on the lips of Heyward, there had arisen such a tumult of yells and cries as served to drive the swift currents of his own blood back from its bounding course into the fountains of his heart. It seemed, for near a minute, as if the demons of hell had possessed themselves of the air about them, and were venting their savage humors in barbarous sounds. The cries came from no particular direction, though it was evident they filled the woods, and as the appalled listeners easily imagined, the caverns of the falls, the rocks, the bed of the river, and the upper air. David raised his tall person in the midst of the infernal din, with a hand on either ear, exclaiming — "Whence comes this discord? Has hell broke loose, that man should utter sounds like these?"

The bright flashes and the quick reports of a dozen rifles, from the opposite banks of the stream, followed this incautious exposure of his person, and left the unfortunate singing-master senseless on that rock where he had been so long slumbering. The Mohicans boldly sent back the intimidating yell of their enemies, who raised a shout of savage triumph at the fall of Gamut. The flash of rifles was then quick and close between them, but either party was too well skilled to leave even a limb exposed to hostile aim.

Duncan listened with intense anxiety for the strokes of the paddle, believing that flight was now their only refuge. The river glanced by with its ordinary velocity, but the canoe was nowhere to be seen on its dark waters. He had just fancied they were cruelly deserted by the scout, as a stream of flame issued from the rock beneath him, and a fierce yell, blended with a shriek of agony, announced that the messenger of death, sent from the fatal weapon of Hawk-eye, had found a victim. At this slight repulse the assailants instantly withdrew, and gradually the place became as still as before the sudden tumult.

Duncan seized the favorable moment to spring to the body of Gamut, which he bore within the shelter of the narrow chasm that protected the sisters In another minute the whole party was collected in this spot of comparative safety.

"The poor fellow has saved his scalp," said Hawk-eye, coolly passing his hand over the head of David; "but he is a proof that a man may be born with too long a tongue! 'Twas downright madness to show six feet of flesh and blood, on a naked rock, to the raging savages. I only wonder he has escaped with life."

"Is he not dead?" demanded Cora, in a voice whose husky tones showed how powerfully natural horror struggled with her assumed firmness. "Can we do aught to assist the wretched man?"

"No, no! the life is in his heart yet, and after he has slept awhile he will come to himself, and be a wiser man for it, till the hour of his real time shall come," returned Hawk-eye, casting another oblique glance at the insensible body, while he filled his charges with admirable nicety. "Carry him in, Uncas, and lay him on the sassafras. The longer his nap lasts the better it will be for him, as I doubt whether he can find a proper cover for such a shape on these rocks; and singing won't do any good with the Iroquois."

"You believe, then, the attack will be renewed?" asked Heyward.

"Do I expect a hungry wolf will satisfy his craving with a mouthful! They have lost a man, and 'tis their fashion, when they meet a loss, and fail in the surprise, to fall back; but we shall have them on again, with new expedients to circumvent us, and master our scalps. Our main hope," he continued, raising his rugged countenance, across which a shade of anxiety just then passed like a darkening cloud, "will be to keep the rock until Munro can send a party to our help! God send it may be soon, and under a leader that knows the Indian customs!"

"You hear our probable fortunes, Cora," said Duncan; "and you know we have everything to hope from the anxiety and experience of your father. Come, then, with Alice, into this cavern, where you, at least, will be safe from the murderous rifles of our enemies, and where you may bestow a care suited to your gentle natures on our unfortunate comrade."

The sisters followed him into the outer cave, where David was beginning, by his sighs, to give symptoms of returning consciousness; and then commending the wounded man to their attention, he immediately prepared to leave them.

"Duncan!" said the tremulous voice of Cora, when he had reached the mouth of the cavern. He turned, and beheld the speaker, whose color had changed to a deadly paleness, and whose lip quivered, gazing after him, with an expression of interest which immediately recalled him to her side. "Remember, Duncan, how necessary your safety is to our own — how you bear a father's sacred trust — how much depends on your discretion and care — in short," she added, while the telltale blood stole over her features, crimsoning her very temples, "how very deservedly dear you are to all of the name of Munro."

"If anything could add to my own base love of life," said Heyward, suffering his unconscious eyes to wander to the youthful form of the silent Alice, "it would be so kind an assurance. As major of the 60th, our honest host will tell you I must take my share of the fray; but our task will be easy; it is merely to keep these blood-hounds at bay for a few hours."

Without waiting for a reply, he tore himself from the presence of the sisters, and joined the scout and his companions, who still lay within the protection of the little chasm between the two caves.

"I tell you, Uncas," said the former, as Heyward joined them, "you are wasteful of your powder, and the kick of the rifle disconcerts your aim! Little powder, light lead, and a long arm, seldom fail of bringing the death screech from a Mingo! At least, such has been my experience with the creature. Come, friends; let us to our covers, for no man can tell when or where a Maqua will strike his blow."

The Indians silently repaired to their appointed stations, which were fissures in the rocks, whence they could command the approaches to the foot of the falls. In the center of the little island, a few short and stunted pines had found root, forming a thicket, into which Hawk-eye darted with the swiftness of a deer, followed by the active Duncan. Here they secured themselves, as well as circumstances would permit, among the shrubs and fragments of stone that were scattered about the place. Above them was a bare, rounded rock, on each side of which the water played its gambols, and plunged into the abysses beneath, in the manner already described. As the day had now dawned, the opposite shores no longer presented a confused outline, but they were able to look into the woods, and distinguish objects beneath the canopy of gloomy pines.

A long and anxious watch succeeded, but without any further evidences of a renewed attack; and Duncan began to hope that their fire had proved more fatal than was supposed, and that their enemies had been effectually repulsed. When he ventured to utter this impression to his companion, he was met by Hawk-eye with an incredulous shake of the head.

"You know not the nature of a Maqua, if you think he is so easily beaten back without a scalp!" he answered. "If there was one of the imps yelling this morning, there were forty! and they know our number and quality too well to give up the chase so soon. Hist! look into the water above, just where it breaks over the rocks. I am no mortal, if the risky devils haven't swam down upon the very pitch, and, as bad luck would have it, they have hit the head of the island. Hist! man, keep close! or the hair will be off your crown in the turning of a knife!"

Heyward lifted his head from the cover, and beheld what he justly considered a prodigy of rashness and skill. The river had worn away the edge of the soft rock in such a manner as to render its first pitch less abrupt and perpendicular than is usual at waterfalls. With no other guide than the ripple of the stream where it met the head of the island, a party of their insatiable foes had ventured into the current, and swam down upon this point, knowing the ready access it would give, if successful, to their intended victims. As Hawk-eye ceased speaking, four human heads could be seen peering above a few logs of drift wood that had lodged on these naked rocks, and which had probably suggested the idea of the practicability of the hazardous undertaking.

At the next moment, a fifth form was seen floating over the green edge of the fall, a little from the line of the island. The savage struggled powerfully to gain the point of safety, and, favored by the glancing water, he was already stretching forth an arm to meet the grasp of his companions, when he shot away again with the whirling current, appeared to rise into the air, with uplifted arms, and starting eyeballs, and fell with a sudden plunge into that deep and yawning abyss over which he hovered. A single, wild, despairing shriek rose from the cavern, and all was hushed again, as the grave.

The first generous impulse of Duncan was to rush to the rescue of the hapless wretch; but he felt himself bound to the spot by the iron grasp of the immovable scout.

"Would ye bring certain death upon us, by telling the Mingoes where we lie?" demanded Hawk-eye, sternly; "'tis a charge of powder saved, — and ammunition is as precious now as breath to a worried deer! Freshen the priming of your pistols — the mist of the falls is apt to dampen the brimstone — and stand firm for a close struggle, while I fire on their rush."

He placed a finger in his mouth, and drew a long, shrill whistle, which was answered from the rocks that were guarded by the Mohicans. Duncan caught glimpses of heads above the scattered drift wood, as this signal rose on the air, but they disappeared again as suddenly as they had glanced upon his sight. A low, rustling sound next drew his attention behind him, and turning his head, he beheld Uncas within a few feet, creeping to his side. Hawk-eye spoke to him in Delaware, when the young chief took his position with singular caution and undisturbed coolness. To Heyward, this was a moment of feverish and impatient suspense; though the scout saw fit to select it as a fit occasion to read a lecture to his more youthful associates on the art of using fire-arms with discretion.

"Of all we'pons," he commenced, "the long-barrelled, true-grooved, soft-metalled rifle is the most dangerous in skilful hands, though it wants a strong arm, a quick eye, and great judgment in charging, to put forth all its beauties. The gun-smiths can have but little insight into their trade, when they make their fowling-pieces and short horse-men's" —

He was interrupted by the low but expressive "hugh" of Uncas.

"I see them, boy, I see them!" continued Hawk-eye; "they are gathering for the rush, or they would keep their dingy backs below the logs. Well, let them," he added, examining his flint; "the leading man certainly comes on to his death, though it should be Montcalm himself!"

At that moment the woods were filled with another burst of cries, and at the signal four savages sprang from the cover of the drift wood. Heyward felt a burning desire to rush forward to meet them, so intense was the delirious anxiety of the moment; but he was restrained by the deliberate examples of the scout and Uncas. When their foes, who leaped over the black rocks that divided them, with long bounds, uttering the wildest yells, were within a few rods, the rifle of Hawk-eye slowly rose among the shrubs, and poured out its fatal contents. The foremost Indian bounded like a stricken deer, and fell headlong among the clefts of the island.

"Now, Uncas," cried the scout, drawing his long knife, while his quick eyes began to flash with ardor, "take the last of the screeching imps; of the other two we are sartin!"

He was obeyed; and but two enemies remained to be overcome. Heyward had given one of his pistols to Hawk-eye, and together they rushed down a little declivity towards their foes; they discharged their weapons at the same instant, and equally without success.

"I know'd it! and I said it!" muttered the scout, whirling the despised little implement over the falls with bitter disdain. "Come on, ye bloody-minded hell-hounds! ye meet a man without a cross!"

The words were barely uttered, when he encountered a savage of gigantic stature, and of the fiercest mien. At the same moment, Duncan found himself engaged with the other, in a similar contest of hand to hand. With ready skill, Hawk-eye and his antagonist each grasped that uplifted arm of the other which held the dangerous knife. For near a minute they stood looking one another in the eye, and gradually exerting the power of their muscles for the mastery. At length the toughened sinews of the white man prevailed over the less practised limbs of the native. The arm of the latter slowly gave way before the increasing force of the scout who, suddenly wresting his armed hand from the grasp of his foe, drove the sharp weapon through his naked bosom to the heart. In the meantime, Heyward had been pressed in a more deadly struggle. His slight sword was snapped in the first encounter. As he was destitute of any other means of defense, his safety now depended entirely on bodily strength and resolution.

Though deficient in neither of these qualities, he had met an enemy every way his equal. Happily he soon succeeded in disarming his adversary, whose knife fell on the rock at their feet; and from this moment it became a fierce struggle who should cast the other over the dizzy height into a neighboring cavern of the falls. Every successive struggle brought them nearer to the verge, where Duncan perceived the final and conquering effort must be made. Each of the combatants threw all his energies into that effort, and the result was, that both tottered on the brink of the precipice. Heyward felt the grasp of the other at his throat, and saw the grim smile the savage gave, under the revengeful hope that he hurried his enemy to a fate similar to his own, as he felt his body slowly yielding to a resistless power, and the young man experienced the passing agony of such a moment in all its horrors. At that instant of extreme danger, a dark hand and glancing knife appeared before him; the Indian released his hold, as the blood flowed freely from around the severed tendons of his wrist; and while Duncan was drawn backward by the saving arm of Uncas, his charmed eyes were still riveted on the fierce and disappointed countenance of his foe, who fell sullenly and disappointed down the irrecoverable precipice.

"To cover! to cover!" cried Hawk-eye, who just then had dispatched his enemy; "to cover, for your lives! the work is but half ended!"

The young Mohican gave a shout of triumph, and, followed by Duncan, he glided up the acclivity they had descended to the combat, and sought the friendly shelter of the rocks and shrubs.

JAMES FENIMORE COOPER.

UNSELFISHNESS.

I FOUND the Battery unoccupied, save by children, whom the weather made as merry as birds. Every thing seemed moving to the vernal tune of

> "Oh, Brignall banks are wild and fair,
> And Greta woods are green." — *Scott's Rokeby.*

To one who was chasing her hoop, I said, smiling, "You are a nice little girl." She stopped, looked up in my face, so rosy and happy, and, laying her hand on her brother's shoulder, exclaimed, earnestly, "And he is a nice little boy, too!" It was a simple, childish act, but it brought a warm gush into my heart. Blessings on all unselfishness! on all that leads us in love to prefer one another! Here lies the secret of universal harmony; this is the diapason which would bring us all into tune. Only by losing ourselves can we find ourselves. How clearly does the divine voice within us proclaim this, by the hymn of joy it sings, whenever we witness an unselfish deed or hear an unselfish thought. Blessings on that loving little one! She made the city seem a garden to me. I kissed my hand to her, as I turned off in quest of the Brooklyn ferry. The sparkling waters swarmed with boats, some of which had taken a big ship by the hand, and were leading her out to sea, as the prattle of childhood often guides wisdom into the deepest and broadest thought.

LYDIA MARIA CHILD.

PHILIP AND LEIGH.

PHILIP had spoken again.

"You know perfectly well that you can," he had said. "My life is in your hands."

Leigh's heart beat fast; and she nervously pulled in pieces a honeysuckle blossom, sacrificing the fragrant, unoffending flower in her troubled mood.

"Mr. Ogden, may I speak very frankly to you? I think there should be no disguise between us, whatever may come, and I know you will not misunderstand me; and you will pardon me if what I am about to say seems strange?"

"Do not hesitate to say anything you wish. I cannot misunderstand."

"In all these days in which you have been so good, and have given me time to think, it seems to me I ought to feel sure of myself, and I am not, Mr. Ogden. I am so sorry, but I feel troubled, full of doubt."

"Why should you not feel so? It is no light thing I ask of you," Philip said gently. Then, after a moment, "Could you tell me what especially makes you troubled?"

"I would like to tell you if I can. I wish to show you what is in my heart. It seems to me the only way," she hesitated. Again the innocent honeysuckle vine suffered, as Leigh's unconscious hands ruthlessly showered leaf and flower upon the steps. Abruptly she began. "Mr. Ogden, it is so different from my theories. All girls have theories, you know. I cannot deny that I care more for you than I ever cared for any one before," she said slowly, and so low, that Philip scarcely heard the words that were so dear to him. "Wait," she went on, with a little imperious gesture, as Philip eagerly began to speak, — "wait. I care for you more, but how can I be sure that I care for you enough? How can I?" And the earnestness of her voice deepened as she asked the question, and looked straight into the eyes of the man that loved her. "You have been good to me. You have

THE HONEYSUCKLE GREW ALL ABOUT.

cared for me constantly in little kind ways. Mrs. Browning says, 'These things have their weight with girls,'" and a faint smile trembled about Leigh's lips. "I suppose she knew. You have been with me weeks and weeks. I have grown used to you, and now you tell me that you love me; and in return I give much regard, a grateful affection perhaps, but is it love? It is not like the love that I have dreamed of!" she exclaimed passionately.

Philip wondered if there were another woman in the world so true as the one who stood before him, trying to let him read her very heart as if it were an open book, and whose face and attitude and voice by sudden, eloquent little changes each moment seemed to reveal every phase of the feeling which stirred her so deeply.

He did not speak, for he saw that she had more to say to him.

"Let me speak more plainly." And she carefully chose her words, and endeavored to be quite calm. "Your presence makes me very happy. I think I would like you to come very often to my sister's home, yet I do not feel that for you I would, if you asked me to-day, give up that home, and all the pleasant things in my old life," Leigh went on bravely, though she was evidently making a mighty effort. "I have always believed no woman ought to marry a man if she feels she can under any circumstances be happy without him. Am I talking strangely? Forgive me. Do not be angry with me. I do care very much for you, and should miss you if you did not come to my home; and I should think of you often at first, but after a time I think I might be quite happy without you." Then, with a tremulous voice, suggestive of the deepest emotion, and also of a nervous desire to laugh, she said, "A woman, if she really loves a man, ought to be willing to go and live in a log-cabin with him out on the prairies; and I do not love you enough for that. I know I do not. Do not think me speaking lightly," she said pleadingly. "It is so hard to tell you exactly what I mean, and I am so sad at heart. But when you offer me so royal a gift as your love, when you place all that you have, and all that you are, at my feet, I must at least give you absolute truth in return. You see how I trust you. I am trying to tell you every thought."

"I know that you trust me," Philip said, taking in his own her two trembling hands and holding them firmly, "and I believe that I can teach you to love me. Leigh, you must love me a little, or you could not let me hold these dear hands in mine, nor touch them with my lips. See, I kiss them over and over, and you do not draw them away. Already you give me far more than I deserve, and for the rest I can wait very, very patiently."

Leigh was touched indescribably by the quiet tenderness of his manner.

"But," she said, "is this right? What if the day comes when I look you in the face and say I do not love you? What would you think of me then?"

"I should think what I think now, — that your true heart had revealed itself to me in all honor."

"But I ought to know; it is weak to hesitate. I cannot bear to think that I may be deceiving you."

"You cannot deceive me. Let your heart be quite at rest. Do not question yourself and be troubled any longer, for whatever comes, you will not have deceived me for a moment. But, dear, I think you will love me. Do you forgive me for feeling so sure?"

"Mr. Ogden, will I seem foolish if I ask you how do I know but some day I may experience a stronger, deeper love than that which I feel for you? I have not seen everybody."

Philip smiled at her unconscious admission and at the utter simplicity of her manner.

"Dear, you will honor me above all the world, if you will give me the happiness of assuming that risk." Then he said more gravely, "I know well that I am no hero. You will meet many a person more like the ideal man you may have dreamed of loving; but I love you with my whole soul, Leigh."

"When you speak so, you place me in a different atmosphere. It is as if I were quite promised to you," Leigh said in a pained, low voice. "I have always been so decided in everything, and I have felt so distressed in the last few days because of my doubts. Love, real love, never hesitates so. Are you sure that you understand? I cannot feel that I wish to lose you utterly; yet, Mr. Ogden, you are very far from being all the world to me. Do you think you understand?"

"Everything, everything, and what you tell me makes me profoundly happy, and I love you a thousand times more for every noble word you have said to-night. I have unspeakable faith in your perfect truth toward me. Whatever you do will be sweet and right."

"I shall feel differently now. You are so good it rests me."

"You have given me such happiness, such blessed hope!"

"Ah, but please do not be happy quite yet! I do not know."

"I know," said Philip, under his breath. . . .

"And are you glad to see me, dear? And are you quite 'sure of yourself' now? And is it like your 'theories'?"

"I was very, very glad, but I think you took an unfair advantage in surprising me, and some day I will have my revenge."

"And will you go out on the prairies and live in a log-cabin with me, if ever I ask you? Will you, Leigh?"

"No, sir, never, if you persist in remembering all the idle words I ever said, and wickedly repeating them to me."

"But would you, Leigh?" he persisted.

"I am really disappointed in you already. I never dreamed you would develop into a tease like Tom. Do you know, I've read that success ruins some natures?"

"But would you?"

She hesitated; then, "I will go to the very end of the world with you one day if you should wish," she said in low, earnest tones.

BLANCHE WILLIS HOWARD.

TO MISS MITFORD.

How do you find yourself? I heard you were poorly. What are you about? I was happy to hear of ———'s safe arrival again, and I shall be most happy to see him, though tell him he will find no more "Solomons" towering up as a background to our conversations. Nothing but genteel-sized drawing-room pocket-history — Alexander in a nutshell; Bucephalus no bigger than a Shetland pony, and my little girl's doll a giantess to my Olympias.

The other night I paid my butcher; one of the miracles of these times, you will say. Let me tell you I have all my life been seeking for a butcher whose respect for genius predominated over his love of gain. I could not make out, before I dealt with this man, his excessive desire that I should be his customer; his sly hints as I passed his shop that he had "a bit of South Down, very fine; a sweetbread, perfection; and a calf's foot that was all jelly without bone!" The other day he called, and I had him sent up into the painting-room. I found him in great admiration of "Alexander." "Quite alive, sir!" "I am glad you think so," said I. "Yes, sir, but, as I have said often to my sister, you could not have painted that picture, sir, if you had not eat my meat, sir!" "Very true, Mr. Sowerby." "Ah! sir, I have a fancy for *genus*, sir!" "Have you, Mr. Sowerby?" "Yes, sir; Mrs. Siddons, sir, has eat my meat, sir; never was such a woman for

AT MILKING-TIME.

chops, sir!"—and he drew up his beefy, shiny face, clean shaved, with a clean blue cravat under his chin, a clean jacket, a clean apron, and a pair of hands that would pin an ox to the earth if he was obstreperous—"Ah! sir, she was a wonderful crayture!" "She was, Mr. Sowerby." "Ah! sir, when she used to act that there character, you see (but Lord, such a head! as I say to my sister)—that there woman, sir, that murders a king between 'm!" "Oh! Lady Macbeth." "Ah, sir, that's it—Lady Macbeth—I used to get up with the butler behind her carriage when she acted, and, as I used to see her looking quite wild, and all the people quite frightened, Ah, ha! my lady, says I, if it wasn't for my meat, though, you wouldn't be able to do that!" "Mr. Sowerby, you seem to be a man of feeling. Will you take a glass of wine?" After a bow or two, down he sat, and by degrees his heart opened. " You see, sir, I have fed Mrs. Siddons, sir; John Kemble, sir; Charles Kemble, sir ; Stephen Kemble, sir ; and Madame Catalani, sir; Morland the painter, and, I beg your pardon, sir, and you, sir." " Mr. Sowerby, you do me honor." " Madame Catalani, sir, was a wonderful woman for sweetbreads ; but the Kemble family, sir, the gentlemen, sir, rump-steaks and kidneys in general was their taste ; but Mrs. Siddons, sir, she liked chops, sir, as much as you do, sir," etc., etc., I soon perceived that the man's ambition was to feed genius. I shall recommend you to him ; but is he not a capital fellow? but a little acting with his remarks would make you roar with laughter. Think of Lady Macbeth eating chops ! Is this not a peep behind the curtain ? . . .

BENJAMIN ROBERT HAYDON.

IN PRAISE OF POETRY.

Now therein — (that is to say, the power of at once teaching and enticing to do well) — now therein, of all sciences — I speak still of human and according to human conceit — is our poet the monarch. For he doth not only show the way, but giveth so sweet a prospect into the way, as will entice any man to enter into it. Nay, he doth, as if your journey should lie through a fair vineyard, at the very first give you a cluster of grapes, that, full of that taste, you may long to pass further. He beginneth not with obscure definitions, which must blur the margent with interpretations, and load the memory with doubtfulness ; but he cometh to you with words set in delightful proportion, either accompanied with, or prepared for, the well-enchanting skill of music ; and with a tale, forsooth, he cometh unto you with a tale which holdeth children from play, and old men from the chimney-corner; and pretending no more, doth intend the winning of the mind from wickedness to virtue, even as the child is often brought to take most wholesome things, by hiding them in such other as have a pleasant taste. For even those hard-hearted evil men, who think virtue a school name, and know no other good but *indulgere genio*, and therefore despise the austere admonitions of the philosopher, and feel not the inward reason they stand upon, yet will be content to be delighted ; which is all the good-fellow poet seems to promise ; and so steal to see the form of goodness — which, seen, they cannot but love ere themselves be aware, as if they had taken a medicine of cherries. By these, therefore, examples and reasons, I think it may be manifest that the poet, with that same hand of delight, doth draw the mind more effectually than any other art doth. And so a conclusion not unfitly ensues, that as virtue is the most excellent resting-place for all worldly learning to make an end of, so poetry, being the most familiar to teach it, and most princely to move towards it, in the most excellent work is the most excellent workman.

Since, then, poetry is of all human learning the most ancient, and of most fatherly antiquity, as from whence other learnings have taken their beginnings; — Since it is so universal that no learned nation doth despise it, no barbarous nation is without it; — Since both Roman and Greek gave such divine names unto it, the one of prophesying, the other of making; and that, indeed, that name of making is fit for it, considering that whereas all other arts retain themselves within their subject, and receive, as it were, their being from it, — the poet, only, bringeth his own stuff, and doth not learn a conceit out of the matter, but maketh matter for a conceit; — Since, neither his description nor end containing any evil, the thing described cannot be evil; — Since his effects be so good as to teach goodness and delight the learners of it; — Since therein (namely, in moral doctrine, the chief of all knowledge) he doth not only far pass the historian, but, for instructing, is well nigh comparable to the philosopher, and for moving, leaveth him behind; — Since the Holy Scripture (wherein there is no uncleanness) hath whole parts in it poetical, and that even our Saviour Christ vouchsafed to use the flowers of it; — Since all its kinds are not only in their united forms, but in their severed dissections fully commendable; — I think — (and I think I think rightly) — the laurel crown appointed for triumphant captains, doth worthily, of all other learnings, honor the poet's triumph.

Sir Philip Sidney.

WEATHERING A GALE.

THE FOOTPRINT ON THE SHORE.

It happened one day about noon, going towards my boat, I was exceedingly surprised with the print of a man's naked foot on the shore, which was very plain to be seen in the sand ; I stood like one thunder-struck, or as if I had seen an apparition : I listened, I looked round me, I could hear nothing, nor see any thing ; I went up to a rising ground to look farther : I went up the shore, and down the shore, but it was all one, I could see no other impression but that one : I went to it again to see if there were any more, and to observe if it might not be my fancy ; but there was no room for that, for there was exactly the very print of a foot, toes, heel, and every part of a foot. How it came thither I knew not, nor could in the least imagine. But after innumerable fluttering thoughts, like a man perfectly confused, and out of myself, I came home to my fortification, not feeling, as we say, the ground I went on, but terrified to the last degree, looking behind me at every two or three steps, mistaking every bush and tree, and fancying every stump at a distance to be a man ; nor is it possible to describe how many various shapes an affrighted imagination represented things to me in ; how many wild ideas were formed every moment in my fancy, and what strange, unaccountable whimsies came into my thoughts by the way.

When I came to my castle, for so I think I called it ever after this, I fled into it like one pursued; whether I went over by the ladder, at first contrived, or went in at the hole in the rock, which I called a door, I cannot remember; for never frighted hare fled to cover, or fox to earth, with more terror of mind than I to this retreat.

How strange a chequer-work of Providence is the life of man! And by what secret differing springs are the affections hurried about, as differing circumstances present! To-day we love what to-morrow we hate; to-day we seek what to-morrow we shun; to-day we desire what to-morrow we fear; nay, even tremble at the apprehensions of. This was exemplified in me at this time in the most lively manner imaginable; for I, whose only affliction was, that I seemed banished from human society, that I was alone, circumscribed by the boundless ocean, cut off from mankind, and condemned to what I call a silent life; that I was as one whom Heaven thought not worthy to be numbered among the living, or to appear among the rest of his creatures; that to have seen one of my own species would have seemed to me a raising me from death to life, and the greatest blessing that Heaven itself, next to the supreme blessing of salvation, could bestow; I say, that I should now tremble at the very apprehensions of seeing a man, and was ready to sink into the ground, at but the shadow, or silent appearance of a man's having set his foot on the island!

Such is the uneven state of human life; and it afforded me a great many curious speculations afterwards, when I had a little recovered my first surprise. I considered that this was the station of life the infinitely wise and good providence of God had determined for me; that as I could not foresee what the ends of divine wisdom might be in all this, so I was not to dispute his sovereignty, who, as I was his creature, had an undoubted right by creation to govern and dispose of me absolutely as he thought fit; and who, as I was a creature who had offended him, had likewise a judicial right to condemn me to what punishment he thought fit; and that it was my part to submit to bear his indignation, because I had sinned against him.

I then reflected, that God, who was not only righteous, but omnipotent, as he had thought fit thus to punish and afflict me, so he was able to deliver me; that if he did not think fit to do it, it was my unquestioned duty to resign myself absolutely and entirely to his will: and, on the other hand, it was my duty also to hope in him, pray to him, and quietly to attend the dictates and directions of his daily providence.

These thoughts took me up many hours, days, nay, I may say, weeks and months; and one particular effect of my cogitations on this occasion I cannot omit; viz., one morning early, lying in my bed, and filled with thoughts about my danger from the appearance of savages, I found it discomposed me very much; upon which those words of the Scripture came into my thoughts, "Call upon me in the day of trouble, and I will deliver thee, and thou shalt glorify me."

Upon this, rising cheerfully out of my bed, my heart was not only comforted, but I was guided and encouraged to pray earnestly to God for deliverance. When I had done praying, I took up my Bible, and, opening it to read, the first words that presented to me, were, "Wait on the Lord, and be of good courage, and he shall strengthen thy heart: Wait, I say, on the Lord." It is impossible to express the comfort this gave me; and in return, I thankfully laid down the book, and was no more sad, at least, not on that occasion.

In the middle of these cogitations, apprehensions, and reflections, it came into my thoughts one day, that all this might be a mere chimera of my own, and that this foot might be the print of my own foot, when I came on shore from my boat: this cheered me up a little too, and I began to persuade myself it was all a delusion; that it was nothing else but my own foot; and why might not I come that way from the boat, as well as I was going that way to the boat? Again, I considered also, that I could by no means tell for certain where I had trod, and where I had not; and that if at last this was only the print of my own foot, I had played the part of those fools, who strive to make stories of specters and apparitions, and then are themselves frighted at them more than anybody else.

Now I began to take courage, and to peep abroad again; for I had not stirred out of my castle for three days and nights, so that I began to starve for provision; for I had little or nothing within doors, but some barley-cakes and water. Then I knew that my goats wanted to be milked too, which usually was my evening diversion; and the poor creatures were in great pain and inconvenience for want of it; and indeed it almost spoiled some of them, and almost dried up their milk.

Heartening myself, therefore, with the belief, that this was nothing but the print of one of my own feet (and so I might be truly said to start at my own shadow), I began to go abroad again, and went to my country-house to milk my flock; but to see with what fear I went forward, how often I looked behind me, how I was ready, every now and then, to lay down my basket, and run for my life; it would have made any one have thought I was haunted with an evil conscience, or that I had been lately most terribly frighted; and so indeed I had.

However, as I went down thus two or three days, and having seen nothing, I began to be a little bolder, and to think there was really nothing in it but my own imagination. But I could not persuade myself fully of this, till I should go down to the shore again, and see this print of a foot, and measure it by my own, and see if there was any similitude or fitness, that I might be assured it was my own foot. But when I came to the place first, it appeared evidently to me, that when I laid up my boat, I could not possibly be on shore anywhere thereabouts. Secondly, when I came to measure the mark with my own foot, I found my foot not so large by a great deal. Both these things filled my head with new imaginations, and gave me the vapors again to the highest degree; so that I shook with cold, like one in an ague; and I went home again, filled with the belief, that some man or men had been on shore there; or, in short, that the island was inhabited, and I might be surprised before I was aware; and what course to take for my security, I knew not. O what ridiculous resolutions men take, when possessed with fear! It deprives them of the use of those means which reason offers for their relief.

Daniel De Foe.

THE RIGHTS OF MAN.

We hold these truths to be self-evident: that all men are created equal; that they are endowed by their Creator with certain unalienable rights; that among these are life, liberty, and the pursuit of happiness; that, to secure these rights, governments are instituted among men, deriving their just powers from the consent of the governed; that, whenever any form of government becomes destructive of these ends, it is the right of the people to alter or to abolish it, and to institute a new government, laying its foundation on such principles, and organizing its powers in such form, as to them shall seem most likely to effect their safety and happiness. Prudence, indeed, will dictate that governments long established should not be changed for light and transient causes; and, accordingly, all experience hath shown that mankind are more disposed to suffer, while evils are sufferable, than to right themselves by abolishing the forms to which they are accustomed. But when a long train of abuses and usurpations, pursuing invariably the same object, evinces a design to reduce them under absolute despotism, it is their right, it is their duty, to throw off such government, and to provide new guards for their future security.

Thomas Jefferson.

TO WILLIAM ROBERTSON.

. . . Do you ask me about my course of life? I can only say that I eat nothing but ambrosia, drink nothing but nectar, breathe nothing but incense, and tread on nothing but flowers. Every man I meet, and still more every lady, would think they were wanting in the most indispensable duty, if they did not make to me a long and elaborate harangue in my praise. What happened last week, when I had the honor of being presented to the Dauphin's children, at Versailles, is one of the most curious scenes I ever yet passed through. The Duc de B., the eldest, a boy of ten years old, stepped forth and told me how many friends and admirers I had in this country, and that he reckoned himself in the number from the pleasure he had received from the reading of many passages in my works. When he had finished, his brother, the Count of P., who is two years younger, began his discourse, and informed me that I had been long and impatiently expected in France, and that he, himself, expected soon to have great satisfaction from the reading of my fine History. But, what is more curious, when I was carried thence to the Count d'A., who is but four years of age, I heard him mumble something, which, though he had forgot it in the way, I conjectured, from some scattered words, to have been also a panegyric dictated to him. . . .

David Hume.

"STAY."

"I intend," John said, "as soon as I am able, to leave Norton Bury, and go abroad for some time."

"Where?"

"To America. It is the best country for a young man who has neither money, nor kindred, nor position, — nothing, in fact, but his own right hand with which to carve out his own fortunes — as I will, if I can."

She murmured something about this being "quite right."

"I am glad you think so." But his voice had resumed that formal tone which ever and anon mingled strangely with its low, deep tenderness. "In any case, I must quit England. I have reasons for so doing."

"What reasons?"

The question seemed to startle John — he did not reply at once.

"If you wish, I will tell you, in order that, should I ever come back — or if I should not come back at all, you who were kind enough to be my friend will know I did not go away from mere youthful recklessness, or love of change."

He waited, apparently for some answer — but it came not, and he continued, —

"I am going, because there has befallen me a great trouble, which, while I stay here, I cannot get free from, or overcome. I do not wish to sink under it — I had rather, as you said, 'do my work in the world,' as a man ought. No man has a right to say unto his Maker: 'My burden is heavier than I can bear.' Do you not think so?"

"I do."

"Do you not think I am right in thus meeting, and trying to conquer an inevitable ill?"

"Is it inevitable?"

"Hush!" John answered wildly. "Don't reason with me — you cannot judge — you do not know. It is enough that I must go. If I stay I shall become unworthy of myself, unworthy of — Forgive me, I have no right to talk thus; but you called me 'Friend,' and I would like you to think kindly of me always. Because — because" — And his voice shook — broke down utterly. "God love thee, and take care of thee, wherever I may go!"

"John, stay!"

It was but a low, faint cry, like that of a little bird. But he heard it — felt it. In the silence of the dark she crept up to him, like a young bird to its mate, and he took her into the shelter of his love forever more. At once, all was made clear between them; for whatever the world might say, they were in the sight of heaven equal, and she received as much as she gave.

Miss Mulock.

THE TRUE TRACK.

Go with me, if you please, to the next station-house, and look off upon that line of railroad. It is as straight as an arrow. Out run the iron lines, glittering in the sun, — out, as far as we can see, until, converging almost to a single thread, they pierce the sky. What were those rails laid in that way for? It is a road, is it? Try your cart or your coach there. The axletrees are too narrow, and you go bumping along upon the sleepers. Try a wheelbarrow. You cannot keep it on the rail. But that road was made for something. Now go with me to the locomotive-shop. What is this? We are told it is a locomotive. What is a locomotive? Why, it is a carriage moved by steam. But it is very heavy. The wheels would sink into a common road to the axle. That locomotive can never run on a common road; and the man is a fool who built it. Strange that men will waste time and money in that way! But stop a moment. Why wouldn't those wheels just fit those rails? We measure them, and then we go to the track and measure its gauge. That solves the difficulty. Those rails were intended for the locomotive, and the locomotive for the rails. They are good for nothing apart. The locomotive is not even safe anywhere else. If it should get off, after it is once on, it would run into rocks and stumps, and bury itself in sands or swamps beyond recovery.

Young man, you are a locomotive. You are a thing that goes by a power planted inside of you. You are made to go. In fact, considered as a machine, you are very far superior to a locomotive. The maker of the locomotive is man; your maker is man's Maker. You are as different from a horse, or an ox, or a camel, as a locomotive is different from a wheelbarrow, a cart, or a coach. Now, do you suppose that the being who made you — manufactured your machine, and put into it the motive power — did not make a special road for you to run upon? My idea of religion is that it is a railroad for a human locomotive, and that just so sure as it undertakes to run upon a road adapted only to animal power, will it bury its wheels in the sand, dash itself among rocks, and come to inevitable wreck. If you don't believe this, try the other thing. Here are forty roads: suppose you choose one of them, and see where you come out. Here is the dramshop road. Try it. Follow it, and see how long it will be before you come to a stump and a smash-up. Here is the road of sensual pleasure. You are just as sure to bury your wheels in the dirt as you try it. Your machine is too heavy for that track altogether. Here is the winding, uncertain path of frivolity. There are morasses on each side of it, and, with the headway that you are under, you will be sure, sooner or later, to pitch into one of them. Here is the road of philosophy, but it runs through a country from which the light of Heaven is shut out; and while you may be able to keep your machine right side up, it will only be by feeling your way along in a clumsy, comfortless kind of style, and with no certainty of ever arriving at the heavenly station-house. Here is the road of skepticism. That is covered with fog, and a fence runs across it within ten rods. Don't you see that your machine was never intended to run on those roads? Don't you know that it never was, and don't you know that the only track under heaven upon which it can run safely is the religious track? Don't you know that just as long as you keep your wheels on that track, wreck is impossible? Don't you know that it is the only track on which wreck is not certain? I know it, if you don't; and I tell you that on that track, which God has laid down expressly for your soul to run upon, your soul will find free play for all its wheels, and an unobstructed and happy progress. It is straight and narrow, but it is safe and solid, and furnishes the only direct route to the heavenly city. Now, if God made your soul, and made religion for it, you are a fool if you refuse to place yourself on the track. You cannot prosper anywhere else, and your machine will not run anywhere else.

JOSIAH GILBERT HOLLAND.

SPIRITUAL EMANCIPATION.

The current skepticism in regard to the tendencies of human nature proceeds upon the fallacy that a man's true wealth, the wealth he covets or prizes, is external to himself, consisting in the abundance of the things he possesses. The skeptic says that if you leave men free from police restraint, however well you may educate them, there will be no security for property. Of course, then, he believes that man values these outward possessions which we call property, above all things. There is no sheerer fallacy current than this. For the undue value men set upon this sort of possession now grows out of its scarcity, grows out of the fact that so many are utterly destitute of it. Appetite is never excessive, never furious, save where it has been starved. The frantic hunger we see it so often exhibiting under every variety of criminal form, marks only the hideous starvation to which society subjects it. It is not a normal, but a morbid state of the appetite, growing exclusively out of the unnatural compression which is imposed upon it by the exigencies of our immature society. Every appetite and passion of man's nature is good and beautiful, and destined to be fully enjoyed, and a scientific society or fellowship among men would ensure this result, without allowing any compromise of the individual dignity, especially without allowing that fierce and disgusting abandonment to them which disfigures so many of our eminent names in church and state, and which infallibly attests the uncleanness of our present morality.

Remove, then, the existing bondage of humanity, remove those factitious restraints which keep appetite and passion on the perpetual lookout for escape, like steam from an over-charged boiler, and their force would instantly become conservative instead of destructive.

For man is destined by the very necessity of his creation, for nothing but the obedience of his inward and divine self-hood, for the obedience of God within him. Even while he is utterly unconscious of his true or inmost self-hood, the aim of his whole existence, the end of all his struggle and toil is to realize it; and when it does dawn upon him, it sheds a complete calm upon the turbid sea of his outward relations.

The effect is irresistible. You cannot arouse a man to self-respect, to a sense of his proper humanity, to a consciousness of the divinity which constitutes his being, without rendering him superior to outward accident. He is no longer the sport of passion, of conscience, or of appetite. The master of the house has come at last, and his servants render him a prompt and joyous obedience. No more in a mere symbolic, but in a very real sense, the Lord has entered his holy temple: all the earth, the entire realm of the outward and finite, spontaneously keeps silence before Him.

HENRY JAMES.

A SUDDEN HURRICANE.

The evening, which had been beautiful before, had undergone a change. The moon was obscured, and gigantic shadows, dense and winged, hurried with deep-toned cries along the heavens, as if in angry pursuit. Occasionally, in sudden gusts, the winds moaned heavily among the pines; a cooling freshness impregnated the atmosphere, and repeated flashes of sharpest lightning imparted to the prospect a splendor which illuminated, while increasing the perils of that path which our adventurers were now pursuing. Large drops, at moments, fell from the driving clouds, and everything promised the coming on of one of those sudden and severe thunder-storms, so common to the early summer of the South.

Singleton looked up anxiously at the wild confusion of sky and forest around him. The woods seemed to apprehend the danger, and the melancholy sighing of their branches appeared to indicate an instinctive consciousness, which had its moral likeness to the feeling in the bosom of the observer. How many of these mighty pines were to be prostrated under that approaching tempest! how many beautiful vines, which had clung to them like affections that only desire an object to fasten upon, would share in their ruin! How could Singleton overlook the analogy between the fortune of his family and friends, and that which his imagination depicted as the probable destiny of the forest?

"We shall have it before long, Humphries, for you see the black horns yonder in the break before us. I begin to feel the warm breath of the hurricane already, and we must look out for some smaller woods. I like not these high pines in a

A SUDDEN STORM.

storm like this, so use your memory, man, and lead on to some thicket of scrubby oaks — if you can think of one near at hand. Ha! — we must speed — we have lingered too long. Why did you not hurry me? You should have known how difficult it was for me to hurry myself in such a situation."

This was spoken by Singleton, at moments when the gusts permitted him to be heard, and when the irregularity of the route suffered his companion to keep beside him. The lieutenant answered promptly : —

"That was the very reason why I did not wish to hurry you, major. I knew you hadn't seen your folks for a mighty long spell, and so I couldn't find it in my heart to break in upon you, though I felt dub'ous that the storm would be soon upon us."

"A bad reason for a soldier. Friends and family are scarcely desirable at such a time as this, since we can seldom see them, or only see their suffering. Ha! — that was sharp!"

"Yes, sir, but at some distance. We are coming to the stunted oaks now, which are rather squat, and not so likely to give as the pines. There ain't so much of 'em, you see. Keep a lookout, sir, or the branches will pull you from your horse. The road here is pretty much overgrown, and the vines crowd thick upon it."

"A word in season!" exclaimed Singleton, as he drew back before an overhanging branch which had been bent by the wind, and was thrust entirely across his path. A few moments were spent in rounding the obstruction, and the storm grew heavier; the winds no longer labored among the trees, but rushed along with a force which flattened their elastic tops, so that it either swept clean through them or laid them prostrate forever. A stronger hold, a positive straining in their effort, became necessary now, with both riders, in order to secure themselves firmly in their saddles; while their horses, with uplifted ears, and an occasional snort, in this manner, not less than by the shiver of their whole frames, betrayed their own apprehensions, and, as it were, appealed to their masters for protection.

"The dumb beast knows where to look, after all, major; he knows that man is most able, you see, to take care of him, though man wants his keeper too. But the beast don't know that. He's like the good soldier that minds his own captain, and looks to him only, though the captain himself has a general from whom he gets his orders. Now, say what you will, major, there's reason in the horse — the good horse, I mean, for some horses that I've straddled in my time have shown themselves mighty foolish and unreasonable."

Humphries stroked the neck of his steed fondly, and coaxed him by an affectionate word, as he uttered himself thus, with no very profound philosophy. He seemed desirous of assuring the steed that he held him of the better class, and favored him accordingly. Singleton assented to the notion of his companion, who did not, however, see the smile which accompanied his answer.

"Yes, yes, Humphries, the horse knows his master, and is the least able or willing of all animals to do without him. I would we had our nags in safety now: I would these five miles were well over."

"It's a tough ride; but that's so much the better, major, the less apt we are to be troubled with the tories."

"I should rather plunge through a crowd of them, now, in a charge against superior cavalry, than take it in such a night as this, when the wind lifts you, at every bound, half out of your saddle, and, but for the lightning, which comes quite too nigh to be at all times pleasant, your face would make momentary acquaintance with boughs and branches, vines and thorns, that give no notice and leave their mark at every brush. A charge were far less difficult."

"Almost as safe, sir, that's certain, and not more unpleasant. But let us hold up, major, for a while, and push for the thicket. We shall now have the worst of the hurricane. See the edge of it yonder — how black! and now — only hear the roaring!"

"Yes, it comes. I feel it on my cheek. It sends a breath like fire before it, sultry and thick, as if it had been sweeping all day over beds of the hottest sand. Lead the way, Humphries."

"Here, sir, — follow close and quick. There's a clump of forest, with nothing but small trees, lying to the left — now, sir, that flash will show it to you — there we can be snug till the storm passes over. It has a long body and it shakes mightily, but it goes too fast to stay long in its journey, and a few minutes, sir — a few minutes is all we want. Mind the vine there, sir; and there, to your left, is a gully, where an old tree's roots have come up. Now, major, the sooner we dismount and squat with our horses the better."

They had now reached the spot to which Humphries had directed his course — a thick undergrowth of small timber — of field pine, the stunted oak, black-jack, and hickory — few of sufficient size to feel the force of the tempest, or prove very conspicuous conductors of the lightning. Obeying the suggestion and following the example of his companion, Singleton dismounted, and the two placed themselves and their horses as much upon the sheltered side of the clump as possible, yet sufficiently far to escape any danger from its overthrow.

Here they awaited the coming of the tempest. The experienced woodman alone could have spoken for its approach. A moment's pause had intervened, when the suddenly aroused elements seemed as suddenly to have sunk into grim repose. A slight sighing of the wind only, as it wound sluggishly along the distant wood, had its warning, and the dense blackness of the embodied storm was only evident at moments when the occasional rush of the lightning made visible its gloomy terrors.

"It's making ready for a charge, major: it's just like a good captain, sir, that calls in his scouts and sentries, and orders all things to keep quiet, and without beat of drum gets all fixed to spring out from the bush upon them that's coming. It won't be long now, sir, before we get it; but just now it's still as the grave. It's waiting for its outriders — them long streaky white clouds it sent out an hour ago, like so many scouts. They're a-coming up now, and when they all get up together — then look out for the squall. Quiet now, Mossfoot — quiet now, creature — don't be frightened — it's not a-going to hurt you, old fellow — not a bit."

Humphries patted his favorite while speaking, and strove to soothe and quiet the impatience which both horses exhibited. This was in that strange pause of the storm which is its most remarkable feature in the South — that singular interregnum of the winds, when, after giving repeated notice of their most terrific action, they seem almost to forget their purpose, and for a few moments appear to slumber in their inactivity.

But the pause was only momentary, and was now at an end. In another instant, they heard the rush and the roar, as of a thousand wild steeds of the desert ploughing the sands ; then followed the mournful howling of the trees — the shrieking of the lashed winds, as if, under the influence of some fierce demon who enjoyed his triumph, they plunged through the forest, wailing at their own destructive progress, yet compelled unswervingly to hurry forward. They twisted the pine from its place, snapping it as a reed, while its heavy fall to the ground which it had so long sheltered, called up, even amid the roar of the tempest, a thousand echoes from the forest. The branches of the wood were prostrated like so much heather, wrested and swept from the tree which yielded them without a struggle to the blast ; and the crouching horses and riders below were in an instant covered with a cloud of fragments. These were the precursors merely ; then came the arrowy flight and form of the hurricane itself — its actual bulk — its embodied power, pressing along through the forest in a gyratory progress, not fifty yards wide, never distending in width, yet capriciously winding from right to left, and left to right, in a zigzag direction, as if a playful spirit thus strove to mix with all the terrors of destruction the sportive mood of the most idle fancy. In this progress, the whole wood in its path underwent prostration — the tall, proud pine, the deep-rooted and unbending oak, the small cedar and the pliant shrub, torn, dismembered of their fine proportions ; some, only by a timely yielding to the pressure, passed over with little injury, as if too much scorned by the assailant for his wrath. The larger trees in the neighborhood of the spot where our partisans had taken shelter, shared the harsher fortune generally, for they were in the very track of the tempest. Too sturdy and massive to yield, they withheld their homage, and were either snapped off relentlessly and short, or were torn and twisted up from their very roots. The poor horses, with eyes staring in the direction of the storm, with ears erect, and manes flying in the wind, stood trembling in every joint, too much terrified, or too conscious of their helplessness, to attempt to fly. All around the crouching party the woods for several seconds absolutely flattened. Huge trees were prostrated, and their branches were clustering thickly, and almost forming a prison around them ; leaving it doubtful, as the huge terror rolled over their heads, whether they could ever make their escape from the enclosure. Rush after rush of the trooping winds went over them, keeping them immovable in their crowded shelter and position — each succeeding troop wilder and weightier than the last, until at length a sullen, bellowing murmur, which before they had not heard, announced the greater weight of the hurricane to be overthrowing the forests in the distance.

The chief danger had overblown. Gradually the warm, oppressive breath passed off ; the air again grew suddenly cool, and a gush of heavy drops came falling from the heavens, as if they too had been just released from the intolerable pressure which had burdened earth. Moaning pitifully, the prostrated trees and shrubs, those which had survived the storm, though shorn by its scythes, gradually and seemingly with painful effort, once more elevated themselves to their old position. Their sighings, as they did so, were almost human to the ears of our crouching warriors, whom their movement in part released. Far and near, the moaning of the forest around them was strangely, but not unpleasantly, heightened in its effect upon their senses, by the distant and declining roar of the past and far-travelling hurricane, as ploughing the deep woods and laying waste all in its progress, it rushed on to a meeting with the kindred storms that gather about the gloomy Cape Hatteras, and stir and foam along the waters of the Atlantic.

WILLIAM GILMORE SIMMS.

ITALIAN LIFE.

ONE TYPE OF ITALIAN BEAUTY.

. . . The manners of Italy are so much changed since we were here last, the alteration is scarcely credible. They say it has been by the last war. The French, being masters, introduced all their customs, which were eagerly embraced by the ladies, and I believe will never be laid aside; yet the different governments make different manners in every state. You know, though the republic is not rich, here are many private families vastly so, and live at a great superfluous expense: all the people of the first quality keep coaches as fine as the Speaker's, and some of them two or three, though the streets are too narrow to use them in the town; but they take the air in them, and their chairs carry them to the gates. The liveries are all plain: gold or silver being forbidden to be worn within the walls, the habits are all obliged to be black, but they wear exceedingly fine lace and linen; and in their country-houses, which are generally in the fauxbourg, they dress very richly, and have extremely fine jewels. Here is nothing cheap but houses. A palace fit for a prince may be hired for fifty pounds per annum: I mean unfurnished. All games of chance are strictly prohibited, and it seems to me the only law they do not try to evade: they play at quadrille, picquet, etc., but not high. Here are no regular public assemblies. I have been visited by all of the first rank, and invited to several fine dinners, particularly to the wedding of one of the House of Spinola, where there were ninety-six sat down to table, and I think the entertainment one of the finest I ever saw. There was, the night following, a ball and supper for the same company, with the same profusion. They tell me that all their great marriages are kept in the same public manner. Nobody keeps more than two horses, all their journeys being post; the expense of them, including the coachman, is (I am told) fifty pounds per annum. A chair is very nearly as much; I give eighteen francs a week for mine. The senators can converse with no strangers during the time of their magistracy, which lasts two years. The number of servants is regulated, and almost every lady has the same, which is two footmen, a gentleman usher, and a page, who follow her chair.

Lady Mary Wortley Montagu.

PROGRESS.

The Greeks had the very largest ideas upon the training of man, and produced specimens of our kind with gifts that have never been surpassed. But the nature of man, such as they knew it, was scarcely at all developed; nay, it was maimed, in its supreme capacity — in its relations towards God. Hence, as in the visions of the prophet, so upon the roll of history, the imposing fabrics of ancient civilization never have endured. Greece has bequeathed to us her ever living tongue, and the immortal productions of her intellect. Rome made ready for Christendom the elements of polity and law; but the brilliant assemblage of endowments which constitutes civilization, having no root in itself, could not bear the shocks of time and vicissitude; it came and it went; it was seen and it was gone.

We now watch with a trembling hope, the course of that later and Christian civilization which arose out of the ashes of the old heathen world, and ask ourselves whether, like the gospel itself, so that which the gospel has wrought beyond itself in the manners, arts, laws, and institutions of men, is in such manner salted with perpetual life, that the gates of hell shall not prevail against it? Will the civilization, which was springing upwards from the days of Charlemagne, and which now over the face of Europe and America, seeking to present to us in bewildering conflict the mingled signs of decrepitude and of vigor, perish like its older types, and like them be known thereafter only in its fragments; or does it bear a charmed life, and will it give shade from the heat and shelter from the storm to all generations of man?

In any answer to such a question, it would perhaps be easier to say what would not than what would be involved. But some things we may observe which may be among the material of a reply.

The arts of war are now so allied with those of peace, that barbarism, once so terrible, is reduced to physical impotence; and what civilized man has had the wit to create, he has also the strength to defend. Thus one grand destructive agency is paralyzed. Time, indeed, is the great destroyer; but his power, too, is greatly neutralized by printing, by commerce which lays the foundation of friendship among nations, by ease of communication which binds men together, by that diffusion of intelligence which multiplies the natural guardians of civilizations. These are perhaps not merely isolated phenomena. Perhaps they are but witnesses, and but a few among many witnesses, to the vast change which has been wrought since the advent of our Lord in the state of man. Perhaps they re-echo to us the truth that apart from sound and sure relations to its Maker, the fitful efforts of mankind must needs be worsted in the conflict with chance and change; but that when by the dispensation of Christianity the order of our moral nature was restored, when the rightful King had once more taken his place upon his throne, then indeed, civilization might come to have a meaning and a vitality such as had before been denied it. Then, at length, it had obtained the key to all the mysteries of the nature of man, to all the anomalies of its condition. Then it had obtained the ground plan of that nature in all its fulness, which before had been known only in remnants or in fragments; fragments of which, even as now in the toppling remains of some ancient church or castle, the true grandeur and the ethereal beauty were even the more conspicuous because of the surrounding ruins. But fragments still, and fragments only, until, by the bringing of life and immortality to light the parts of our nature were reunited, its harmony was re-established, the riddle of life, heretofore unsolved, was at length read as a discipline, and so obtained its just interpretation. All that had before seemed idle conflict, wasted energy, barren effort, was seen to be but the preparation for a glorious future; and death itself instead of extinguishing the last hopes of man, became the means and the pledge of his perfection.

WILLIAM E. GLADSTONE.

PERSONAL INFLUENCE.

On went the talk and laughter. Two or three of the little boys in the long dormitory were already in bed, sitting up with their chins on their knees. The light burned clear, the noise went on. It was a trying moment for Arthur, the poor little lonely boy; however, this time he didn't ask Tom what he might or might not do, but dropped on his knees by his bedside, as he had done every day from his childhood, to open his heart to Him who heareth the cry and beareth the sorrows of the tender child, and the strong man in agony. . . .

There were many boys in the room by whom that little scene was taken to heart before they slept. But sleep seemed to have deserted the pillow of poor Tom. For some time his excitement, and the flood of memories which chased one another through his brain kept him from thinking or resolving. His head throbbed, his heart leapt, and he could hardly keep himself from springing out of bed and rushing about the room. Then the thought of his own mother came across him, and the promise he had made at her knee, years ago, never to forget to kneel by his bedside, and give himself up to his Father, before he laid his head on the pillow, from which it might never rise; and he lay down gently and cried as if his heart would break. He was only fourteen years old.

It was no light act of courage in those days for a little fellow to say his prayers publicly, even at Rugby. A few years later, when Arnold's manly piety had begun to leaven the school, the tables turned; before he died, in the school-house, at least, and I believe in the other houses, the rule was the other way. But poor Tom had come to school in other times. The first few nights after he came he did not kneel down because of the noise, but sat up in bed till the candle was out, and then stole out and said his prayers, in fear lest some one should find him out. So did many another poor little fellow. Then he began to think that he might just as well say his prayers in bed, and then it didn't matter whether he was kneeling, or sitting, or lying down. And so it had come to pass with Tom, as with all who will not confess their Lord before men; and for the last year he had probably not said his prayers in earnest a dozen times.

Poor Tom! the first and bitterest feeling which was likely to break his heart was the sense of his own cowardice. The vice of all others which he loathed was brought in and burned in on his own soul. He had lied to his mother, to his conscience, to his God. How could he bear it? And then the poor little weak boy, whom he had pitied and almost scorned for his weakness, had done that which he, braggart as he was, dared not do. The first dawn of comfort came to him in swearing to himself that he would stand by that boy through thick and thin, and cheer him, and help him, and bear his burdens, for the good deed done that night. Then he resolved to write home next day and tell his mother all, and what a coward her son had been. And then peace came to him, as he resolved lastly, to bear his testimony next morning. The morning would be harder than the night to begin with, but he felt that he could not afford to let one chance slip. Several times he faltered, for the devil showed him first all his old friends calling him "Saint" and "Square-toes," and a dozen hard names, and whispered to him that his motives would be misunderstood, and he would only be misunderstood, and he would only be left alone with the new boy; whereas it was his duty to keep all means of influence, that he might do good to the largest number. And then came the more subtle temptation, "Shall I not be showing myself braver than others by doing this? Have I any right to begin it now? Ought I not rather to pray in my own study, letting other boys know that I do so, and trying to lead them to it, while in public at least I should go on as I have done?" However, his good angel was too strong that night, and he turned on his side and slept, tired of trying to reason but resolved to follow the impulse which had been so strong, and in which he had, found peace.

RUGBY SCHOOL.

Next morning he was up and washed and dressed, all but his jacket and waist-coat, just as the ten minutes' bell began to ring, and then in the face of the whole room knelt down to pray. Not five words could he say — the bell mocked him; he was listening for every whisper in the room — what were they all thinking of him? He was ashamed to go on kneeling ashamed to rise from his knees. At last, as it were from his inmost heart, a still small voice seemed to breathe forth the words of the publican, "God be merciful unto me a sinner!" He repeated them over and over, clinging to them as for his life, and rose from his knees comforted and humbled, and ready to face the whole world. It was not needed: two other boys beside Arthur had already followed his example, and he went down to the great school with a glimmering of another lesson in his heart — the lesson that he who has conquered his own coward spirit has conquered the whole outward world; and that other one which the old prophet learned in the cave in Mount Horeb, that however we may fancy ourselves alone on the side of good, the King and Lord of men is nowhere without his witnesses; for in every society there are those who have not bowed the knee to Baal.

THOMAS HUGHES.

MR. TARBOX AND ZOSÉPHINE.

SHE nodded and began to move slowly on, he following.
"I'm not betraying any one's confidence," persisted he; "but I can't help but have a care for you. Not that you need it, or anybody's. You can take care of yourself if any man or woman can. Every time your foot touches the ground it says so as plain as words. That's what first caught my fancy. You haven't got to have somebody to take care of you. Oh, Josephine! that's just why I want to take care of you so bad! I can take care of myself, and I used to like to do it; I was just that selfish and small; but love's widened me. I can take care of myself; but, oh! what satisfaction is there in it? Is there any? Now, I ask you! It may do for you, for you're worth taking care of; but I want to take care of something I needn't be ashamed to love!" He softly stole her hand as they went. She let it stay, yet looked away from him, up through the darkling branches, and distressfully shook her head.

"Don't, Josephine!— don't do that! I want you to take care of me. You could do better, I know, if love wasn't the count; but when it comes to loving you, I'm the edition deloox! I know you've an aspiring nature, but so have I; and I believe with you to love and you loving me, and counselling and guiding me, I could climb high. Oh, Josephine! it isn't this poor Tarbox I'm asking you to give yourself to; it's the Tarbox that is to be; it's the coming Tarbox! Why, it's even a good business move! If it wasn't, I wouldn't say a word! You know I can, and will, take the very best care of everything you've got; and I know you'll take the same of mine. It's a good move, every way. Why, here's everything just fixed for it! Listen to the mocking-bird! See him yonder, just at the right of the stile. See! Oh, Josephine! don't you see he isn't

> ' Still singing where the weeping-willows wave;'

he's on the myrtle; the myrtle, Josephine, and the crape-myrtle at that! — widowhood, unwidowed! — Now he's on the fence, — but he'll not stay there, — and you mustn't either!" The suitor smiled at his own ludicrousness, yet for all that looked beseechingly in earnest. He stood still again, continuing to hold her hand. She stole a furtive glance here and there for possible spectators. He smiled again.

"You don't see anybody; the world waives its claim." But there was such distress on her face that his smile passed away, and he made a new effort to accommodate his suit to her mood.

"Josephine, there's no eye on us except it's overhead. Tell me this; if he that was yours until ten years ago was looking down now and could speak to us, don't you believe he'd say yes?"

"Oh! I dunno. Not to-day! Not dis day!" The widow's eyes met his gaze

of tender inquiry and then sank to the ground. She shook her head mournfully.
"Naw, naw ; not dis day. 'Tis to-day 'Thanase was kill' !"

Mr. Tarbox relaxed his grasp, and Zoséphine's hand escaped. She never had betrayed to him so much distress as filled her face now. "De man what kill' him git away. You t'ink I git marrie' while dat man alive ? Ho-o-o ! You t'ink I let Marguerite see me do dat ? Ah, naw !" She waved him away and turned to leave the spot, but he pressed after, and she paused once more. . . .

"But I tell you, yes ! you, Josephine ! I'm poor sort enough yet ; but I could have done things once that I can't do now. There was a time when if some miserable outlaw stood, or even seemed, maybe, to stand between me and my chances for happiness, I could have handed him over to human justice, so called, as easy as wink ; but now ? No, never any more ! Josephine, I know that man whose picture I've just looked at. I could see you avenged. I could lay my hands, and the hands of the law, on him inside of twenty-four hours. You say you can't marry till the law has laid its penalties on him, or at least while he lives and escapes them. Is that right ?"

Zoséphine had set her face to oppose his words only with unyielding silence, but the answer escaped her :

"Yass, 'tis so. 'Tis ri-ight !"

"No, Josephine. I know you feel as if it were ; but you don't think so. No, you don't ; I know you better in this matter than you know yourself, and you don't think it's right. You know justice belongs to the State, and that when you talk to yourself about what you owe to justice it means something else, that you're too sweet and good to give the right name to and still want it. You don't want it; you don't want revenge, and here's the proof ; for, Josephine, you know, and I know, that if I — even without speaking — with no more than one look of the eye — should offer to buy your favor at that price, even ever so lawfully, you'd thank me for one minute and then loathe me to the end of your days."

Zoséphine's face had lost its hardness. It was drawn with distress. With a gesture of repulsion and pain she exclaimed :

"I di'n' mean — I di'n' mean — Ah !"

"What ? private revenge ? No, of course you didn't ! But what else would it be ? Oh, Josephine ! don't I know you didn't mean it ? Didn't I tell you so ? But I want you to go farther ; I want you to put away forever the feeling. I want to move and stand between you and it, and say — whatever it costs me to say it — God forbid ! I do say it ; I say it now. I can't say more ; I can't say less ; and somehow — I don't know how — wherever you learned it — I've learned it from you."

Zoséphine opened her lips to refuse ; but they closed and tightened upon each other, her narrowed eyes sent short flashes out upon his, and her breath came and went long and deep without sound. But at his last words she saw — the strangest thing — to be where she saw it — a tear — tears — standing in his eyes ; saw them a moment ; and then could see them no more for her own. Her lips relaxed, her form drooped, she lifted her face to reply, but her mouth twitched ; she could not speak.

"I'm not so foolish as I look," he said, trying to smile away his emotion. "If the State chose to hunt him out and put him to trial and punishment, I don't say I'd stand in the way; that's the State's business; that's for the public safety. But it's too late — you and Bonaventure have made it too late — for me to help any one, least of all the one I love, to be revenged." He saw his words were prevailing, and followed them up. "Oh! you don't need it any more than you really want it, Josephine. You mustn't ever look toward it again. I throw myself and my love across the path. Don't walk over us. Take my hand; give me yours; come another way; and if you'll let such a poor excuse for a teacher and guide help you, I'll help you all I can to learn to say 'Forgive us our trespasses.' You can begin now, by forgiving me. I may have thrown away my last chance with you, but I can't help it; it's my love that spoke. And if I have spoiled all, and if for the tears you're shedding I've got to pay with the greatest disappointment of my life, still I've had the glory and the sanctification of loving you. If I must say, I can say,

> "'Tis better to have loved and lost,
> Than never to have loved at all.'

Must I? Are you going to make me say that?"

Zoséphine, still in tears, silently and with drooping head pushed her way across the stile and left him standing on the other side. He sent one pleading word after her:

"Isn't it most too late to go the rest of the way alone?"

She turned, lifted her eyes to his for an instant, and nodded. In a twinkling he was at her side. She glanced at him again and said quite contentedly:

"Yass; 'tis so," and they went the short remnant of the way together.

GEORGE W. CABLE.

AN ENCOUNTER WITH A PANTHER.

As soon as I had effected my dangerous passage, I screened myself behind a cliff, and gave myself up to reflection. While occupied with these reflections, my eyes were fixed upon the opposite steeps. The tops of the trees, waving to and fro in the wildest commotion, and their trunks occasionally bending to the blast, which, in these lofty regions, blew with a violence unknown in the tracts below, exhibited an awful spectacle. At length my attention was attracted by the trunk which lay across the gulf, and which I had converted into a bridge. I perceived that it had already swerved somewhat from its original position; that every blast broke or loosened some of the fibers by which its roots were connected with the opposite bank; and that, if the storm did not speedily abate, there was imminent danger of its being torn from the rock and precipitated into the chasm. Thus my retreat would be cut off, and the evils from which I was endeavoring to rescue another would be experienced by myself.

I believed my destiny to hang upon the expedition with which I should recross this gulf. The moments that were spent in these deliberations were critical, and I shuddered to observe that the trunk was held in its place by one or two fibers, which were already stretched almost to breaking.

To pass along the trunk, rendered slippery by the wet and unsteadfast by the wind, was eminently dangerous. To maintain my hold in passing, in defiance of the whirlwind, required the most vigorous exertions. For this end, it was necessary to discommode myself of my cloak and of the volume which I carried in the pocket of my coat.

Just as I had disposed of these encumbrances, and had risen from my seat, my attention was again called to the opposite steep by the most unwelcome object that at this time could possibly occur. Something was perceived moving among the bushes and rocks, which, for a time, I hoped was nothing more than a raccoon or opossum, but which presently appeared to be a panther. His gray coat, extended claws, fiery eyes, and a cry which he at that moment uttered, and which, by its resemblance to the human voice, is peculiarly terrific, denoted him to be the most ferocious and untamable of that detested race. The industry of our hunters has nearly banished animals of prey from these precincts. The fastnesses of Norwalk, however, could not but afford refuge to some of them. Of late I had met them so rarely that my fears were seldom alive, and I trod without caution the ruggedest and most solitary haunts. Still, however, I had seldom been unfurnished in my rambles with the means of defense.

The unfrequency with which I had lately encountered this foe, and the encumbrance of provision, made me neglect, on this occasion, to bring with me my usual arms. The beast that was now before me, when stimulated by hunger, was accustomed to assail whatever could provide him with a banquet of blood. He would set upon the man and the deer with equal and irresistible ferocity. His sagacity was equal to his strength, and he seemed able to discover when his antagonist was armed and prepared for defense.

My past experience enabled me to estimate the full extent of my danger. He sat on the brow of the steep, eying the bridge, and apparently deliberating whether he should cross it. It was probable that he had scented my footsteps thus far, and, should he pass over, his vigilance could scarcely fail of detecting my asylum.

Should he retain his present station, my danger was scarcely lessened. To pass over in the face of a famished tiger was only to rush upon my fate. The falling of the trunk, which had lately been so anxiously deprecated, was now with no less solicitude desired. Every new gust, I hoped, would tear asunder its remaining bands, and, by cutting off all communication between the opposite steeps, place me in security. My hopes, however, were destined to be frustrated. The fibers of the prostrate tree were obstinately tenacious of their hold, and presently the animal scrambled down the rock and proceeded to cross it.

Of all kinds of death, that which now menaced me was the most abhorred. To die by disease or by the hand of a fellow-creature was propitious and lenient in comparison with being rent to pieces by the fangs of this savage. To perish in this obscure retreat by means so impervious to the anxious curiosity of my friends, to lose my portion of existence by so untoward and ignoble a destiny, was insupportable. I bitterly deplored my rashness in coming hither unprovided for an encounter like this.

The evil of my present circumstances consisted chiefly in suspense. My death was unavoidable, but my imagination had leisure to torment itself by anticipations. One foot of the savage was slowly and cautiously moved after the other. He struck his claws so deeply into the bark that they were with difficulty withdrawn. At length he leaped upon the ground. We were now separated by an interval of scarcely eight feet. To leave the spot where I crouched was impossible. Behind and beside me the cliff rose perpendicularly, and before me was this grim and terrible visage. I shrunk still closer to the ground, and closed my eyes.

From this pause of horror I was aroused by the noise occasioned by a second spring of the animal. He leaped into the pit in which I had so deeply regretted that I had not taken refuge, and disappeared. My rescue was so sudden, and so much beyond my belief or my hope, that I doubted for a moment whether my senses did not deceive me. This opportunity of escape was not to be neglected. I left my place and scrambled over the trunk with a precipitation which had like to have proved fatal. The tree groaned and shook under me, the wind blew with unexampled violence, and I had scarcely reached the opposite steep when the roots were severed from the rock, and the whole fell thundering to the bottom of the chasm.

My trepidations were not speedily quieted. I looked back with wonder on my hairbreadth escape, and on that singular concurrence of events which had placed me in so short a period in absolute security. Had the trunk fallen a moment earlier, I should have been imprisoned on the hill or thrown headlong. Had its fall been delayed another moment, I should have been pursued; for the beast now issued from his den, and testified his surprise and disappointment by tokens the sight of which made my blood run cold.

He saw me, and hastened to the verge of the chasm. He squatted on his hind legs, and assumed the attitude of one preparing to leap. My consternation was excited afresh by these appearances. It seemed at first as if the rift was too wide for any power of muscles to carry him in safety over; but I knew the unparalleled agility of this animal, and that his experience had made him a better judge of the practicability of this exploit than I was.

Still there was hope that he would relinquish this design as desperate. This hope was quickly at an end. He sprung, and his fore-legs touched the verge of the rock on which I stood. In spite of vehement exertions, however, the surface was too smooth and too hard to allow him to make good his hold. He fell, and a piercing cry uttered below, showed that nothing had obstructed his descent to the bottom.

CHARLES BROCKDEN BROWN.

TO A CHILD.

So you are at Sandgate! Of course, wishing for your old play-fellow, M—— H—— (he can play — it's work to me), to help you to make little puddles in the Sand, and swing on the Gate. But perhaps there are no sand and gate at Sandgate, which, in that case, nominally tells us a fib. But there must be little crabs somewhere, which you can catch, if you are nimble enough, so like spiders, I wonder they do not make webs. The large crabs are scarcer.

If you do catch a big one with strong claws — and like experiments, — you can shut him up in a cupboard with a loaf of sugar, and you can see whether he will break it up with his nippers. Besides crabs, I used to find jelly-fish on the beach, made, it seemed to me, of sea-calves' feet, and no sherry.

The mermaids eat them, I suppose, at their wet water-parties, or salt *soirées*.

I suppose you never gather any sea-flowers, but only sea-weeds. The truth is, Mr. David Jones never rises from his bed, and so has a garden full of weeds, like Dr. Watts' Sluggard. . . .

I have heard that you bathe in the sea, which is very refreshing, but it requires care; for if you stay under water too long you may come up a mermaid, who is only half a lady, with a fish's tail, — which she can boil if she likes. You had better try this with your doll, whether it turns her into half a "dollfin."

I hope you like the sea. I always did when I was a child, which was about two years ago. Sometimes it makes such a fizzing and foaming, I wonder some of our London cheats do not bottle it up, and sell it for ginger-pop. . . .

Sometime ago exactly, there used to be, about the part of the coast where you are, large white birds with black-tipped wings, that went flying and screaming over the sea, and now and then plunged down into the water after a fish. Perhaps they catch their sprats now with nets or hooks and lines. Do you ever see such birds? We used to call them "gulls,' — but they didn't mind it! Do you ever see any boats or vessels? And don't you wish, when you see a ship, that somebody was a sea-captain instead of a doctor, that he might bring you home a pet lion, or calf elephant, ever so many parrots, or a monkey, from foreign parts? I knew a little girl who was promised a baby whale by her sailor brother, and who *blubbered* because he did not bring it. I suppose there are no whales at Sandgate, but you might find a seal about the beach; or, at least, a stone for one. The sea stones are not pretty when they are dry, but look beautiful when they are wet, and we can always keep sucking them!

If you can find one, pray pick me up a pebble for a seal. I prefer the red sort, like Mrs. Jenkins' brooch and ear-rings, which she calls " red chamelion." Well, how happy you must be! Childhood is such a joyous, merry time; and I often wish I was two or three children! But I suppose I can't be; or else I would be Jeanie, and May, and Dunnie Elliot. And wouldn't I pull off my three pairs of shoes and socks, and go paddling in the sea up to my six knees! And oh! how I could climb up the downs, and roll down the ups on my three backs and stomachs! . . .

THOMAS HOOD.

THE PREMIER GLADSTONE.

CARLTON HOUSE TERRACE is one of the historic spots in London. It is a long, stately row of mansions flanking St. James Park. At the foot of broad Waterloo Place stands the lofty column to the Duke of York. As he died heavily in debt, the wags say "The Duke was put up on top of the column to get him out of the reach of his creditors." In the second or third house from the monument resides Britain's ruler, the Premier Gladstone. Technically, the ruler of the realm dwells in Windsor Palace. But Major Jack Downing tells us that when General Jackson — on his visit to Downingville — got tired of shaking hands with the crowd, he (the Major) hid behind him, and poking his arm under Old Hickory's shoulder, he "shuck hands for the Gineral." So the hand of royalty in England is really the hand of William E. Gladstone slipped under the regal robes. I had the honor of two very delightful interviews with the Premier last summer. As the "Alabama question" was just at its most exciting point, Mr. Gladstone was quite ready to converse freely with any American who was supposed to be familiar with the state of public sentiment on this side of the water. He very kindly invited me to visit him. He received me with cordial freedom, and in the half-hour's chat he opened his mind to me with that transparent sincerity which belongs to the character of a Christian statesman. As I rose to leave, saying to him, "Your time belongs to the British Empire and not to an American traveler," he very cordially said, "Come and breakfast with me on Thursday." Breakfast is the familiar meal in English home-life, as "tea-drinking" is with us. I went at ten o'clock on a June morning, and found the Premier standing out on his rear balcony, overlooking cool, verdant St. James Park. Mr. Gladstone is in excellent preservation; his walk is alert, and his broad shoulders have never stooped under the load of official responsibilities. One secret of his vigorous health is that he is a capital sleeper. "I never," said he to me, "allow the cares of State to get inside of my bed-chamber door." He says that he does not remember that he was ever kept awake for half an hour by anxiety but once. And that was at the country-seat of his brother-in-law, Lord Lyttelton, where he had been chopping down a tree just at twilight. He did not quite finish the job, and the fear that the tree might blow down before morning worried him out of a little sleep. I am afraid that President Lincoln knew but little of such quiet slumbers during his stormy administration.

At the breakfast-table of Mr. Gladstone I met the venerable Dean Ramsay, of Edinburgh, and the Rev. Newman Hall, who is on quite intimate terms of friendship with the Premier. Mrs. Gladstone — who was the daughter of a substantial country gentleman, and with whom Mr. Gladstone fell in love in his student days — is a warm-hearted lady, whose beauty of character and manners surpass her beauty of person. She is an untiring worker in several schemes of active philanthropy. A son was at the table, and a noble-looking daughter. Another son is in the Church of England pulpit. And what a charming hour of chat was that at the

Prime Minister's breakfast! A package of private despatches from the Geneva Arbitrators was quietly laid aside unopened until the coffee and toast and strawberries were disposed of. The Presidential campaign in America seemed to interest Mr. Gladstone deeply, and he inquired, "Have you read Mr. Sumner's speech against the President? It is an extraordinary speech. If his charges are unjust, they ought never to have been made. If they are just, it seems to me that impeachment is inevitable. It would be thought so here. We do not quite understand your freedom of Congressional criticisms."

But politics were soon ruled out for a playful discussion of American humor, especially of the negro type. Mr. Gladstone enjoyed hugely some stories of plantation preaching, and said afterwards that he had not laughed so heartily in many a day. Negro wit (like negro music) is so indigenous to our soil that it is fresher to foreign ears than to our own. As the hour came for a morning session of Parliament, we withdrew. Mr. Gladstone's last words to me were, "I cannot tell what Providence may order, but no power on earth can hinder the peaceful settlement of our controversy with your country, and the complete triumph of the treaty." He was a true prophet. And let us rejoice that during all that long controversy, the sagacious brain, and the noble Christian heart of William E. Gladstone, guided the diplomacy of the British Empire.

THEODORE L. CUYLER.

THE FOURTH OF JULY.

YESTERDAY the greatest question was decided which ever was debated in America, and a greater, perhaps, never was nor will be decided among men. A resolution was passed, without one dissenting colony, "that these United Colonies are, and of right ought to be, free and independent States, and as such they have, and of right ought to have, full power to make war, conclude peace, establish commerce, and to do all other acts and things which other States may rightfully do." You will see, in a few days, a Declaration setting forth the causes which have impelled us to this mighty revolution, and the reasons which will justify it in the sight of God and man. A plan of confederation will be taken up in a few days.

When I look back to the year 1761, and recollect the argument concerning writs of assistance in the superior court, which I have hitherto considered as the commencement of this controversy between Great Britain and America, and run through the whole period, from that time to this, and recollect the series of political events, the chain of causes and effects, I am surprised at the suddenness as well as greatness of this revolution. Britain has been filled with folly, and America with wisdom; at least, this is my judgment. Time must determine. It is the will of Heaven that the two countries should be sundered forever. It may be the will of Heaven that America should suffer calamities still more wasting, and distresses yet more dreadful. If this is to be the case, it will have this good effect at least. It will inspire us with many virtues which we have not, and correct many errors, follies, and vices which threaten to disturb, dishonor, and destroy us. The furnace of affliction produces refinement in states as well as individuals. And the new governments we are assuming in every part will require a purification from our vices, and an augmentation of our virtues, or they will be no blessings. The people will have unbounded power, and the people are extremely addicted to corruption and venality, as well as the great. But I must submit all my hopes and fears to an overruling Providence, in which, unfashionable as the faith may be, I firmly believe.

JOHN ADAMS.

THE PRIVATE CHARACTER OF WEBSTER.

To appreciate the variety and accuracy of his knowledge, and even the true compass of his mind, you must have had some familiarity with his friendly written correspondence, and you must have conversed with him with some degree of freedom. There, more than in senatorial or forensic debate, gleamed the true riches of his genius, as well as the goodness of his large heart, and the kindness of his noble nature. There, with no longer a great part to discharge, no longer compelled to weigh and measure propositions, to tread the dizzy heights which part the antagonisms of the Constitution, to put aside allusions and illustrations which crowded on his mind in action, but which the dignity of a public appearance had to reject; in the confidence of hospitality (which ever he dispensed as a prince who also was a friend), his memory — one of his most extraordinary faculties, quite in proportion to all the rest — swept free over the readings and labors of more than half a century; and then, allusions, direct and ready quotations, a passing mature criticism, sometimes only a recollection of the mere emotions which a glorious passage or interesting event had once excited, darkening for a moment the face and filling the eye, often an instructive exposition of a current maxim of philosophy or politics, the history of an invention, the recital of some incident casting a new light on some transaction or some institution, — this flow of unstudied conversation, quite as remarkable as any other exhibition of his mind, better than any other, perhaps, at once opened an unexpected glimpse of his various acquirements, and gave you to experience, delightedly, that the "mild sentiments have their eloquence as well as the stormy passions."

There must be added, next, the element of an impressive character, inspiring regard, trust, and admiration, not unmingled with love. It had, I think, intrinsically a charm such as belongs only to a good, noble, and beautiful nature. In its combination with so much fame, so much force of will, and so much intellect, it filled and fascinated the imagination and heart. It was affectionate in childhood and youth, and it was more than ever so in the few last months of his long life.

It is the universal testimony that he gave to his parents in largest measure, honor, love, obedience; that he eagerly appropriated the first means which he could command to relieve the father from the debts contracted to educate his brother and himself; that he selected his first place of professional practice that he might soothe the coming on of his old age; that all through life he neglected no occasion — sometimes when leaning on the arm of a friend, alone, with faltering voice, sometimes in the presence of great assemblies, where the tide of general emotion made it graceful — to express his "affectionate veneration of him who reared and defended the log cabin in which his elder brothers and sisters were born, against savage violence and destruction, cherished all the domestic virtues beneath its roof, and through the fire and blood of some years of revolutionary war, shrank from no danger, no toil, no sacrifice, to serve his country, and to raise his children to a condition better than his own."

Equally beautiful was his love of all his kindred and all his friends. When I hear him accused of selfishness, and a cold, bad nature, I recall him lying sleepless all night, not without tears of boyhood, conferring with Ezekiel how the darling desire of both hearts should be compassed, and he, too, admitted to the precious privileges of education; courageously pleading the cause of both brothers in the morning; prevailing by the wise and discerning affection of the mother; suspending his studies of the law, and registering deeds and teaching school to earn the means, for both, of availing themselves of the opportunity which the parental self-sacrifice had placed within their reach; loving him through life, mourning him when dead, with a love and a sorrow very wonderful, passing the sorrow of woman; I recall the husband, the father of the living and of the early departed, the friend, the counsellor of many years, and my heart grows too full and liquid for the refutation of words.

His affectionate nature, craving ever friendship, as well as the presence of kindred blood, diffused itself through all his private life, gave sincerity to all his hospitalities, kindness to his eye, warmth to the pressure of his hand; made his greatness and genius unbend themselves to the playfulness of childhood, flowed out in graceful memories indulged of the past or the dead, of incidents when life was young and promised to be happy, — gave generous sketches of his rivals, — the high contention now hidden by the handful of earth, — hours passed fifty years ago with great authors, recalled for the vernal emotions which then they made to live and revel in the soul. And from these conversations of friendship, no man — no man, old or young — went away to remember one word of profaneness, one allusion of indelicacy, one impure thought, one unbelieving suggestion, one doubt cast on the reality of virtue, of patriotism, of enthusiasm, of the progress of man, — one doubt cast on righteousness, or temperance, or judgment to come.

Every one of his tastes and recreations announced the same type of character.

His love of agriculture, of sports in the open air, of the outward world in starlight and storms, and sea and boundless wilderness, — partly a result of the influences of the first fourteen years of his life, perpetuated like its other affections and its other lessons of a mother's love, — the Psalms, the Bible, the stories of the wars, — partly the return of an unsophisticated and healthful nature, tiring for a space of the idle business of political life, its distinctions, its artificialities, to employments, to sensations which interest without agitating the universal race alike, as God has framed it, in which one feels himself only a man, fashioned from the earth, set to till it, appointed to return to it, yet made in the image of his Maker, and with a spirit that shall not die, — all displayed a man whom the most various intercourse with the world, the longest career of strife and honors, the consciousness of intellectual supremacy, the coming in of a wide fame, constantly enlarging, left, as he was at first, natural, simple, manly, genial, kind.

RUFUS CHOATE.

THE SABBATH.

I FEEL by experience the eternal obligation, because of the eternal necessity, of the Sabbath. The soul withers without it; it thrives in proportion to the fidelity of its observance. Nay, I can believe the stern rigor of the Puritan Sabbath had a grand effect upon the soul. Fancy a man thrown in upon himself, with no permitted music, no relaxation, nor literature, nor secular conversation — nothing but his Bible, his own soul and God's silence! What hearts of iron this system must have made. How different from our stuffed arm-chair religion and "gospel of comfort!" as if to be made comfortable were the great end of religion. I am persuaded, however, that the Sabbath must rest not on an enactment, but on the necessities of human nature. It is necessary not because it is commanded; but it is commanded because it is necessary. If the Bible says, "Eat the herb of the field," self-sustenance does not become a duty in consequence of the enactment, but the enactment is only a statement of the law of human nature. And so with the Sabbath. . . . As to the enactment, a greater part is indisputably dispensed with. The day, the mode of observance, the manner of computing twenty-four hours from twelve to twelve, or from sunset to sunset. If these be ceremonial, who is to prove that the number one in seven is not ceremonial, too, and that it might not be changed for one in ten? If all this is got rid of, and "no manner of work" is construed to permit hot dinners and fly-driving on the Sabbath, then it is only an arbitrary distinction to call any other part, or even the whole of it, of moral or eternal instead of ceremonial obligation. You cannot base it on a law; but you can show that the law was based on an eternal fitness. There I think it never can be dislodged.

FREDERICK W. ROBERTSON.

JOAN.

It was a retired nook where evergreens were growing, and where the violet fragrance was more powerful than anywhere else, for the rich, moist earth of one bed was blue with them. Joan was standing near these violets, — he saw her as he turned into the walk, — a motionless figure in heavy brown drapery.

She heard him and started from her reverie. In another half-dozen steps he was at her side.

"Don't look as if I had alarmed you," he said. "It seems such a poor beginning to what I have come to say."

Her hand trembled so that one or two of the loose violets she held fell at his feet. She had a cluster of their fragrant bloom fastened in the full knot of her hair. The dropping of the flowers seemed to help her to recover herself. She drew back a little, a shade of pride in her gesture, though the color dyed her cheeks and her eyes were downcast.

"I cannot — I cannot listen," she said.

The slight change which he noted in her speech touched him unutterably. It was not a very great change; she spoke slowly and uncertainly, and the quaint Northern burr still held its own, and here and there a word betrayed her effort.

"No, no," he said, "you will listen. You gave me back my life. You will not make it worthless. If you cannot love me" — his voice shaking — "it would have been less cruel to have left me where you found me, — a dead man, — for whom all pain was over."

He stopped. The woman trembled from head to foot. She raised her eyes from the ground and looked at him, catching her breath.

"Yo' are askin' me to be yore wife!" she said. "Me!"

"I love you," he answered. "You, and no other woman!"

She waited a moment and then turned suddenly away from him, and leaned against the tree under which they were standing, resting her face upon her arm. Her hand clung among the ivy leaves and crushed them. Her old speech came back in the quick, hushed cry she uttered. "I conna turn yo' fro' me," she said. "Oh! I conna!"

"Thank God! Thank God!" he cried.

He would have caught her to his breast, but she held up her hand to restrain him.

"Not yet," she said, "not yet. I conna turn you fro' me, but theer's summat I must ask. Give me th' time to make myself worthy — give me th' time to work an' strive; be patient with me, until th' day comes when I can come to yo' an' know I need not shame yo'. They say I am na slow at learnin' — wait and see how I can work for th' mon — for th' mon I love."

FRANCES HODGSON BURNETT.

A QUESTION OF LOVING.

WHEN Gabriel had gone about two hundred yards along the down, he heard a "hoi-hoi!" uttered behind him, in a piping note of more treble quality than that in which the exclamation usually embodies itself when shouted across a field. He looked round, and saw a girl racing after him, waving a white handkerchief.

Oak stood still—and the runner drew nearer. It was Bathsheba Everdene. Gabriel's color deepened; hers was already deep, not, as it appeared, from emotion, but from running.

"Farmer Oak—I"—she said, pausing for want of breath, pulling up in front of him with a slanted face, and putting her hand to her side.

"I have just called to see you," said Gabriel, pending her further speech.

"Yes—I know that," she said, panting like a robin, her face moist and red with her exertions, like a peony petal, before the sun dries off the dew, "I didn't know you had come to ask to have me, or I should have come in from the garden instantly. I ran after you to say—that my aunt made a mistake in sending you away from courting me"—

Gabriel expanded. "I'm sorry to have made you run so fast, my dear," he said, with a grateful sense of favors to come. "Wait a bit till you've found your breath."

"It was quite a mistake,—aunt's telling you I had a young man already," Bathsheba went on. "I haven't a sweetheart at all—and I never had one, and I thought that, as times go with women, it was such a pity to send you away thinking I had several."

"Really and truly I am glad to hear that!" said Farmer Oak, smiling one of his long special smiles, and blushing with gladness. He held out his hand to take hers, which, when she had eased her side by pressing it there, was prettily extended upon her bosom to still her loud-beating heart. Directly he seized it she put it behind her, so that it slipped through his fingers like an eel.

"I have a nice snug little farm," said Gabriel, with half a degree less assurance than when he had seized her hand.

"Yes; you have."

"A man has advanced me money to begin with, but still, it will soon be paid off, and, though I am only an every-day sort of a man, I have got on a little since I was a boy." Gabriel uttered "a little" in a tone to show her that it was the complacent form of "a great deal." He continued: "When we are married, I am quite sure I can work twice as hard as I do now."

He went forward and stretched out his arm again. Bathsheba had overtaken him at a point beside which stood a low stunted holly-bush, now laden with red berries. Seeing his advance take the form of an attitude threatening a possible enclosure, if not compression, of her person, she edged off round the bush.

"Why, Farmer Oak," she said, over the top, looking at him with rounded eyes, "I never said I was going to marry you."

BATHSHEBA.

"Well — that is a tale!" said Oak, with dismay. "To run after anybody like this, and then say you don't want me!"

"What I meant to tell you was only this," she said eagerly, and yet half-conscious of the absurdity of the position she had made for herself; "that nobody has got me yet as a sweetheart, instead of my having a dozen, as my aunt said; I hate to be thought any one's property in that way, though possibly I shall be had some day. Why, if I'd wanted you I shouldn't run after you like this; 'twould have been the forwardest thing! But there was no harm in hurrying to correct a piece of false news that had been told you."

"Oh, no — no harm at all." But there is such a thing as being too generous in expressing a judgment impulsively, and Oak added, with a more appreciative sense of all the circumstances, "Well, I am not quite certain it was no harm."

"Indeed, I hadn't time to think before starting whether I wanted to marry or not, for you'd have been gone over the hill."

"Come," said Gabriel, freshening again; "think a minute or two. I'll wait awhile, Miss Everdene. Will you marry me? Do, Bathsheba. I love you far more than common!"

"I'll try to think," she observed, rather more timorously; "if I can think out of doors; but my mind spreads away so."

"But you can give a guess."

"Then give me time." Bathsheba looked thoughtfully into the distance, away from the direction in which Gabriel stood.

"I can make you happy," said he to the back of her head, across the bush. "You shall have a piano in a year or two — farmer's wives are getting to have pianos now — and I'll practice up the flute right well to play with you in the evenings."

"Yes; I should like that."

"And have one of those little ten-pound gigs for market — and nice flowers and birds — cocks and hens, I mean, because they are useful," continued Gabriel, feeling balanced between poetry and prose. . . .

She was silent awhile. He regarded the red berries between them over and over again, to such an extent, that holly seemed in his after-life to be a cipher signifying a proposal of marriage. Bathsheba decisively turned to him.

"No; 'tis no use," she said. "I don't want to marry you."

"Try."

"I have tried hard all the time I've been thinking; for a marriage would be very nice in one sense. People would talk about me, and think I had won my battle. And I should feel triumphant, and all that. But a husband" —

"Well?"

"Why, he'd always be there, as you say; whenever I looked up, there he'd be."

"Of course he would — I, that is."

"Well, what I mean is that I shouldn't mind being a bride at a wedding, if I could be one without having a husband. But since a woman can't show off in that way by herself, I sha'n't marry — at least yet."

"That's a terrible wooden story."

At this elegant criticism of her statement, Bathsheba made an addition to her dignity by a slight sweep away from him.

"Upon my heart and soul, I don't know what a maid can say stupider than that," said Oak. "But, dearest," he continued, in a palliative tone, "don't be like it!" Oak sighed deep honest sigh — none the less so in that, being like the sigh of a pine plantation, it was rather noticeable as a disturbance of the atmosphere. "Why won't you have me?" he said appealingly, creeping round the holly to reach her side.

"I cannot," she said, retreating.

"But why?" he persisted, standing still at last, in despair of ever reaching her, and facing over the bush.

"Because I don't love you."

"Yes, but " —

She contracted a yawn to an inoffensive smallness, so that it was hardly ill-mannered at all. "I don't love you," she said.

"But I love you — and, as for myself, I am content to be liked." . . .

"No — no — I cannot. Don't press me any more — don't. I don't love you — so 'twould be ridiculous," she said, with a laugh.

No man likes to see his emotions the sport of a merry-go-round of skittishness.

"Very well," said Oak, firmly, with the bearing of one who was going to give his days and nights to Ecclesiastes forever. "Then I'll ask you no more."

THOMAS HARDY.

MILTON AT THE ORGAN.

REFORM.

Methinks I see in my mind a noble and puissant nation rousing herself like a strong man after sleep, and shaking her invincible locks; methinks I see her as an eagle muing her mighty youth, and kindling her dazzled eyes at the full midday beam; purging and unscaling her long abused sight at the fountain itself of heavenly radiance; while the whole noise of timorous and flocking birds, with those also that love the twilight, flutter about, amazed at what she means, and in their envious gabble would prognosticate a year of sects and schisms.

Error supports custom, custom countenances error; and these two between them would persecute and chase away all truth and solid wisdom out of human life, were it not that God, rather than man, once in many ages calls together the prudent and religious counsels of men, deputed to repress the encroachments, and to work off the inveterate blots and obscurities wrought upon our minds by the subtle insinuating of error and custom; who, with the numerous and vulgar train of their followers, make it their chief design to envy and cry down the industry of free reasoning under the terms of humor and innovation; as if the womb of teeming Truth were to be closed up, if she presumed to bring forth aught that sorts not with their unchewed notions and suppositions.

John Milton.

COUNTRY HOSPITALITY.

THOSE inferior duties of life, which the French call *les petites morales*, or the smaller morals, are with us distinguished by the name of good manners or breeding. This I look upon, in the general notion of it, to be a sort of artificial good sense, adapted to the meanest capacities, and introduced to make mankind easy in their commerce with each other. Low and little understandings, without some rules of this kind, would be perpetually wandering into a thousand indecencies and irregularities in behavior; and in their ordinary conversation, fall into the same boisterous familiarities that one observes among them where intemperance has quite taken away the use of their reason. In other instances, it is odd to consider, that for want of common discretion, the very end of good breeding is wholly perverted; and civility, intended to make us easy, is employed in laying chains and fetters upon us, in debarring us of our wishes, and in crossing our most reasonable desires and inclinations.

This abuse reigns chiefly in the country, as I found to my vexation when I was last there, in a visit I made to a neighbor about two miles from my cousin. As soon as I entered the parlor, they put me into the great chair that stood close by a huge fire, and kept me there by force until I was almost stifled. Then a boy came in a great hurry to pull off my boots, which I in vain opposed, urging that I must return soon after dinner. In the mean time the good lady whispered her eldest daughter, and slipped a key into her hand; the girl returned instantly with a beer-glass half-full of *aqua mirabilis* and sirup of gillyflowers. I took as much as I had a mind for, but madam vowed I should drink it off; for she was sure it would do me good after coming out of the cold air; and I was forced to obey, which absolutely took away my stomach. When dinner came in, I had a mind to sit at a distance from the fire; but they told me it was as much as my life was worth, and sat me with my back just against it. Although my appetite was quite gone, I was resolved to force down as much as I could, and desired the leg of a pullet. "Indeed, Mr. Bickerstaff," says the lady, "you must eat a wing, to oblige me;" and so put a couple upon my plate. I was persecuted at this rate during the whole meal: as often as I called for small beer, the master tipped the wink, and the servant brought me a brimmer of October.

Some time after dinner, I ordered my cousin's man, who came with me, to get ready the horses; but it was resolved I should not stir that night; and when I seemed pretty much bent upon going, they ordered the stable door to be locked, and the children hid my cloak and boots. The next question was, What should I have for supper? I said, I never eat anything at night; but was at last, in my own defense, obliged to name the first thing that came into my head. After three hours, spent chiefly in apologies for my entertainment, insinuating to me, "That this was the worst time of the year for provisions; that they were at a great distance from any market; that they were afraid I should be starved; and that they knew they kept me to my loss;" the lady went, and left me to her husband; for they took special care I should never be alone. As soon as her back was turned, the little misses ran backward and forward every moment, and constantly as they came in, or went out, made a courtesy directly at me, which, in good manners, I was forced to return with a bow, and "your humble servant, pretty miss." Exactly at eight, the mother came up, and discovered, by the redness of her face, that supper was not far off. It was twice as large as the dinner, and my persecution doubled in proportion. I desired at my usual hour to go to my repose, and was conducted to my chamber by the gentleman, his lady, and the whole train of children. They importuned me to drink something before I went to bed; and, upon my refusing, at last left a bottle of stingo, as they call it, for fear I should wake and be thirsty in the night.

I was forced in the morning to rise and dress myself in the dark, because they would not suffer my kinsman's servant to disturb me at the hour I desired to be called. I was now resolved to break through all measures to get away; and, after sitting down to a monstrous breakfast of cold beef, mutton, neat's tongues, venison pasty, and stale beer, took leave of the family. But the gentleman would needs see me part of the way, and carry me a short cut through his own ground, which he told me would save half a mile's riding. This last piece of civility had like to have cost me dear, being once or twice in danger of my neck by leaping over his ditches, and at last forced to alight in the dirt, when my horse, having slipped his bridle, ran away, and took us up more than an hour to recover him again.

JONATHAN SWIFT.

THE AMERICAN INDIAN.

IT is a sad commentary on civilization that it has but degraded where it should have exalted. The Indian has been the white man's foil where he should have proved his friend. Christianity failed to christianize him not because of any inherent weakness in the grandest of religious faiths, but because of the lack of real Christianity itself in the exponents of that faith. Appetite which should have been cultivated into gentle living was used to make him a brute; association which should have uplifted him made him an outcast and a vagrant; statesmanship which should have constituted him the peer of his white brethren has alternately persecuted and petted, domineered over and degraded him. And the race that took from him his land has taken from him also his ambition, his manhood and his life.

The American Indian has reason to be proud of his race. His has been a record which even dead civilizations might well have envied. Evolved from savagery through years of partial progress, he became as bold a warrior as ever Homer sung, as eloquent an orator as Greek or Roman knew. His barbaric virtues could shame the sloth and license of Tiberius' day, his simple manliness could put to blush the servile manners of Justinian's court. His rude manufactures and yet ruder art have, rude as they were, still furnished suggestions upon which modern invention can scarcely improve, and his governmental policy of a league of freemen is that toward which all the world is tending.

His manners and his methods will compare favorably with those of any barbaric people. With no more brutality than the Huns of Attila, no greater ferocity than the sea-wolves of Olaf the Viking, and no deeper strain of vindictiveness than the Goths of Alaric, the American Indian has been eliminated as a factor in a fusing civilization where these bloodier compeers have been accepted as the bases of refined nationalities.

The Indian knew no law but that of simple justice, no dealings other than those of simple honesty, no order more binding than that of simple equality. His mind, hampered by the superstition that always inheres in an out-of-door race, was still no greater slave to the supernatural than is that of the agricultural peasantry of any land, and the spell of the scalp-lock, or the magic of the "fetich" was not so very far removed from the slavish manipulation of the myriad gods of Rome, the mystic "unicorn-horn" of the bloody Torqumada, the dread of the "evil eye" among the peasantry of England, or the fancied "overlooking" which led to such a tragic farce upon the slope of Witches' Hill.

All this may appear to practical folk as an heroic and over-drawn estimate of a very ordinary and limited intelligence. But it is an estimate that is borne out by facts, and is one, moreover, that the justice of the conquerors should allow to the conquered. The shame of it all lies in the knowledge that a civilization which might have moulded has only marred, and that a promising barbarism that in time might have developed into a completed native civilization has been smothered and contemptuously blotted out by the followers of a Master whose greatest precept was: Love one another. But it is never too late to be just.

ELBRIDGE S. BROOKS.

BURR AND BLENNERHASSET.

LET us put the case between Burr and Blennerhasset. Let us compare the two men and settle this question of precedence between them. It may save a good deal of troublesome ceremony hereafter.

Who Aaron Burr is, we have seen in part already. I will add that, beginning his operations in New York, he associates with him men whose wealth is to supply the necessary funds. Possessed of the main-spring, his personal labor contrives all the machinery. Pervading the continent from New York to New Orleans, he draws into his plan, by every allurement which he can contrive, men of all ranks and descriptions. To youthful ardor he presents danger and glory; to ambition, rank and titles and honors; to avarice, the mines of Mexico. To each person whom he addresses he presents the object adapted to his taste. His recruiting-officers are appointed. Men are engaged throughout the continent. Civil life is, indeed, quiet upon its surface, but in its bosom this man has contrived to deposit the materials which, with the slightest touch of his match, produce an explosion to shake the continent. All this his restless ambition has contrived; and, in the autumn of 1806, he goes forth for the last time to apply this match. On this occasion he meets with Blennerhasset.

Who is Blennerhasset? A native of Ireland, a man of letters, who fled from the storms of his own country to find quiet in ours. His history shows that war is not the natural element of his mind; if it had been, he never would have exchanged Ireland for America. So far is an army from furnishing the society natural and proper to Mr. Blennerhasset's character, that, on his arrival in America, he retired even from the population of the Atlantic States, and sought quiet and solitude in the bosom of our Western forests. But he carried with him taste and science and wealth; and, lo! the desert smiled. Possessing himself of a beautiful island in the Ohio, he rears upon it a palace and decorates it with every romantic embellishment of fancy. A shrubbery that Shenstone might have envied blooms around him. Music that might have charmed Calypso and her nymphs is his. An extensive library spreads its treasures before him. A philosophical apparatus offers to him all the secrets and mysteries of nature. Peace, tranquillity, and innocence shed their mingled delights around him. And, to crown the enchantment of the scene, a wife, who is said to be lovely even beyond her sex, and graced with every accomplishment that can render it irresistible, had blessed him with her love and made him the father of several children. The evidence would convince you that this is but a faint picture of the real life. In the midst of all this peace, this innocent simplicity and this tranquillity, this feast of the mind, this pure banquet of the heart, the destroyer comes; he comes to change this paradise into a hell. Yet the flowers do not wither at his approach. No monitory shuddering through the bosom of their unfortunate possessor warns him of the ruin that is coming upon him. A stranger presents himself. Introduced to their civilities by the high rank which he had lately held in his country, he soon finds his way to their hearts by the dignity and elegance of his demeanor, the light and beauty of his conversation, and the seductive and fascinating power of his address. The conquest was not difficult. Innocence is ever simple and credulous. Conscious of no design itself, it suspects none in others. It wears no guard before its breast. Every door and portal and avenue of the heart is thrown open, and all who choose it enter. Such was the state of Eden when the serpent entered its bowers. The prisoner, in a more engaging form, winding himself into the open and unpractised heart of the unfortunate Blennerhasset, found but little difficulty in changing the native character of that heart and the objects of its affection. By degrees he infuses into it the poison of his own ambition. He breathes into it the fire of his own courage, — a daring and desperate thirst for glory, an ardor panting for great enterprises, for all the storm and bustle and hurricane of life. In a short time the whole man is changed, and every object of his former delight is relinquished. No more he enjoys the tranquil scene: it has become flat and insipid to his taste. His books are abandoned. His retort and crucible are thrown aside. His shrubbery blooms and breathes its fragrance upon the air in vain: he likes it not. His ear no longer drinks the rich melody of music: it longs for the trumpet's clangor and the cannon's roar. Even the prattle of his babes, once so sweet, no longer affects him; and the angel-smile of his wife, which hitherto touched his bosom with ecstasy so unspeakable, is now unseen and unfelt. Greater objects have taken possession of his soul. His imagination has been dazzled by visions of diadems, of stars and garters and titles of nobility. He has been taught to burn with restless emulation at the names of great heroes and conquerors. His enchanted island is destined soon to relapse into a wilderness; and in a few months we find the beautiful and tender partner of his bosom, whom he lately "permitted not the winds of " summer "to visit too roughly," we find her shivering at midnight on the wintry banks of the Ohio, and mingling her tears with the torrents that froze as they fell. Yet this unfortunate man, thus deluded from his interest and his happiness, thus seduced from the paths of innocence and peace, thus confounded in the toils that were deliberately spread for him, and overwhelmed by the mastering spirit and genius of another, — this man, thus ruined and undone and made to play a subordinate part in this grand drama of guilt and treason, — this man is to be called the principal offender, while he by whom he was thus plunged in misery is comparatively innocent, — a mere accessory! Is this reason? Is it law? Is it humanity? Sir, neither the human heart nor the human understanding will bear a perversion so monstrous and absurd! so shocking to the soul! so revolting to reason! Let Aaron Burr, then, not shrink from the high destination which he has courted; and, having already ruined Blennerhasset in fortune, character, and happiness forever, let him not attempt to finish the tragedy by thrusting that ill-fated man between himself and punishment.

WILLIAM WIRT.

ISABELLA OF SPAIN.

HER person was of the middle height, and well proportioned. She had a clear, fresh complexion, with light blue eyes and auburn hair, — a style of beauty exceedingly rare in Spain. Her features were regular, and universally allowed to be uncommonly handsome. The illusion which attaches to rank, more especially when united with engaging manners, might lead us to suspect some exaggeration in the encomiums so liberally lavished on her. But they would seem to be in a great measure justified by the portraits that remain of her, which combine a faultless symmetry of features with singular sweetness and intelligence of expression.

Her manners were most gracious and pleasing. They were marked by natural dignity and modest reserve, tempered by an affability which flowed from the kindliness of her disposition. She was the last person to be approached with undue familiarity; yet the respect which she imposed was mingled with the strongest feelings of devotion and love. She showed great tact in accommodating herself to the peculiar situation and character of those around her. She appeared in arms at the head of her troops, and shrunk from none of the hardships of war. During the reforms introduced into the religious houses, she visited the nunneries in person, taking her needlework with her, and passing the day in the society of the inmates. When traveling in Galicia, she attired herself in the costume of the country, borrowing for that purpose the jewels and other ornaments of the ladies there, and returning them with liberal additions. By this condescending and captivating deportment, as well as by her higher qualities, she gained an ascendency over her turbulent subjects which no king of Spain could ever boast.

She spoke the Castilian with much elegance and correctness. She had an easy fluency of discourse, which, though generally of a serious complexion, was occasionally seasoned with agreeable sallies, some of which have passed into proverbs. She was temperate even to abstemiousness in her diet, seldom or never tasting wine, and so frugal in her table, that the daily expenses of herself and family did not exceed the moderate sum of forty ducats. She was equally simple and economical in her apparel. On all public occasions, indeed, she displayed a royal magnificence; but she had no relish for it in private; and she freely gave away her clothes and jewels as presents to her friends. Naturally of a sedate, though cheerful temper, she had little taste for the frivolous amusements which make up so much of a court life; and, if she encouraged the presence of minstrels and musicians in her palace, it was to wean her young nobility from the coarser and less intellectual pleasures to which they were addicted.

Among her moral qualities, the most conspicuous, perhaps, was her magnanimity. She betrayed nothing little or selfish in thought or action. Her schemes were vast, and executed in the same noble spirit in which they were conceived. She never employed doubtful agents or sinister measures, but the most direct and open policy. She scorned to avail herself of advantages offered by the perfidy of others. Where she had once given her confidence, she gave her hearty and steady support; and she was scrupulous to redeem any pledge she had made to those who ventured in her cause, however unpopular. She sustained Ximenes in all his obnoxious but salutary reforms. She seconded Columbus in the prosecution of his arduous enterprise, and shielded him from the calumny of his enemies. She did the same good service to her favorite, Gonsalvo de Cordova; and the day of her death was felt, and, as it proved, truly felt, by both, as the last of their good fortune. Artifice and duplicity were so abhorent to her character, and so averse from her domestic policy, that, when they appear in the foreign relations of Spain, it is certainly not imputable to her. She was incapable of harboring any petty distrust or latent malice; and, although stern in the execution and exaction of public justice, she made the most generous allowance, and even sometimes advances, to those who had personally injured her.

But the principle which gave a peculiar coloring to every feature of Isabella's mind was piety. It shone forth from the very depths of her soul with a heavenly radiance, which illuminated her whole character. Fortunately, her earliest years had been passed in the rugged school of adversity, under the eye of a mother who implanted in her serious mind such strong principles of religion as nothing in afterlife had power to shake. At an early age, in the flower of youth and beauty, she was introduced to her brother's court; but its blandishments, so dazzling to a young imagination, had no power over hers, for she was surrounded by a moral atmosphere of purity,—

"Driving far off each thing of sin and guilt."

Such was the decorum of her manners that, though encompassed by false friends and open enemies, not the slightest reproach was breathed on her fair name in this corrupt and calumnious court.

WILLIAM HICKLING PRESCOTT.

THE LEGEND OF THE DATE-TREE.

IN a lot situated at the corner of Orleans and Dauphine streets, in the city of New Orleans, there is a tree which nobody looks at without curiosity and without wondering how it came there. For a long time it was the only one of its kind known in the state, and from its isolated position, it has always been cursed with sterility. It reminds one of the warm climes of Africa or Asia, and wears the aspect of a stranger of distinction driven from his native country. Indeed, with its sharp and thin foliage, sighing mournfully under the blast of one of our November northern winds, it looks as sorrowful as an exile. Its enormous trunk is nothing but an agglomeration of knots and bumps, which each passing year seems to have deposited there as a mark of age, and as a protection against the blows of time and of the world. Inquire for its origin, and every one will tell you that it has stood there from time immemorial. A sort of vague but impressive mystery is attached to it, and it is as superstitiously respected as one of the old oaks of Dodona. Bold would be the axe that should strike the first blow at that foreign patriarch ; and if it were prostrated to the ground by a profane hand, what native of the city would not mourn over its fall, and brand the act as an unnatural and criminal deed ? So, long live the date-tree of Orleans street — that time-honored descendant of Asiatic ancestors!

In the beginning of 1727, a French vessel of war landed at New Orleans a man of haughty mien, who wore the Turkish dress, and whose whole attendance was a single servant.

He was received by the governor with the highest distinction, and was conducted by him to a small but comfortable house with a pretty garden, then existing at the corner of Orleans and Dauphine streets, and which, from the circumstance of its being so distant from other dwellings, might have been called a rural retreat, although situated in the limits of the city. There, the stranger, who was understood to be a prisoner of state, lived in the greatest seclusion ; and although neither he nor his attendant could be guilty of indiscretion, because none understood their language, and although Governor Périer severely rebuked the slightest inquiry, yet it seemed to be the settled conviction in Louisiana, that the mysterious stranger was a brother of the Sultan, or some great personage of the Ottoman empire, who had fled from the anger of the viceregent of Mohammed, and who had taken refuge in France. The Sultan had peremptorily demanded the fugitive, and the French government, thinking it derogatory to its dignity to comply with that request, but at the same time not wishing to expose its friendly relations with the Moslem monarch, and perhaps desiring, for political purposes, to keep in hostage the important guest it had in its hands, had recourse to the expedient of answering that he had fled to Louisiana, which was so distant a country that it might be looked upon as the grave, where, as it was suggested, the fugitive might be suffered to wait in peace for actual death, without danger or offence to the Sultan. Whether this story be true or not is now a matter of so little consequence, that it would not repay the trouble of a strict historical investigation.

The year 1727 was drawing to its close, when on a dark, stormy night, the howling and barking of the numerous dogs in the streets of New Orleans were observed to be fiercer than usual, and some of that class of individuals who pretend to know everything, declared that, by the vivid flashes of the lightning, they had seen, swiftly and stealthily gliding toward the residence of the unknown, a body of men who wore the scowling appearance of malefactors and ministers of blood. There afterward came also a report that a piratical-looking Turkish vessel had been hovering a few days previous in the bay of Barataria. Be it as it may, on the next morning the house of the stranger was deserted. There were no traces of mortal struggle to be seen; but in the garden, the earth had been dug, and there was the unmistakable indication of a recent grave. Soon, however, all doubts were removed by the finding of an inscription in Arabic characters, engraved on a marble tablet, which was subsequently sent to France. It ran thus: "The justice of heaven is satisfied, and the date-tree shall grow on the traitor's tomb. The sublime Emperor of the faithful, the supporter of the faith, the omnipotent master and Sultan of the world, has redeemed his vow. God is great, and Mohammed is his prophet. Allah!" Some time after this event, a foreign-looking tree was seen to peep out of the spot where a corpse must have been deposited in that stormy night, when the rage of the elements yielded to the pitiless fury of man, and it thus explained in some degree this part of the inscription, "the date-tree shall grow on the traitor's grave."

Who was he, or what had he done, who had provoked such relentless and far-seeking revenge? Ask Nemesis, or — at that hour when evil spirits are allowed to roam over the earth, and magical invocations are made — go, and interrogate the tree of the dead.

CHARLES ÉTIENNE ARTHUR GAYARRÉ.

A SCENE IN THE FORECASTLE.

I HAD scarcely been aboard of the ship twenty-four hours, when a circumstance occurred, which, although noways picturesque, is so significant of the state of affairs, that I cannot forbear relating it.

In the first place, however, it must be known, that among the crew was a man so excessively ugly, that he went by the ironical appellation of "Beauty." He was the ship's carpenter; and for that reason was sometimes known by his nautical cognomen of "Chips." There was no absolute deformity about the man; he was symmetrically ugly. But ill-favored as he was in person, Beauty was none the less ugly in temper; but no one could blame him; his countenance had soured his heart. Now Jermin and Beauty were always at sword's points. The truth was, the latter was the only man in the ship whom the mate had never decidedly got the better of; and hence the grudge he bore him. As for Beauty, he prided himself upon talking up to the mate, as we shall soon see.

Toward evening there was something to be done on deck, and the carpenter who belonged to the watch was missing. "Where's that skulk, Chips?" shouted Jermin down the forecastle scuttle.

"Taking his ease, d'ye see, down here on a chest, if you want to know," replied

UNDER SAIL.

that worthy himself, quietly withdrawing his pipe from his mouth. This insolence flung the fiery little mate into a mighty rage; but Beauty said nothing, puffing away with all the tranquillity imaginable. Here, it must be remembered that, never mind what may be the provocation, no prudent officer ever dreams of entering a ship's forecastle on a hostile visit. If he wants to see anybody who happens to be there, and refuses to come up, why he must wait patiently until the sailor is willing. The reason is this. The place is very dark; and nothing is easier than to knock one descending on the head, before he knows where he is, and a very long while before he ever finds out who did it.

Nobody knew this better than Jermin, and so he contented himself with looking down the scuttle and storming. At last Beauty made some cool observation which set him half wild.

A SCENE IN THE FORECASTLE.

"Tumble on deck," he then bellowed — "come, up with you, or I'll jump down and make you." The carpenter begged him to go about it at once.

No sooner said than done: prudence forgotten, Jermin was there; and by a sort of instinct, had his man by the throat before he could well see him. One of the men now made a rush at him, but the rest dragged him off, protesting that they should have fair play.

"Now, come on deck," shouted the mate, struggling like a good fellow to hold the carpenter fast.

"Take me there," was the dogged answer, and Beauty wriggled about in the nervous grasp of the other like a couple of yards of boa-constrictor.

His assailant now undertook to make him up into a compact bundle, the more easily to transport him. While thus occupied, Beauty got his arms loose, and threw him over backward. But Jermin quickly recovered himself, when for a time they had it every way, dragging each other about, bumping their heads against the projecting beams, and returning each other's blows the first favorable opportunity that offered.

Unfortunately, Jermin at last slipped and fell; his foe seating himself on his chest and keeping him down. Now this was one of those situations in which the voice of counsel, or reproof, comes with peculiar unction. Nor did Beauty let the opportunity slip. But the mate said nothing in reply, only foaming at the mouth and struggling to rise.

Just then a thin tremor of a voice was heard from above. It was the captain, who, happening to ascend to the quarter-deck at the commencement of the scuffle, would gladly have returned to the cabin, but was prevented by the fear of ridicule. As the din increased, and it became evident that his officer was in serious trouble, he thought it would never do to stand leaning over the bulwarks, so he made his appearance on the forecastle, resolved, as his best policy, to treat the matter lightly.

"Why, why," he began, speaking pettishly, and very fast, "what's all this about? Mr. Jermin, Mr. Jermin — carpenter, carpenter; what are you doing down there? Come on deck; come on deck."

Whereupon Doctor Long Ghost cries out in a squeak, "Ah! Miss Guy, is that you? Now, my dear, go right home, or you'll get hurt."

"Pooh, pooh! you, sir, whoever you are, I was not speaking to you; none of your nonsense. Mr. Jermin, I was talking to you: have the kindness to come on deck, sir; I want to see you."

"And how, in the devil's name, am I to get there?" cried the mate furiously. "Jump down here, Captain Guy, and show yourself a man. Let me up, you Chips! unhand me, I say! Oh! I'll pay you for this, some day! Come on, Captain Guy!"

At this appeal, the poor man was seized with a perfect spasm of fidgets. "Pooh, pooh, carpenter; have done with your nonsense! Let him up, sir; let him up! Do you hear? Let Mr. Jermin come on deck!"

"Go along with you, Paper Jack," replied Beauty; "this quarrel's between the mate and me; so go aft, where you belong!"

As the captain once more dipped his head down the scuttle to make answer, from an unseen hand he received, full in the face, the contents of a tin can of soaked biscuit and tea-leaves. The doctor was not far off just then. Without waiting for anything more, the discomfited gentleman, with both hands to his streaming face, retreated to the quarter-deck.

A few moments more, and Jermin, forced to a compromise, followed after, in his torn frock and scarred face, looking for all the world as if he had just disentangled himself from some intricate piece of machinery. For about half an hour both remained in the cabin, where the mate's rough tones were heard high above the low, smooth voice of the captain.

Of all his conflicts with the men, this was the first in which Jermin had been worsted; and he was proportionably enraged. Upon going below — as the steward afterward told us — he bluntly informed Guy, that, for the future, he might look out for his ship himself; for his part, he was done with her, if that was the way he allowed his officers to be treated. After many high words, the captain finally assured him that the first fitting opportunity the carpenter should be cordially flogged; though, as matters stood, the experiment would be a hazardous one. Upon this Jermin reluctantly consented to drop the matter for the present; and he soon drowned all thoughts of it in a can of flip, which Guy had previously instructed the steward to prepare, as a sop to allay his wrath.

HERMAN MELVILLE.

CAPTAIN CUTTLE'S ISLAND.

It happened by evil chance to be one of Mrs. MacStinger's great cleaning days. On these occasions Mrs. MacStinger was knocked up by the policeman at a quarter before three in the morning, and rarely succumbed before twelve o'clock next night. The chief object of this institution appeared to be, that Mrs. MacStinger should move all the furniture into the back-garden at early dawn, walk about the house in pattens all day, and move the furniture back again after dark. These ceremonies greatly fluttered those doves, the young MacStingers, who were not only unable at such times to find any resting-place for the soles of their feet, but generally came in for a good deal of pecking from the maternal bird during the progress of the solemnities.

At the moment when Florence and Susan Nipper presented themselves at Mrs. MacStinger's door, that worthy but redoubtable female was in the act of conveying Alexander MacStinger, aged two years and three months, along the passage for forcible deposition in a sitting posture on the street pavement; Alexander being black in the face with holding his breath after punishment, and a cool paving-stone being usually found to act as a powerful restorative in such cases.

The feelings of Mrs. MacStinger, as a woman and a mother, were outraged by the look of pity for Alexander which she observed in Florence's face. Therefore, Mrs. MacStinger asserting those finest emotions of our nature, in preference to weakly gratifying her curiosity, shook and buffeted Alexander, both before and during the application of the paving-stone, and took no further notice of the strangers.

"I beg your pardon, ma'am," said Florence when the child had found his breath again, and was using it. "Is this Captain Cuttle's house?"

"No," said Mrs. MacStinger.

"Not Number Nine?" asked Florence, hesitating.

"Who said it wasn't Number Nine?" said Mrs. MacStinger.

Susan Nipper instantly struck in, and begged to inquire what Mrs. MacStinger meant by that, and if she knew whom she was talking to.

Mrs. MacStinger, in retort, looked at her all over. "What do you want with Captain Cuttle, I should wish to know?" said Mrs. MacStinger.

"Should you? Then I'm sorry that you won't be satisfied," returned Miss Nipper.

"Hush, Susan! If you please!" said Florence. "Perhaps you can have the goodness to tell us where Captain Cuttle lives, ma'am, as he don't live here."

"Who says he don't live here?" retorted the implacable MacStinger. "I said it wasn't Cap'en Cuttle's house — and it a'nt his house — and forbid it that it ever should be his house — for Cap'en Cuttle don't know how to keep a house — and don't deserve to have a house — it's my house — and when I let the upper floor to Cap'en Cuttle, oh, I do a thankless thing, and cast pearls before swine!"

Mrs. MacStinger pitched her voice for the upper windows in offering these remarks, and cracked off each clause sharply by itself, as if from a rifle possessing an infinity of barrels. After the last shot, the captain's voice was heard to say, in feeble remonstrance from his own room, "Steady below!"

"Since you want Cap'en Cuttle, there he is!" said Mrs. MacStinger, with an angry motion of her hand. On Florence making bold to enter without any more parley, and on Susan following, Mrs. MacStinger recommenced her pedestrian exercise in pattens, and Alexander MacStinger (still on the paving-stone), who had stopped in his crying to attend to the conversation, began to wail again, entertaining himself during that dismal performance, which was quite mechanical, with a general survey of the prospect, terminating in the hackney coach.

The captain in his own apartment was sitting with his hands in his pockets, and his legs drawn up under his chair, on a very small, desolate island, lying about midway in an ocean of soap-and-water. The captain's windows had been cleaned, the walls had been cleaned, the stove had been cleaned, and everything, the stove excepted, was wet, and shining with soft soap and sand: the smell of which dry-saltery impregnated the air. In the midst of the dreary scene, the captain, cast away upon his island, looked round on the waste of waters with a rueful countenance, and seemed waiting for some friendly bark to come that way, and take him off.

But when the captain, directing his forlorn visage towards the door, saw Florence appear with her maid, no words can describe his astonishment. Mrs. MacStinger's eloquence having rendered all other sounds but imperfectly distinguishable, he had looked for no rarer visitor than the potboy or the milkman; wherefore, when Florence appeared, and, coming to the confines of the island, put her hand in his, the captain stood up aghast, as if he supposed her, for the moment, to be some young member of the Flying Dutchman's family.

Instantly recovering his self-possession, however, the captain's first care was to place her on dry land, which he happily accomplished with one motion of his arm. Issuing forth, then, upon the main, Captain Cuttle took Miss Nipper round the waist, and bore her to the island also. Captain Cuttle, then, with great respect and admiration, raised the hand of Florence to his lips, and standing off a little (for the island was not large enough for three), beamed on her from the soap-and-water like a new description of Triton.

"You are amazed to see us, I am sure," said Florence with a smile.

The inexpressibly gratified captain kissed his hook in reply and growled, as if a choice and delicate compliment were included in the words, "Stand by! Stand by!"

"But I couldn't rest," said Florence, "without coming to ask you what you think about dear Walter — who is my brother now — and whether there is anything to fear, and whether you will not go and console his poor uncle every day, until we have some intelligence of him?"

At these words Captain Cuttle, as by an involuntary gesture, clapped his hand to his head, on which the hard glazed hat was not, and looked discomfited.

"Have you any fears for Walter's safety?" inquired Florence, from whose face the captain (so enraptured he was with it) could not take his eyes: while she, in her turn, looked earnestly at him, to be assured of the sincerity of his reply.

"No, Heart's Delight," said Captain Cuttle, "I am not afeard. Wal'r is a lad as'll go through a good deal o' hard weather. Wal'r is a lad as'll bring as much success to that 'ere brig as a lad is capable on. Wal'r," said the captain, his eyes glistening with the praise of his young friend, and his hook raised to announce a beautiful quotation, "is what you may call a out'ard and visible sign of a in'ard and spirited grasp, and when found make a note of."

Florence, who did not quite understand this, though the captain evidently thought it full of meaning, and highly satisfactory, mildly looked to him for something more.

"I am not afeard, my Heart's Delight," resumed the captain. "There's been most uncommon bad weather in them latitudes, there's no denyin', and they have drove and drove, and been beat off, maybe t'other side the world. But the ship's a good ship, and the lad's a good lad; and it ain't easy, thank the Lord," the captain made a little bow, "to break up hearts of oak, whether they're in brigs or buzzums. Here we have 'em both ways, which is bringing it up with a round turn, and so I ain't a bit afeard as yet."

"As yet?" repeated Florence.

"Not a bit," returned the captain, kissing his iron hand; "and afore I begin to be, my Heart's Delight, Wal'r will have wrote home from the island, or from some port or another, and made all taut and ship-shape. And with regard to old Sol Gills," here the captain became solemn, "who I'll stand by, and not desert until death doe us part, and when the stormy winds do blow, do blow, do blow — overhaul the catechism," said the captain parenthetically, "and there you'll find them expressions — if it would console Sol Gills to have the opinion of a seafaring man as has got a mind equal to any undertaking that he puts it alongside of, and as was all but smashed in his 'prenticeship, and of which the name is Bunsby, that 'ere man shall give him such an opinion in his own parlour as'll stun him. Ah!"

"THE STORMY WINDS DO BLOW."

said Captain Cuttle vauntingly, "as much as if he'd gone and knocked his head again a door!"

"Let us take this gentleman to see him, and let us hear what he says," cried Florence. "Will you go with us now? We have a coach here."

Again the captain clapped his hand to his head, on which the hard glazed hat was not, and looked discomfited. But at this instant a most remarkable phenomenon occurred. The door opening without any note of preparation, and apparently of itself, the hard glazed hat in question skimmed into the room like a bird, and alighted heavily at the captain's feet. The door then shut as violently as it had opened, and nothing ensued in explanation of the prodigy.

Captain Cuttle picked up his hat, and, having turned it over with a look of interest and welcome, began to polish it on his sleeve. While doing so, the captain eyed his visitors intently, and said in a low voice:

"You see I should have bore down on Sol Gills yesterday and this morning, but she — she took it away and kept it. That's the long and short of the subject."

"Who did, for goodness' sake?" asked Susan Nipper.

"The lady of the house, my dear," returned the captain in a gruff whisper, and making signals of secrecy. "We had some words about the swabbing of these here planks, and she — in short," said the captain, eying the door, and relieving himself, with a long breath, "she stopped my liberty."

"Oh! I wish she had me to deal with!" said Susan, reddening with the energy of the wish. "I'd stop her!"

"Would you, do you think, my dear?" rejoined the captain, shaking his head doubtfully, but regarding the desperate courage of the fair aspirant with obvious admiration. "I don't know. It's difficult navigation. She's very hard to carry on with, my dear. You never can tell how she'll head, you see. She's full one minute, and round upon you next. And when she is a Tartar," said the captain, with the perspiration breaking out on his forehead —— There was nothing but a whistle emphatic enough for the conclusion of the sentence, so the captain whistled tremulously. After which he again shook his head, and recurring to his admiration of Miss Nipper's devoted bravery, timidly repeated, "Would you, do you think, my dear?"

Susan only replied with a bridling smile, but that was so very full of defiance, that there is no knowing how long Captain Cuttle might have stood entranced in its contemplation, if Florence in her anxiety had not again proposed their immediately resorting to the oracular Bunsby. Thus reminded of his duty, Captain Cuttle put on the glazed hat firmly, took up another knobby stick, with which he had supplied the place of that one given to Walter, and offering his arm to Florence, prepared to cut his way through the enemy.

It turned out, however, that Mrs. MacStinger had already changed her course and that she headed, as the captain had remarked she often did, in quite a new direction. For, when they got downstairs, they found that exemplary woman beating the mats on the door-steps, with Alexander, still upon the paving-stone, dimly looming through a fog of dust; and so absorbed was Mrs. MacStinger in her household occupation, that when Captain Cuttle and his visitors passed, she beat the harder, and neither by word nor gesture showed any consciousness of their vicinity. The captain was so well pleased with this easy escape — although the effect of the door-mats on him was like a copious administration of snuff, and made him sneeze until the tears ran down his face — that he could hardly believe his good fortune; but more than once, between the door and the hackney coach, looked over his shoulder, with an obvious apprehension of Mrs. MacStinger's giving chase yet.

CHARLES DICKENS.

BARBERRY ISLAND.

Barberry Island is the Island of Calm Delights. That is, you feel sure of it if you land from the quiet little cove on the western side, whence the grass-grown main street of the village takes up its gentle way.

The houses on each side of the road lack the staring white and green paint and smart and thrifty air that characterize New England houses generally; they were evidently smart and shining once, but that was before Barberry Island went to sleep. Now and then there is a garden with old-fashioned flowers in it; but the drowsy poppies and the melancholy mourning brides look much more at home than the sturdy London pride and the pert little bachelors' buttons, which are almost strangled by weeds. The church is as guiltless of paint as the dwelling-houses, and looks as it were tottering to its fall. The carpet is threadbare and moth-eaten, the walls and roof stained and cracked, and the sounding board over the pulpit must arouse the most dismal apprehensions in the mind of the minister who stands under it. On some pews, and on the Bibles and hymn books in their racks, dust has been allowed to gather thickly; whether this is from a melancholy sense of the fitness of things or a desire to economize labor, we can only guess.

A fence suddenly arrests our footsteps. There is a gate, fastened only by a feeble and frayed rope, but there is little temptation to venture beyond it; marshy land, beset with bogs and pitfalls, lies on the other side. The grassy road has come to an end — because, as Deacon Manley informed us, "there didn't seem to be no pertickler reason for its goin' no further."

It is just beyond the burying ground that the road comes to an end, as if, having brought its travelers to that peaceful bourne, its mission was accomplished.

The cemetery looks neglected. Evidently Luke Hadlock was not the only "corpse" whose relatives had "died out." A tangle of grass and weeds and running vines links the graves together; the paths are "past finding out." There are many more stones than graves.

In an obscure corner we come suddenly upon a new grave. It causes a shock of surprise, for the ancient dates on the stones have beguiled us into the fancy that nobody dies on Barberry Island nowadays. And on the Island of Calm Delights, so far from the jars and turmoils of the world, there wouldn't, as Deacon Manley would say, "seem to be no pertikler reason" why people should not live on forever. The headstone has been very recently set, but the date upon it is of a year ago. "Jonas Battles. Aged 43. Drowned off Dead Man's Point, Aug. 13, 187—." Probably a fisherman whose boat was overtaken by a sudden squall; such accidents are common enough along this rocky coast. But a sudden recollection strikes us that we have heard of a man who saved a young girl from drowning here last summer, and lost his own life.

Only half-remembering the details, and feeling an interest in Jonas Battles and his fate, we go over to Deacon Manley's store to make inquiries. The Deacon is a living chronicle of all the events which have transpired on Barberry Island for the last fifty years.

His stock in trade is small in extent, but great in variety. Every article has its price plainly marked upon it, and in a conspicuous position is this notice: "Customers will please make change for themselves in the drawer." The door stands wide open, and there is neither proprietor nor clerk about the premises. The drawer, filled with money, is unlocked. Apparently the millennium has begun on Barberry Island.

SOPHIE SWETT.

THE TOWN PUMP.

NOON, by the north clock! Noon, by the east! High noon, too, by these hot sunbeams, which fall, scarcely aslope, upon my head, and almost make the water bubble and smoke in the trough under my nose. Truly, we public characters have a tough time of it! And, among all the town officers, chosen at March meeting, where is he that sustains, for a single year, the burden of such manifold duties as are imposed, in perpetuity, upon the Town Pump? The title of "town treasurer" is rightfully mine, as guardian of the best treasure that the town has. The overseers of the poor ought to make me their chairman, since I provide bountifully for the pauper, without expense to him that pays taxes. I am at the head of the fire-department, and one of the physicians to the board of health. As a keeper of the peace, all water-drinkers will confess me equal to the constable. I perform some of the duties of the town clerk, by promulgating public notices when they are posted on my front. To speak within bounds, I am the chief person of the municipality, and exhibit, moreover, an admirable pattern to my brother officers, by the cool, steady, upright, downright, and impartial discharge of my business, and the constancy with which I stand to my post. Summer or winter, nobody seeks me in vain; for, all day long, I am seen at the busiest corner, just above the market, stretching out my arms to rich and poor alike; and at night, I hold a lantern over my head, both to show where I am, and keep people out of the gutters.

At this sultry noontide, I am cupbearer to the parched populace, for whose benefit an iron goblet is chained to my waist. Like a dramseller on the mall at muster-day, I cry aloud to all and sundry, in my plainest accents, and at the very tiptop of my voice. Here it is, gentlemen! Here is the good liquor! Walk up, walk up, gentlemen, walk up, walk up! Here is the superior stuff! Here is the unadulterated ale of father Adam, — better than Cognac, Hollands, Jamaica, strong beer, or wine of any price; here it is by the hogshead or the single glass, and not a cent to pay! Walk up, gentlemen, walk up, and help yourselves!

It were a pity if all this outcry should draw no customers. Here they come. A hot day, gentlemen! Quaff, and away again, so as to keep yourselves in a nice cool sweat. You, my friend, will need another cupful, to wash the dust out of your throat, if it be as thick there as it is on your cow-hide shoes. I see that you have trudged half a score of miles to-day, and, like a wise man, have passed by the taverns, and stopped at the running brooks and well-curbs. Otherwise, betwixt heat without and fire within, you would have been burnt to a cinder, or melted down to nothing at all, in the fashion of a jelly-fish. Drink, and make room for that other fellow, who seeks my aid to quench the fiery fever of last night's potations, which he drained from no cup of mine. Welcome, most rubicund sir! You

NATHANIEL HAWTHORNE

and I have been great strangers, hitherto; nor, to confess the truth, will my nose be anxious for a closer intimacy, till the fumes of your breath be a little less potent. Mercy on you, man! the water absolutely hisses down your red-hot gullet, and is converted quite to steam, in the miniature tophet which you mistake for a stomach. Fill again, and tell me, on the word of an honest toper, did you ever, in cellar, tavern, or any kind of a dram-shop, spend the price of your children's food for a swig half so delicious? Now, for the first time these ten years, you know the flavor of cold water. Good-by; and, whenever you are thirsty, remember that I keep a constant supply, at the old stand. Who next? Oh, my little friend, you are let loose from school, and come hither to scrub your blooming face, and drown the memory of certain taps of the ferrule, and other school-

boy troubles, in a draught from the Town Pump. Take it, pure as the current of your young life. Take it, and may your heart and tongue never be scorched with a fiercer thirst than now! There, my dear child, put down the cup, and yield your place to this elderly gentleman, who treads so tenderly over the paving-stones, that I suspect he is afraid of breaking them. What! he limps by, without so much as thanking me, as if my hospitable offers were meant only for people who have no wine-cellars. Well, well, sir, — no harm done, I hope! Go draw the cork, tip the decanter; but, when your great toe shall set you a-roaring, it will be no affair of mine. If gentlemen love the pleasant titillation of the gout, it is all one to the Town Pump. This thirsty dog, with his red tongue lolling out, does not scorn my hospitality, but stands on his hind legs and laps eagerly out of the trough. See how lightly he capers away again! Jowler, did your worship ever have the gout?

Your pardon, good people! I must interrupt my stream of eloquence, and spout forth a stream of water, to replenish the trough for this teamster and his two yoke of oxen, who have come from Topsfield, or somewhere along that way. No part of my business is pleasanter than the watering of cattle. Look! how rapidly they lower the water-mark on the sides of the trough, till their capacious stomachs are moistened with a gallon or two apiece, and they can afford time to breathe it in, with sighs of calm enjoyment. Now they roll their quiet eyes around the brim of their monstrous drinking-vessel. An ox is your true toper. . . .

Ahem! Dry work, this speechifying; especially to an unpractised orator. I never conceived till now what toil the temperance lecturers undergo for my sake. Hereafter they shall have the business to themselves. Do, some kind Christian, pump a stroke or two, just to wet my whistle. Thank you, sir. My dear hearers, when the world shall have been regenerated by my instrumentality, you will collect your useless vats and liquor-casks into one great pile, and make a bonfire in honor of the Town Pump. And when I shall have decayed, like my predecessors, then, if you revere my memory, let a marble fountain, richly sculptured, take my place upon the spot. Such monuments should be erected everywhere, and inscribed with the names of the distinguished champions of my cause. . . .

One o'clock! Nay, then, if the dinner-bell begins to speak, I may as well hold my peace. — Here comes a pretty young girl of my acquaintance, with a large stone pitcher for me to fill. May she draw a husband, while drawing her water, as Rachel did of old! Hold out your vessel, my dear! There it is, full to the brim; so now run home, peeping at your sweet image in the pitcher as you go; and forget not, in a glass of my own liquor, to drink — "Success to the Town Pump!"

NATHANIEL HAWTHORNE.

ADAM AND DINAH.

ADAM looked at her; it was so sweet to look at her eyes, which had now a self-forgetful questioning in them — for a moment he forgot that he wanted to say anything, or that it was necessary to tell her what he meant.

"Dinah," he said suddenly, taking both her hands between his, "I love you with my whole heart and soul. I love you next to God who made me."

Dinah's lips became pale, like her cheeks, and she trembled violently under the shock of painful joy. Her hands were cold as death between Adam's. She could not draw them away, because he held them fast.

"Don't tell me you can't love me, Dinah. Don't tell me we must part, and pass our lives away from one another."

The tears were trembling in Dinah's eyes, and they fell before she could answer. But she spoke in a quiet, low voice. "Yes, dear Adam, we must submit to another Will. We must part."

"Not if you love me, Dinah, — not if you love me," Adam said passionately. "Tell me — tell me if you can love me better than a brother."

Dinah was too entirely reliant on the Divine Will to attempt to achieve any end by a deceptive concealment. She was recovering now from the first shock of emotion, and she looked at Adam with simple, sincere eyes as she said, —

"Yes, Adam, my heart is drawn strongly toward you; and of my own will, if I had no clear showing to the contrary, I could find my happiness in being near you, and ministering to you continually. I fear I should forget to rejoice and weep with others; nay, I fear I should forget the Divine Presence, and seek no love but yours." . . .

Adam went on presently with his pleading: —

"And you can do almost as much as you do now; I won't ask you to go to church with me of a Sunday. You shall go where you like among the people, and teach 'em; for though I like church best, I don't put my soul above yours, as if my words was better for you t' follow than your own conscience. And you can help the sick just as much, and you'll have more means o' making 'em a bit comfortable; and you'll be among all your own friends as love you, and can help 'em, and be a blessing to 'em, till their dying day. Surely, Dinah, you'd be as near to God as if you were living lonely and away from me."

Dinah made no answer for some time. Adam was still holding her hands, and looking at her with almost trembling anxiety, when she turned her grave, loving eyes on his, and said in rather a sad voice, —

"Adam, there is truth in what you say; and there's many of God's servants who have greater strength than I have, and find their hearts enlarged by the cares of husband and kindred. But I have not faith that it would be so with me, for since my affections have been set above measure on you, I have had less peace and joy in God; I have felt, as it were, a division in my heart. And think how it is with me, Adam: that life I have led is like a land I have trodden in blessedness since my childhood; and if I long for a moment to follow the voice which calls me to another land that I know not, I cannot but fear that my soul might hereafter yearn for that early blessedness which I had forsaken; and where doubt enters, there is not perfect love. I must wait for clearer guidance; I must go from you, and we must submit ourselves entirely to the Divine Will. We are sometimes required to lay our natural, lawful affections on the altar."

Adam dared not plead again, for Dinah's was not the voice of caprice or insincerity. But it was very hard for him; his eyes got dim as he looked at her.

"But you may come to feel satisfied — to feel that you may come to me again, and we may never part, Dinah?"

"We must submit ourselves, Adam. With time, our duty will be made clear. It may be, when I have entered on my former life, I shall find all these new thoughts and wishes vanish, and become as things that were not. Then I shall know that my calling is not toward marriage. But we must wait."

. . . .

He came within three paces of her, and then said, "Dinah!" She started without looking round, as if she connected the sound with no place. "Dinah!" Adam said again. He knew quite well what was in her mind. She was so accustomed to think of impressions as purely spiritual monitions, that she looked for no material visible accompaniment of the voice.

But this second time she looked round. What a look of yearning love it was that the mild gray eyes turned on the strong, dark-eyed man! She did not start again at the sight of him; she said nothing, but moved toward him so that his arm could clasp her round.

And they walked on so in silence, while the warm tears fell. Adam was content, and said nothing. It was Dinah who spoke first.

"Adam," she said, "it is the Divine Will. My soul is so knit to yours that it is but a divided life I live without you. And this moment, now you are with me, and I feel that our hearts are filled with the same love, I have a fulness of strength to bear and do our Heavenly Father's will that I had lost before."

Adam paused, and looked into her sincere, loving eyes.

"Then we'll never part any more, Dinah, till death parts us."

And they kissed each other with a deep joy.

GEORGE ELIOT.

INDOLENCE.

STRENUOUS individual application is the price paid for distinction; excellence of any sort being invariably placed beyond the reach of indolence. It is the diligent hand and head alone that maketh rich, — in self-culture, growth in wisdom, and in business. Even when men are born to wealth and high social position, any solid reputation which they may individually achieve can only be obtained by energetic application; for though an inheritance of acres may be bequeathed, an inheritance of knowledge and wisdom cannot. The wealthy man may pay others for doing his work for him, but it is impossible to get his thinking done for him by another, or to purchase any kind of self-culture. Indeed, the doctrine that excellence in any pursuit is only to be achieved by laborious application, holds as true in the case of the man of wealth as in that of Drew and Gifford, whose only school was a cobbler's stall, or Hugh Miller, whose only college was a Cromarty stone quarry.

Although much may be accomplished by means of individual industry and energy, as these and other instances set forth in the following pages serve to illustrate, it must at the same time be acknowledged that the help which we derive from others in the journey of life is of very great importance. The poet Wordsworth has well said that "these two things, contradictory though they may seem, must go together, — manly dependence and manly independence, manly reliance and manly self-reliance." From infancy to old age, all are more or less indebted to others for nurture and culture; and the best and strongest are usually found the readiest to acknowledge such help.

A human character is moulded by a thousand subtle influences; by example and precept; by life and literature; by friends and neighbors; by the world we live in as well as by the spirits of our forefathers, whose legacy of good words and deeds we inherit. But great, unquestionably, though these influences are acknowledged to be, it is nevertheless equally clear that men must necessarily be the active agents of their own well-being and well-doing; and that, however much the wise and the good may owe to others, they themselves must in the very nature of things be their own best helpers.

SAMUEL SMILES.

THE TOURNAMENT.

ABOUT the hour of ten o'clock, the whole plain was crowded with horsemen, horsewomen, and foot passengers, hastening to the tournament; and shortly after, a grand flourish of trumpets announced Prince John and his retinue, attended by many of those knights who meant to take share in the game, as well as others who had no such intention.

About the same time arrived Cedric the Saxon, with the Lady Rowena, unattended, however, by Athelstane. This Saxon lord had arrayed his tall and strong person in armor, in order to take his place among the combatants; and, considerably to the surprise of Cedric, had chosen to enlist himself on the part of the Knight Templar. The Saxon, indeed, had remonstrated strongly with his friend upon the injudicious choice he had made of his party; but he had only received that sort of answer usually given by those who are more obstinate in following their own course, than strong in justifying it.

His best, if not his only reason, for adhering to the party of Brian de Bois-Guilbert, Athelstane had the prudence to keep to himself. Though his apathy of disposition prevented his taking any means to recommend himself to the Lady Rowena, he was, nevertheless, by no means insensible to her charms, and considered his union with her as a matter already fixed beyond doubt, by the assent of Cedric and her other friends. It had therefore been with smothered displeasure that the proud though indolent Lord of Coningsburg beheld the victor of the preceding day select Rowena as the object of that honor which it became his privilege to confer. In order to punish him for a preference which seemed to interfere with his own suit, Athelstane, confident of his strength, and to whom his flatterers, at least, ascribed great skill in arms, had determined not only to deprive the Disinherited Knight of his powerful succor, but, if an opportunity should occur, to make him feel the weight of his battle-axe.

THE TOURNAMENT.

De Bracy, and other knights attached to Prince John, in obedience to a hint from him, had joined the party of the challengers, John being desirous to secure, if possible, the victory to that side. On the other hand, many other knights, both English and Norman, natives and strangers, took part against the challengers, the more readily that the opposite band was to be led by so distinguished a champion as the Disinherited Knight had approved himself.

As soon as Prince John observed that the destined Queen of the day had arrived upon the field, assuming that air of courtesy which sat well upon him when he was pleased to exhibit it, he rode forward to meet her, doffed his bonnet, and, alighting from his horse, assisted the Lady Rowena from her saddle, while his followers uncovered at the same time, and one of the most distinguished, dismounted to hold her palfrey.

"It is thus," said Prince John, "that we set the dutiful example of loyalty to the Queen of Love and Beauty, and are ourselves her guide to the throne which she must this day occupy. — Ladies," he said, "attend your Queen, as you wish in your turn to be distinguished by like honors."

So saying, the Prince marshalled Rowena to the seat of honor opposite his own, while the fairest and most distinguished ladies present crowded after her to obtain places as near as possible to their temporary sovereign.

No sooner was Rowena seated than a burst of music, half-drowned by the shouts of the multitude, greeted her new dignity. Meantime, the sun shone fierce and bright upon the polished arms of the knights of either side, who crowded the opposite extremities of the lists, and held eager conference together concerning the best mode of arranging their line of battle, and supporting the conflict.

The heralds then proclaimed silence until the laws of the tourney should be rehearsed. These were calculated in some degree to abate the dangers of the day; a precaution the more necessary, as the conflict was to be maintained with sharp swords and pointed lances.

The champions were therefore prohibited to thrust with the sword, and were confined to striking. A knight, it was announced, might use a mace or battle-axe at pleasure, but the dagger was a prohibited weapon. A knight unhorsed might renew the fight on foot with any other on the opposite side in the same predicament; but mounted horsemen were in that case forbidden to assail him. When any knight could force his antagonist to the extremity of the lists, so as to touch the palisade with his person or arms, such opponent was obliged to yield himself vanquished, and his armor and horse were placed at the disposal of the conqueror. A knight thus overcome was not permitted to take further share in the combat. If any combatant was struck down, and unable to recover his feet, his squire or page might enter the lists, and drag his master out of the press; but in that case the knight was adjudged vanquished, and his arms and horse declared forfeited. The combat was to cease as soon as Prince John should throw down his leading staff, or truncheon; another precaution usually taken to prevent the unnecessary effusion of blood by the too long endurance of a sport so desperate. Any knight breaking the rules of the tournament, or otherwise transgressing the rules of honorable chivalry, was liable to be stript of his arms, and, having his shield reversed, to be placed in that posture astride upon the bars of the palisade, and exposed to public derision, in punishment of his unknightly conduct. Having announced these precautions, the heralds concluded with an exhortation to each good knight to do his duty, and to merit favor from the Queen of Beauty and Love.

This proclamation having been made, the heralds withdrew to their stations. The knights, entering at either end of the lists in long procession, arranged themselves in a double file, precisely opposite to each other, the leader of each party being in the center of the foremost rank, a post which he did not occupy until each had carefully arranged the ranks of his party, and stationed every one in his place.

It was a goodly, and at the same time an anxious sight, to behold so many gallant champions, mounted bravely, and armed richly, stand ready prepared for an encounter so formidable, seated on their war-saddles like so many pillars of iron, and awaiting the signal of encounter with the same ardor as their generous steeds, which, by neighing and pawing the ground, gave signal of their impatience.

As yet the knights held their long lances upright, their bright points glancing to the sun, and the streamers with which they were decorated fluttering over the plumage of the helmets. Thus they remained while the marshals of the field surveyed their ranks with the utmost exactness, lest either party had more or fewer than the appointed number. The tale was found exactly complete. The marshals then withdrew from the lists, and William de Wyvil, with a voice of thunder, pronounced the signal words — *Laissez aller!* The trumpets sounded as he spoke — the spears of the champions were at once lowered and placed in the rests — the spurs were dashed into the flanks of the horses, and the two foremost ranks of either party rushed upon each other in full gallop, and met in the middle of the lists with a shock, the sound of which was heard at a mile's distance. The rear rank of each party advanced at a slower pace to sustain the defeated, and follow up the success of the victors of their party.

The consequences of the encounter was not instantly seen, for the dust raised by the tramping of so many steeds darkened the air, and it was a minute ere the anxious spectators could see the fate of the encounter. When the fight became visible, half the knights on each side were dismounted, some by the dexterity of their adversary's lance, — some by the superior weight and strength of opponents, which had borne down both horse and man, — some lay stretched on earth as if never more to rise, — some had already gained their feet, and were closing hand to hand with those of their antagonists who were in the same predicament, — and several on both sides, who had received wounds by which they were disabled, were stopping their blood by their scarfs, and endeavoring to extricate themselves from the tumult. The mounted knights, whose lances had been almost all broken by the fury of the encounter, were now closely engaged with their swords, shouting their war-cries, and exchanging buffets, as if honor and life depended on the issue of the combat.

The tumult was presently increased by the advance of the second rank on either side, which, acting as a reserve, now rushed on to aid their companions. The followers of Brian de Bois-Guilbert shouted—"Ha! *Beau-scant! Beau-scant!* *—For the Temple—For the Temple!" The opposite party shouted in answer—*Desdichado! Desdichado!*"—which watchword they took from the motto upon their leader's shield.

The champions thus encountering each other with the utmost fury, and with alternate success, the tide of battle seemed to flow now toward the southern, now toward the northern extremity of the lists, as the one or the other party prevailed. Meantime the clang of the blows, and the shouts of the combatants, mixed fearfully with the sound of the trumpets, and drowned the groans of those who fell, and lay rolling defenseless beneath the feet of the horses. The splendid armor of the combatants was now defaced with dust and blood, and gave way at every stroke of the sword and battle-axe. The gay plumage, shorn from the crests, drifted upon the breeze like snow-flakes. All that was beautiful and graceful in the martial array had disappeared, and what was now visible was only calculated to awake terror or compassion.

Yet such is the force of habit, that not only the vulgar spectators, who are naturally attracted by sights of horror, but even the ladies of distinction, who crowded the galleries, saw the conflict with a thrilling interest certainly, but without a wish to withdraw their eyes from a sight so terrible. Here and there, indeed, a fair cheek might turn pale, or a faint scream might be heard, as a lover, a brother, or a husband, was struck from his horse. But, in general, the ladies around encouraged the combatants, not only by clapping their hands and waving their veils and kerchiefs, but even by exclaiming, "Brave lance! Good sword!" when any successful thrust or blow took place under their observation.

Such being the interest taken by the fair sex in this bloody game, that of the men is the more easily understood. It showed itself in loud acclamations upon every change of fortune, while all eyes were so riveted on the lists, that the spectators seemed as if they themselves had dealt and received the blows which were there so freely bestowed. And between every pause was heard the voice of the heralds, exclaiming, "Fight on, brave knights! Man dies, but glory lives!—Fight on—death is better than defeat!—Fight on, brave knights!—for bright eyes behold your deeds."

Amid the varied fortunes of the combat, the eyes of all endeavored to discover the leaders of each band, who, mingling in the thick of the fight, encouraged their companions both by voice and example. Both displayed great feats of gallantry, nor did either Bois-Guilbert or the Disinherited Knight find in the ranks opposed to them a champion who could be termed their unquestioned match. They repeatedly endeavored to single out each other, spurred by mutual animosity, and aware that the fall of either leader might be considered as decisive of victory. Such, however, was the crowd and confusion, that during the earlier part of the conflict, their efforts to meet were unavailing, and they were repeatedly separated

---

* *Beau-seant* was the name of the Templars' banner, which was half-black, half-white, to intimate, it is said, that they were candid and fair towards Christians, but black and terrible towards infidels.

by the eagerness of their followers, each of whom was anxious to win honor, by measuring his strength against the leader of the opposite party.

But when the field became thin by the number on either side who had yielded themselves vanquished, had been compelled to the extremity of the lists, or been otherwise rendered incapable of continuing the strife, the Templar and the Disinherited Knight at length encountered hand to hand, with all the fury that mortal animosity, joined to rivalry of honor, could inspire. Such was the address of each in parrying and striking, that the spectators broke forth into a unanimous and involuntary shout, expressive of their delight and admiration.

But at this moment the party of the Disinherited Knight had the worst; the gigantic arm of Front-de-Bœuf on the flank, and the ponderous strength of Athelstane on the other, bearing down and dispersing those immediately exposed to them. Finding themselves freed from their immediate antagonists, it seemed to have occurred to both these knights at the same instant, that they would render the most decisive advantage to their party, by aiding the Templar in his contest with his rival. Turning their horses, therefore, at the same moment, the Norman spurred against the Disinherited Knight on the one side, and the Saxon on the other. It was utterly impossible that the object of this unequal and unexpected assault could have sustained it, had he not been warned by a general cry from the spectators, who could not but take interest in one exposed to such disadvantage.

"Beware! Beware! Sir Disinherited!" was shouted so universally, that the knight became aware of his danger; and striking a full blow at the Templar, he reined back his steed in the same moment, so as to escape the charge of Athelstane and Front-de-Bœuf. These knights, therefore, their aim being thus eluded, rushed from opposite sides betwixt the object of their attack and the Templar, almost running their horses against each other ere they could stop their career. Recovering their horses, however, and wheeling them round, the whole three pursued their united purpose of bearing to the earth the Disinherited Knight.

Nothing could have saved him, except the remarkable strength and activity of his noble horse which he had won on the preceding day.

This stood him in the more stead, as the horse of Bois-Guilbert was wounded, and those of Front-de-Bœuf and Athelstane were both tired with the weight of their gigantic masters, clad in complete armor, and with the preceding exertions of the day. The masterly horsemanship of the Disinherited Knight, and the activity of the noble animal which he mounted, enabled him for a few minutes to keep at sword's point his three antagonists, turning and wheeling with the agility of a hawk upon the wing, keeping his enemies as far separate as he could, and rushing now against the one, now against the other, dealing sweeping blows with his sword, without waiting to receive those which were aimed at him in return.

But although the lists rang with the applauses of his dexterity, it was evident that he must at last be overpowered; and the nobles around Prince John implored him with one voice to throw down his warder, and to save so brave a knight from the disgrace of being overcome by odds.

"Not I, by the light of Heaven!" answered Prince John; "this same springal, who conceals his name, and despises our proffered hospitality, hath already gained one prize, and may now afford to let others have their turn." As he spoke thus, an unexpected incident changed the fortune of the day.

There was among the ranks of the Disinherited Knight, a champion in black armor, mounted on a black horse, large of size, tall, and to all appearance powerful and strong, like the rider by whom he was mounted. This knight, who bore on his shield no device of any kind, had hitherto evinced very little interest in the event of the fight, beating off with seeming ease those combatants who attacked him, but neither pursuing his advantages, nor himself assailing any one. In short, he had hitherto acted the part rather of a spectator than of a party in the tournament, a circumstance which procured him among the spectators the name of *Le Noir Faineant*, or the Black Sluggard.

At once this knight seemed to throw aside his apathy, when he discovered the leader of his party so hard bestead; for, setting spurs to his horse, which was quite fresh, he came to his assistance like a thunderbolt, exclaiming in a voice like a trumpet-call, "*Desdichado*, to the rescue!" It was high time, for, while the Disinherited Knight was pressing upon the Templar, Front-de-Bœuf had got nigh to him with his uplifted sword; but ere the blow could descend, the Sable Knight dealt a stroke on his head, which, glancing from the polished helmet lighted with violence scarcely abated on the *chafron* of the steed, and Front-de-Bœuf rolled on the ground, both horse and man equally stunned by the fury of the blow. *Le Noir Faineant* then turned his horse upon Athelstane of Coningsburgh; and his own sword having been broken in the encounter with Front-de-Bœuf, he wrenched from the hand of the bulky Saxon the battle-axe which he wielded, and, like one familiar with the use of the weapon, bestowed him such a blow upon the crest, that Athelstane also lay senseless on the field. Having achieved this double feat, for which he was the more highly applauded that it was totally unexpected from him, the knight seemed to resume the sluggishness of his character, returning calmly to the northern extremity of the lists, leaving his leader to cope as he best could with Brian de Bois-Guilbert. This was no longer matter of so much difficulty as formerly. The Templar's horse had bled much, and gave way under the shock of the Disinherited Knight's charge. Brian de Bois-Guilbert rolled on the field, encumbered with the stirrup, from which he was unable to draw his foot. His antagonist sprung from horseback, waved his fatal sword over the head of his adversary, and commanded him to yield himself; when Prince John, more moved by the Templar's dangerous situation than he had been by that of his rival, saved him the mortification of confessing himself vanquished, by casting down his warder, and putting an end to the conflict.

It was, indeed, only the relics and embers of the fight which continued to burn; for of the few knights who still continued in the lists, the greater part had, by tacit consent, forborne the conflict for some time, leaving it to be determined by the strife of the leaders.

The squires, who had found it a matter of danger and difficulty to attend their masters during the engagement, now thronged into the lists to pay their dutiful attendance to the wounded, who were removed with the utmost care and attention to the neighboring pavilions, or to the quarters prepared for them in the adjoining village.

Thus ended the memorable field of Ashby-de-la-Zouche, one of the most gallantly contested tournaments of that age; for although only four knights, including one who was smothered by the heat of his armor, that died upon the field, yet upwards of thirty were desperately wounded, four or five of whom never recovered. Several more were disabled for life; and those who escaped best carried the marks of the conflict to the grave with them. Hence it is always mentioned in the old records, as the Gentle and Joyous Passage of Arms of Ashby.

It being now the duty of Prince John to name the knight who had done best, he determined that the honor of the day remained with the knight whom the popular voice had termed *Le Noir Faineant.* It was pointed out to the Prince, in impeachment of this decree, that the victory had been in fact won by the Disinherited Knight, who, in the course of the day, had overcome six champions with his own hand, and who had finally unhorsed and struck down the leader of the opposite party. But Prince John adhered to his own opinion, on the ground that the Disinherited Knight and his party had lost the day, but for the powerful assistance of the Knight of the Black Armor, to whom, therefore, he persisted in awarding the prize.

Sir Walter Scott.

THE CULTURE OF THE PURITANS.

Whatever may have taken place later, the Puritanism of the first forty years of the seventeenth century was not tainted with degrading or ungraceful associations of any sort. The rank, the wealth, the chivalry, the genius, the learning, the accomplishments, the social refinements and elegance of the time, were largely represented in its ranks. Not to speak of Scotland, where soon Puritanism had few opponents in the class of the high-born and the educated, the severity of Elizabeth scarcely restrained, in her latter days, its predominance among the most exalted orders of her subjects. The Earls of Leicester, Bedford, Huntingdon and Warwick, Sir Nicholas Bacon, his greater son, Walsingham, Burleigh, Mildmay, Sadler, Knollys, were specimens of a host of eminent men more or less friendly to or tolerant of it. Throughout the reign of James the First, it controlled the House of Commons, composed chiefly of the landed gentry of the kingdom; and if it had less sway among the Peers, this was partly because the number of lay nobles did not largely exceed that of the Bishops, who were mostly creatures of the crown. The aggregate property of that Puritan House of Commons, whose dissolution has been just now related, was computed to be three times as great as that of the Lords. The statesmen of the first period of that Parliament, which by and by dethroned

Charles the First, had been bred in the luxury of the landed aristocracy of the realm; while of the nobility, Manchester, Essex, Warwick, Brooke, Fairfax, and others, and of the gentry, a long roll of men of the scarcely inferior position of Hampden and Waller, commanded and officered its armies and fleets. A Puritan was the first Protestant founder of a college at an English University. Among the clergy, representing mainly the scholarship of the country, nothing is more incontrovertible than that the permanent ascendency of Puritanism was only prevented by the severities of the governments of Elizabeth and her Scottish kinsmen under the several administrations of Parker, Whitgift, Bancroft, and Laud.

It may be easily believed that none of the guests whom the Earl of Leicester placed at his table by the side of his nephew, Sir Philip Sydney, were clowns. But the supposition of any necessary connection between Puritanism and what is harsh and rude in taste and manners will not even stand the test of an observation of the character of men who figured in its ranks, when the lines came to be most distinctly drawn. The Parliamentary general, Devereux, Earl of Essex, was no straitlaced gospeller, but a man formed with every grace of person, mind, and culture, to be the ornament of a splendid court, the model knight — the idol, as long as he was the comrade, of the royal soldiery — the Bayard of the time. The position of Manchester and Fairfax, of Hollis, Fiennes and Pierrepont, was by birthright in the most polished circle of English society. In the memoirs of the young regicide, Colonel Hutchinson, recorded by his beautiful and gentle wife, we may look at the interior of a Puritan household, and see its graces, divine and human, as they shone with a naturally blended luster in the most strenuous and most afflicted times. The renown of English learning owes something to the sect which enrolled the names of Seldon, Lightfoot, Gale, and Owen. Its seriousness and depth of thought had lent their inspiration to the delicate muse of Spenser. Judging between their colleague preachers, Travers, and Hooker, the critical Templars awarded the palm of scholarly eloquence to the Puritan. When the Puritan lawyer Whitelock was ambassador to Queen Christina, he kept a magnificent state, which was the admiration of her court, perplexed as they were by his persistant Puritanical testimony against the practise of drinking healths. For his Latin Secretary, the Puritan Protector employed a man at once equal to the foremost of mankind in genius and learning, and skilled in all manly exercises, proficient in the lighter accomplishments beyond any other Englishman of his day, and caressed in his youth, in France and Italy, for eminence in the studies of their fastidious scholars and artists. The king's camp and court at Oxford had not a better swordsman or amateur musician than John Milton, and his portraits exhibit him with locks as flowing as Prince Rupert's. In such trifles as the fashion of apparel, the usage of the best modern society vindicates, in characteristic particulars, the Roundhead judgment and taste of the century before the last. The English gentleman now, as the Puritan gentleman then, dresses plainly in " sad " colors, and puts his lace and embroidery on his servants.

JOHN GORHAM PALFREY.

BESS AND THE SNAKE.

"He does not come, — he does not come," she murmured, as she stood contemplating the thick copse spreading before her, and forming the barrier which terminated the beautiful range of oaks which constituted the grove. How beautiful was the green and garniture of that little copse of wood! The leaves were thick, and the grass around lay folded over and over in bunches, with here and there a wild flower gleaming from its green and making of it a beautiful carpet of the richest and most various texture. A small tree rose from the center of a clump around which a wild grape gadded luxuriantly; and, with an incoherent sense of what she saw, she lingered before the little cluster, seeming to survey that which, though it seemed to fix her eye, yet failed to fill her thought. Her mind wandered, — her soul was far away; and the objects in her vision were far other than those which occupied her imagination. Things grew indistinct beneath her eye. The eye rather slept than saw. The musing spirit had given holiday to the ordinary senses, and took no heed of the forms that rose, and floated, or glided away, before them. In this way, the leaf detached made no impression upon the sight that was yet bent upon it; she saw not the bird, though it whirled, untroubled by a fear, in wanton circles around her head, — and the black snake, with the rapidity of an arrow, darted over her path without arousing a single terror in the form that otherwise would have shivered at its mere appearance. And yet, though thus indistinct were all things around her to the musing mind of the maiden, her eye was yet singularly fixed, — fastened, as it were, to a single spot, gathered and controlled by a single object, and glazed, apparently, beneath a curious fascination.

Before the maiden rose a little clump of bushes, — bright tangled leaves flaunting wide in glossiest green, with vines trailing over them, thickly decked with blue and crimson flowers. Her eye communed vacantly with these; fastened by a starlike shining glance, — a subtle ray, that shot out from the circle of green leaves, — seeming to be their very eye, — and sending out a fluid luster that seemed to stream across the space between, and find its way into her own eyes. Very piercing and beautiful was that subtle brightness, of the sweetest, strangest power. And now the leaves quivered and seemed to float away, only to return, and the vines waved and swung around in fantastic mazes, unfolding ever-changing varieties of form and color to her gaze; but the star-like eye was ever steadfast, bright and gorgeous gleaming in their midst, and still fastened, with strange fondness, upon her own. How beautiful, with wondrous intensity, did it gleam, and dilate, growing larger and more lustrous with every ray which it sent forth! And her own glance became intense, fixed also; but, with a dreaming sense that conjured up the wildest fancies, terribly beautiful, that took her soul away from her, and wrapt it about as with a spell. She would have fled, she would have flown; but she had not power to move. The will was wanting to her flight. She felt that she could have bent forward to pluck the gem-like thing from the bosom of the leaf in which it seemed to grow, and which it irradiated with its bright white gleam; but ever as she aimed to stretch forth her hand and bend forward, she heard a rush of wings and a shrill scream from the tree above her, — such a scream as the mock-bird makes, when, angrily, it raises its dusky crest and flaps its wings furiously against its slender sides. Such a scream seemed like a warning, and, though yet unawakened to full consciousness, it startled her and forbade her effort. More than once, in her survey of this strange object, had she heard that shrill note, and still had it carried to her ear the same note of warning, and to her mind the same vague consciousness of an evil presence. But the star-like eye was yet upon her own, — a small, bright eye, quick like that of a bird, now steady in its place and observant seemingly only of hers, now darting forward with all the clustering leaves about it, and shooting up towards her, as if wooing her to seize. At another moment, riveted to the vine which lay around it, it would whirl round and round, dazzlingly bright and beautiful, even as a torch waving hurriedly by night in the hands of some playful boy; but, in all this time, the glance was never taken from her own : there it grew, fixed, — a very principle of light, — and such a light, — a subtle, burning, piercing, fascinating gleam, such as gathers in vapor above the old grave, and blinds us as we look, — shooting, darting directly into her eye, dazzling her gaze, defeating its sense of discrimination, and confusing strangely that of perception. She felt dizzy; for, as she looked, a cloud of colors, bright, gay, various colors, floated and hung like so much drapery around the single object that had so secured her attention and spellbound her feet. Her limbs felt momently more and more insecure — her blood grew cold, and she seemed to feel the gradual freeze of vein by vein throughout her person.

At that moment a rustling was heard in the branches of the tree beside her, and the bird, which had repeatedly uttered a single cry above her, as it were of warning, flew away from his station with a scream more piercing than ever. This movement had the effect, for which it really seemed intended, of bringing back to her a portion of the consciousness she seemed so totally to have been deprived of before. She strove to move from before the beautiful but terrible presence, but for a while she strove in vain. The rich, star-like glance still riveted her own, and the subtle fascination kept her bound. The mental energies, however, with the moment of their greatest trial, now gathered suddenly to her aid; and, with a desperate effort, but with a feeling still of most annoying uncertainty and dread, she succeeded partially in the attempt, and threw her arms backwards, her hands grasping the neighboring tree, feeble, tottering, and depending upon it for that support which her own limbs almost entirely denied her. With her movement, however, came the full development of the powerful spell and dreadful mystery before her. As her feet receded, though but a single pace, to the tree against which she now rested, the audibly-articulated ring, like that of a watch when wound up with the verge broken, announced the nature of that splendid yet dangerous presence, in the form of the monstrous rattlesnake, now but a few feet before her, lying coiled at the bottom of a beautiful shrub, with which, to her dreaming eye, many of its own glorious hues had become associated. She was at length conscious enough to perceive and to feel all her danger; but terror had denied her the strength necessary to fly from her dreadful enemy. There still the eye glared beautifully bright and piercing upon her own; and, seemingly in a spirit of sport, the insidious reptile slowly unwound himself from his coil, but only to gather himself up again into his muscular rings, his great flat head rising in the midst, and slowly nodding, as it were, towards her, the eye still peering deeply into her own; — the rattle still slightly ringing at intervals, and giving forth that paralyzing sound, which, once heard, is remembered forever.

The reptile all this while appeared to be conscious of, and to sport with, while seeking to excite, her terrors. Now, with its flat head, distended mouth, and curving neck, would it dart forward its long form towards her, — its fatal teeth, unfolding on either side of its upper jaw, seeming to threaten her with instantaneous death, while its powerful eye shot forth glances of that fatal power of fascination, malignantly bright, which, by paralyzing, with a novel form of terror and of beauty, may readily account for the spell it possesses of binding the feet of the timid, and denying to fear even the privilege of flight. Could she have fled! She felt the necessity; but the power of her limbs was gone! and there still it lay, coiling and uncoiling, its arching neck glittering like a ring of brazed copper, bright and lurid; and the dreadful beauty of its eye still fastened, eagerly contemplating the victim, while the pendulous rattle still rang the death-note, as if to prepare the conscious mind for the fate which is momently approaching to the blow. Meanwhile the stillness became death-like with all surrounding objects. The bird had gone with its scream and rush. The breeze was silent. The vines ceased to wave. The leaves faintly quivered on their stems. The serpent once more lay still; but the eye was never once turned away from the victim. Its corded muscles are all in coil. They have but to unclasp suddenly, and the dreadful folds will be upon her, its full length, and the fatal teeth will strike, and the deadly venom which they secrete will mingle with the life-blood in her veins.

The terrified damsel, her full consciousness restored, but not her strength, feels all the danger. She sees that the sport of the terrible reptile is at an end. She cannot now mistake the horrid expression of its eye. She strives to scream, but the voice dies away, a feeble gurgling in her throat. Her tongue is paralyzed; her lips are sealed; once more she strives for flight, but her limbs refuse their office. She has nothing left of life but its fearful consciousness. It is in her despair that, a last effort, she succeeds to scream, a single wild cry, forced from her by the accumulated agony; she sinks down upon the grass before her enemy, — her eyes, however, still open, and still looking upon those which he directs forever upon them. She sees him approach, — now advancing, now receding, — now swelling in every part with something of anger, while his neck is arched beautifully like that of a wild horse under the curb; until, at length, tired as it were of play, like the cat with its victim, she sees the neck growing larger and becoming completely bronzed as about to strike, — the huge jaws unclosing almost directly above her, the long tubulated fang, charged with venom, protruding from the cavernous mouth, — and she sees no more! Insensibility came to her aid, and she lay almost lifeless under the very folds of the monster.

In that moment the copse parted,—and an arrow, piercing the monster through and through the neck, bore his head forward to the ground, alongside of the maiden, while his spiral extremities, now unfolding in his own agony, were actually, in part, writhing upon her person. The arrow came from the fugitive Occonestoga, who had fortunately reached the spot, in season, on his way to the Block House. He rushed from the copse as the snake fell, and, with a stick, fearlessly approached him where he lay tossing in agony upon the grass. Seeing him advance, the courageous reptile made an effort to regain his coil, shaking the fearful rattle violently at every evolution which he took for that purpose; but the arrow, completely passing through his neck, opposed an unyielding obstacle to the endeavor; and, finding it hopeless, and seeing the new enemy about to assault him, with something of the spirit of the white man under like circumstances, he turned desperately round, and striking his charged fangs, so that they were riveted in the wound they made, into a susceptible part of his own body, he threw himself over with a single convulsion, and, a moment after, lay dead beside the utterly unconscious maiden.

WILLIAM GILMORE SIMMS.

HOME.

The home should not be considered merely as an eating and sleeping place, but as a place where self-respect may be preserved, and comforts secured, and domestic pleasures enjoyed. Three fourths of the petty vices which degrade society, and swell into crimes which disgrace it, would shrink before the influence of self-respect. To be a place of happiness, exercising beneficial influences upon its members, and especially upon the children growing up within it, the home must be pervaded by the spirit of comfort, cleanliness, affection and intelligence. And in order to secure this, the presence of a well-ordered, industrious and educated woman is indispensable. So much depends upon the woman, that we might almost pronounce the happiness or unhappiness of the home to be woman's work. No nation can advance except through the improvement of the nation's homes; and they can only be improved through the instrumentality of woman. They must know how to make homes comfortable; and before they can know, they must have been taught.

Homes are the manufactories of men; and as the homes are, so will the men be. Mind will be degraded by the physical influences around it, decency will be destroyed by constant contact with impurity and defilement, and coarseness of manners, habits and tastes will become inevitable. You cannot rear a kindly nature, sensitive against evil, careful of proprieties, and desirous of moral and intellectual improvement, amidst the darkness, dampness, disorder and discomfort which unhappily characterize so large a portion of the dwellings of the poor in our large towns; and until we can, by some means or other, improve their domestic accommodation, their low moral and social condition must be regarded as inevitable.

SAMUEL SMILES.

THE TOWER OF LONDON.

MR. PUNCH, — *My Dear Sir:* — I skurcely need inform you that your excellent Tower is very pop'lar with pe'ple from the agricultooral districks, and it was chiefly them class which I found waitin at the gates the other mornin.

I saw at once that the Tower was established on a firm basis. In the entire history of firm basisis I don't find a basis more firmer than this one.

"You have no Tower in America?" said a man in the crowd, who had somehow detected my denomination.

"Alars! no," I anserd; "we boste of our enterprise and improovements, and yit we are devoid of a Tower. America oh my onhappy country! thou hast not got no Tower! It's a sweet Boon."

The gates was opened after a while, and we all purchist tickets, and went into a waitin-room.

"My frens," said a pale-faced little man, in black close, "this is a sad day."

"Inasmuch as to how?" I said.

"I mean it is sad to think that so many pe'ple have been killed within these gloomy walls. My frens, let us drop a tear!"

"No," I said, "you must excuse me. Others may drop one if they feel like it; but as for me, I decline. The early managers of this institootion were a bad lot, and their crimes were trooly orful; but I can't sob for those who died four or five hundred years ago. If they was my own relations I couldn't. It's absurd to shed sobs over things which occurd during the rain of Henry the Three. Let us be cheerful," I continnered. "Look at the festiv Warders, in their red flannil jackets. They are cheerful, and why should it not be thusly with us?"

A Warder now took us in charge, and showed us the Trater's Gate, the armers, and things. The Trater's Gate is wide enuff to admit about twenty traters abrest, I should jedge; but beyond this, I couldn't see that it was superior to gates in gen'ral.

Traters, I will here remark, are a onfornit class of pe'ple. If they wasn't, they wouldn't be traters. They conspire to bust up a country — they fail, and they're traters. They bust her, and they become statesmen and heroes.

Take the case of Gloster, afterwards Old Dick the Three, who may be seen at the Tower on horseback, in a heavy tin overcoat — take Mr. Gloster's case. Mr. G. was a conspirator of the basist dye, and if he'd failed, he would have been hung on a sour apple-tree. But Mr. G. succeeded, and became great. He was slewed by Col. Richmond, but he lives in history, and his equestrian figger may be seen daily for a sixpence, in conjunction with other em'nent persons, and no extra charge for the Warder's able and bootiful lectur.

There's one king in this room who is mounted onto a foaming steed, his right hand graspin a barber's pole. I didn't learn his name.

The room where the daggers and pistils and other weppins is kept is interestin. Among this collection of choice cuttlery I notist the bow and arrer which those hot-hedded old chaps used to conduct battles with. It is quite like the bow and arrer used at this day by certain tribes of American Injuns, and they shoot 'em off with such a excellent precision that I almost sigh'd to be an Injun when I was in the Rocky Mountain regin. They are a pleasant lot them Injuns. Mr. Cooper and Dr. Catlin have told us of the red man's wonerful eloquence, and I found it so. Our party was stopt on the plains of Utah by a band of Shoshones, whose chief said :

"Brothers! the pale-face is welcome. Brothers! the sun is sinking in the west, and Wa-na-bucky-she will soon cease speaking. Brothers! the poor red man belongs to a race which is fast becomin extink."

He then whooped in a shrill manner, stole all our blankets and whisky, and fled to the primeval forest to conceal his emotions.

I will remark here, while on the subjeck of Injuns, that they are in the main a very shaky set, with even less sense than the Fenians, and when I hear philanthropists bewailin the fack that every year "carries the noble red man nearer the settin sun," I simply have to say I'm glad of it, tho' it is rough on the settin sun. They call you by the sweet name of Brother one minit, and the next they scalp you with their Thomas-hawks. But I wander. Let us return to the Tower.

At one end of the room where the weppins is kept, is a wax figger of Queen Elizabeth, mounted on a fiery stuffed hoss, whose glass eye flashes with pride, and whose red morocker nostril dilates hawtily, as if conscious of the royal burden he bears. I have associated Elizabeth with the Spanish Armady. She's mixed up with it at the Surrey Theatre, where Troo to the Core is bein acted, and in which a full bally core is introjooced on board the Spanish Admiral's ship, giving the audiens the idee that he intends openin a moosic-hall in Plymouth the moment he conkers that town. But a very interesting drammer is Troo to the Core, notwithstandin the eccentric conduct of the Spanish Admiral ; and very nice it is in Queen Elizabeth to make Martin Truegold a baronet.

The Warder shows us some instroments of tortur, such as thumbscrews, throat-collars, etc., statin that these was conkered from the Spanish Armady, and addin what a crooil pe'ple the Spaniards was in them days — which elissited from a bright-eyed little girl of about twelve summers the remark that she tho't it was rich to talk about the crooilty of the Spaniards usin thumbscrews, when he was in a Tower where so many poor people's heads had been cut off. This made the Warder stammer and turn red.

I was so pleased with the little girl's brightness that I could have kissed the dear child, and I would if she'd been six years older.

I think my companions intended makin a day of it, for they all had sandwiches, sassiges, etc. The sad-lookin man, who had wanted us to drop a tear afore we started to go round, fling'd such quantities of sassige into his mouth that I expected to see him choke hisself to death; he said to me, in the Beauchamp Tower, where the poor prisoners writ their onhappy names on the cold walls, "This is a sad sight."

"It is indeed," I anserd. "You're black in the face. You shouldn't eat sassige in public without some rehearsals beforehand. You manage it orkwardly."

"No," he said, "I mean this sad room."

Indeed, he was quite right. Tho' so long ago all these drefful things happened, I was very glad to git away from this gloomy room, and go where the rich and

THE TOWER OF LONDON.

sparklin Crown Jewils is kept. I was so pleased with the Queen's Crown, that it occurd to me what a agree'ble surprise it would be to send a sim'lar one home to my wife; and I asked the Warder what was the vally of a good, well-constructed Crown like that. He told me, but on cypherin up with a pencil the amount of funs I have in the Jint Stock Bank, I conclooded I'd send her a genteel silver watch instid.

And so I left the Tower. It is a solid and commandin edifis, but I deny that it is cheerful. I bid it adoo without a pang.

I was droven to my hotel by the most melancholly driver of a four-wheeler that I ever saw. He heaved a deep sigh as I gave him two shillings.

"I'll give you six d.'s more," I said, "if it hurts you so."

"It isn't that," he said, with a hart-rendin groan, "it's only a way I have. My mind's upset to-day. I at one time thought I'd drive you into the Thames. I've been readin all the daily papers to try and understand about Governor Eyre, and my mind is totterin. It's really wonderful I didn't drive you into the Thames."

I asked the onhappy man what his number was, so I could redily find him in case I should want him agin, and bad him good-bye. And then I tho't what a frollicsome day I'd made of it. Respectably, etc.,

CHARLES FARRAR BROWNE (Artemus Ward).

TILLY BONES.

It was a delightful morning in the early spring, and on such a morning a woman of horticultural tastes is much happier out of doors than in the house ; but for this, I am inclined to think, I should not have admitted Tilly to my kitchen, for the expression of her countenance was sullen and vengeful *à faire peur.*

However, in other respects she was a comely enough young negro woman, tall, slim, and tidy ; therefore, reflecting that I might wait longer and fare worse, I engaged her services. I had had experience to teach me that a cook may smile and smile, and be a villain, so I hoped that signs might go by contraries in Tilly's case. Nor were my hopes disappointed. Her sullen gloom of countenance did by no means betoken the inward spirit, for Tilly possessed a power of grin and chuckle beyond any young African that ever I saw. She was a genuine plantation darky, untamed by that *bizarre* form of civilization developed in the negro that dwells in cities ; and her quaint originality and keen observation furnished me unfailing amusement. For Tilly was neither shy nor reticent. I found her "garrulously given, a babbler in the land," and I became the repository of her opinions and experiences.

As to her name, her "krizten" name, as she called it, was unmistakably Tilly ; and another name she had which certainly began with a B, but owing to her peculiar utterance this remained to me a mystery forever, and therefore I took the liberty of calling her Bones, partly because this was as near as I could attain to the name she gave, and partly because of her extreme angularity. Inasmuch as she had, from the beginning of our acquaintance, dignified my own name with the prefix of Mac, we could consider ourselves quits on that score. I acquiesced in the accession of a syllable with secret amusement, and she accepted her new appellation with an appreciative grin. Evidently she considered it a distinction to have a name so difficult of pronunciation.

"Hit's a easier name to say, Bones is," she commented, with patronizing indulgence for my inability to catch the proper sound ; "en ef you knows hit, Miz McAnderson, en me knows hit, why, hit's all squay." (This is as near as phonetic spelling can come to Tilly's "square.") " Hit ain't Pawndus's name, Bones ain't ;

but," she continued, consoling, "you ain't got no 'casion fur ter call Pawndus, noway, fur I rekin Pawndus ain't comin' 'roun' whey dey is a boss. So hit doan mek no diffunce; en ef you's minded fur ter say Bones, Miz McAnderson, why I ain't no ways cawntrairy."

Tilly talked a great deal about this "Pawndus." He was her husband, and a very terrible reality at times. He was directly responsible for Tilly's sour and sullen 'havior of the visage, as I soon discovered.

I should explain that in this part of the world servants' rooms are useless appendages to any establishment. Not one woman in a hundred who is willing to "hire out" by the month can be induced to take a room on the employer's

TILLY ON THE PLANTATION.

premises; one and all, they would rather walk any distance through any weather, and Tilly was no exception to the general rule. I offered her a room that had the advantage of being detached from my dwelling-house, a room that was, I have no doubt, infinitely more comfortable than the shanty she occupied on the other side of town. But though I extended the privilege of a domicile to " Pawndus" also, Tilly stoutly refused; and, rain or shine, heat or cold, sick or well, she tramped her mile back and forth, so long as she remained in my service.

She came to her work one morning with a countenance so lowering that I was constrained to inquire what was the matter. To which Tilly, with lips projecting, made answer in this wise:

"Pawndus he" — then followed an inarticulate mumble.

Now I had heard, in a roundabout way, from a lady who had once employed Tilly, that Pawndus was given to wife-beating, and I had asked Tilly about this. But she had indignantly denied it: "Pawndus knowed better, he did, den ter tech her." Nevertheless, on the morning I speak of, I had my suspicions as to what had happened, but I only said:

"I don't understand you, Tilly. Is Pondus sick, or are you sick?"

A faint flicker of a smile, that wondrous smile of hers, that was like a sunburst from behind a thunder-cloud, played over her dusky features as she answered:

"'Speck Pawndus sick 'nuff jez 'bout now." And then her smile vanished, and her lips again stuck out amazingly, as she continued in a grumbling tone: "'N' as fum me, might jez well be sick; my arm dat lame, can't lif pot-lid."

"What is the matter with your arm?"

"Well, Miz McAnderson, fac' is, you see, Pawndus he do git perpetuil, times; 'n' las' night he done let loose on me wid de tongs. Dat's what matter wid my arm."

"I will give you some arnica for it, Tilly. But I thought Pondus did not beat you?"

"No, Miz McAnderson," said Tilly, solemnly shaking her head, " I ain't nuver said Pawndus doan beat me no time. But, you see, w'en I wuz livin' to Miz Ginnie Vine's, Miz Ginnie she sot a heap by me, Miz Ginnie did, en' Mr. Vine he 'lowed efen Pawndus dared to tech me, he'd jez war him clean plum out; 'n' Pawndus"—(here a gleeful chuckle)—"he so skeered Mr. Vine, he done broke hisse'f dat foolishness. Praise de Lawd, I mek sho' he same ez forgit all 'bout beaten me, 'n' I warn't gwine lay no pas' doin's aginst him; but here, now, come las' night, debbil in him done broke out fresh. Tink Mr. Vine ain't yere 'bout hit."

"Does he drink?" I asked.

"No, Miz McAnderson, nuttin' but debbil in him," she responded gloomily.

"Why don't you go down to the mayor, and have him bound over to keep the peace?" I counselled.

"Kee, he! Miz McAnderson! You doan know nuttin'!" cackled Tilly, forgetting her bruises so far as to double up with laughter. "Me go carry complain', caws money; git me 'noder beaten, w'en I go home; den mebbe put Pawndus in jail. Nigger doan mine jail; nuttin' ter do, an' Pawndus outen wuk — dat's what's matter wid dat nigger now — den he come outen jail and beat me 'gin. Tell you, Miz McAnderson, dat sort doin's cawses money, dey do. Now, one time I waz mad wid a gal, en we fout, 'n' I stobbed her. I waz dat mad 'peared like I couldn't see; en den, w'en de blood hit come, de sorry en de skeered tergyedder jez swallered up de mad. Den de sing out 'Perlice!' en yere we go to de mare, en money ter pay. Me'n' dat gal's been good friends sence dat time. No, I ainter gwine ter no mare; I knows better w'at ter do wid dat nigger Pawndus den dat. I done come by Aunt Becky's, dis yer mawnin', en I 'low she'll mek Pawndus 'pent."

"And who is Aunt Becky?" I asked, with some vague notion that Aunt Becky might be in the dread secrets of African sorcery.

"Aunt Becky, she's he maw. Eve'y time Pawndus he whack me, Aunt Becky she whack Pawndus; caze he's her chile, en she's boun' ter raise him right; en I 'low effen he ain't raised yit, she'll keep on spilin' de rods tell he is."

ELIZABETH WHITFIELD BELLAMY.

IN VENICE.

LAST night in my gondola I made a vow I would write you a letter, if it was only to beg you would write to me at Rome. Like the great Marco Polo, however, whose tomb I saw to-day, I have a secret wish to astonish you with my travels, and would take you with me, as you would not go willingly, from London to Paris, and from Paris to the Lake of Geneva, and so on to this city of romantic adventure, the place from which he started. . . . But I must talk to you a little about Venice. I cannot tell you what I felt, when the postilion turned gaily around, and, pointing with his whip, cried out "Venezia!" For there it was, sure enough, with its long line of domes and turrets glittering in the sun. I walk about here all day long in a dream. Is that the Rialto, I say to myself? Is this St. Mark's Place? Do I see the Adriatic? I think if you and I were together here, my dear Moore, we might manufacture something from the ponte dei sospiri, the scala dei giganti, the piombi, the pozzi, and the thousand ingredients of mystery and terror that are here at every turn. Nothing can be more luxurious than a gondola and its little black cabin, in which you can fly about unseen, the gondoliers so silent all the while. They dip their oars as if they were afraid of disturbing you; yet you fly. As you are rowed through one of the narrow streets, often do you catch the notes of a guitar, accompanied by a female voice, through some open window; and at night, on the Grand Canal, how amusing is it to observe the moving lights (every gondola has its light), one now and then shooting across at a little distance, and vanishing into a smaller canal. Oh, if you had any pursuit of love or pleasure, how nervous they would make you, not knowing their contents or their destination! and how infinitely more interesting, as more mysterious, their silence, than the noise of carriage-wheels! Before the steps of the opera house they are drawn up in array with their shining prows of white metal, waiting for the company. One man remains in your boat, while the other stands at the door of your loge. When you come out, he attends you down, and calling "Pietro," or "Giacoma," is answered from the water, and away you go. The gliding motion is delightful, and would calm you after any scene in a casino. The gondolas of the foreign ministers carry the national flag. I think you would be pleased with an Italian theatre. It is lighted only from the

A LEAP FROM THE RIALTO.

stage, and the soft shadows that are thrown over it produce a very visionary effect. Here and there the figures in a box are illuminated from within, and glimmering and partial lights are almost magical. . . . This is indeed a fairy-land, and Venice particularly so. If at Naples you see most with the eye, and at Rome with the memory, surely at Venice you see most with the imagination. . . .

SAMUEL ROGERS.

THE STANDARD OF SPEECH.

WHATEVER predilection the Americans may have for their native European tongues, and particularly the British descendants for the English, yet several circumstances render a future separation of the American tongue from the English, necessary and unavoidable. The vicinity of the European nations, with the uninterrupted communication in peace and the changes of dominion in war, are gradually assimilating their respective languages. The English with others is suffering continual alterations. America, placed at a distance from those nations, will feel in a much less degree, the influence of the assimilating causes ; at the same time, numerous local causes, such as a new country, new associations of people, new combinations of ideas in arts and science, and some intercourse with tribes wholly unknown in Europe, will introduce new words into the American tongue. These causes will produce, in a course of time, a language in North America as different from the future language of England as the modern Dutch, Danish, and Swedish are from the German, or from one another ; like remote branches of a tree springing from the same stock, or rays of light, shot from the same center, and diverging from each other in proportion to their distance from the point of separation.

Whether the inhabitants of America can be brought to a perfect uniformity in the pronunciation of words, it is not easy to predict ; but it is certain that no attempt of the kind has been made, and an experiment, begun and pursued on the right principles, is the only way to decide the question. Schools in Great Britain have gone far towards demolishing local dialects — commerce has also had its influence — and in America these causes, operating more generally, must have a proportional effect.

In many parts of America, people at present attempt to copy the English phrases and pronunciation — an attempt that is favored by their habits, their prepossessions, and the intercourse between the two countries. This attempt has, within the period of a few years, produced a multitude of changes in these particulars, especially among the leading classes of people. These changes make a difference between the language of the higher and common ranks, and indeed between the same ranks in different States, as the rage for copying the English does not prevail equally in every part of North America.

But besides the reasons already assigned to prove this imitation absurd, there is a difficulty attending it which will defeat the end proposed by its advocates; which is, that the English themselves have no standard of pronunciation, nor can they ever have one on the plan they propose. The authors, who have attempted to give us a standard, make the practice of the court and stage in London the sole criterion of propriety in speaking. An attempt to establish a standard on this foundation is both unjust and idle. It is unjust, because it is abridging the nation of its rights. The general practice of a nation is the rule of propriety, and this practice should at least be consulted in so important a matter as that of making laws for speaking. While all men are upon a footing and no singularities are accounted vulgar or ridiculous, every man enjoys perfect liberty. But when a particular set of men, in exalted stations, undertake to say, "we are the standards of propriety and elegance, and if all men do not conform to our practice they shall be accounted vulgar and ignorant," they take a very great liberty with the rules of the language and the rights of civility.

But an attempt to fix a standard on the practice of any particular class of people is highly absurd; as a friend of mine once observed, it is like fixing a light-house on a floating island. It is an attempt to fix that which is in itself variable; at least it must be variable so long as it is supposed that a local practice has no standard but a local practice, that is, no standard but itself. While this doctrine is believed, it will be impossible for a nation to follow as fast as the standard changes — for if the gentlemen at court constitute a standard, they are above it themselves, and their practice must shift with their passions and their whims.

But this is not all. If the practice of a few men in the capital is to be the standard, a knowledge of this must be communicated to the whole nation. Who shall do this? An able compiler perhaps attempts to give this practice in a dictionary; but it is probable that the pronunciation, even at court or on the stage, is not uniform. The compiler therefore must follow his particular friends and patrons, in which case he is sure to be opposed and the authority of his standard called in question; or he must give two pronunciations as the standard, which leaves the student in the same uncertainty as it found him. Both these events have actually taken place in England, with respect to the most approved standards; and of course no one is universally followed.

Besides, if language must vary, like fashions, at the caprice of a court, we must have our standard dictionaries republished with the fashionable pronunciation, at least once in five years; otherwise a gentleman in the country will become intolerably vulgar by not being in a situation to adopt the fashion of the day. The new editions of them will supersede the old, and we shall have our pronunciation to relearn, with the polite alterations, which are generally corruptions.

Such are the consequences of attempting to make a local practice the standard of language in a nation. The attempt must keep the language in perpetual fluctuation, and the learner in uncertainty.

If a standard therefore cannot be fixed on local and variable custom, on what shall it be fixed? If the most eminent speakers are not to direct our practice, where shall we look for a guide? The answer is extremely easy; the rules of the language itself, and the general practice of the nation, constitute propriety in speaking. If we examine the structure of any language, we shall find a certain principle of analogy running through the whole. We shall find in English that similar combinations of letters have usually the same pronunciation, and that words having the same terminating syllable generally have the accent at the same distance from that termination.

These principles of analogy were not the result of design — they must have been the effect of accident, or that tendency which all men feel towards uniformity. But the principles, when established, are productive of great convenience, and become an authority superior to the arbitrary decisions of any man or class of men. There is one exception only to this remark. When a deviation from analogy has become the universal practice of a nation, it then takes place of all rules and becomes the standard of propriety.

The two points, therefore, which I conceive to be the basis of a standard in speaking, are these — universal undisputed practice, and the principle of analogy. Universal practice is generally, perhaps always, a rule of propriety; and in disputed points, where people differ in opinion and practice, analogy should always decide the controversy.

These are authorities to which all men will submit — they are superior to the opinions and caprices of the great, and to the negligence and ignorance of the multitude. The authority of individuals is always liable to be called in question, but the unanimous consent of a nation, and a fixed principle interwoven with the very construction of a language, coeval and coextensive with it, are like the common laws of a land or the immutable rules of morality, the propriety of which every man, however refractory, is forced to acknowledge, and to which most men will readily submit. Fashion is usually the child of caprice and the being of a day; principles of propriety are founded in the very nature of things, and remain unmoved and unchanged, amidst all the fluctuations of human affairs and the revolutions of time.

NOAH WEBSTER.

DRAMATIC REALISM.

In a piece at the Ambigu, called the "Rentrée à Paris," a mere scene in honor of the return of the troops from the Crimea the other day, there is a novelty which I think it worth letting you know of, as it is easily available, either for a serious or a comic interest — the introduction of a supposed electric telegraph. The scene is the railway terminus at Paris, with the electric telegraph-office on the prompt side, and the clerks with their backs to the audience — much more real than if they were, as they infallibly would be, staring about the house — working the needles; and the little bell perpetually ringing. There are assembled to greet the soldiers all the easily and naturally imagined elements of interest — old veteran fathers, young children, agonized mothers, sisters and brothers, girl lovers — each impatient to know of his or her own object of solicitude. Enter to these a certain marquis, full of sympathy for all, who says, "My friends, I am one of you. My brother has no commission yet. He is a common soldier. I wait for him as well as all brothers and sisters here wait for their brothers. Tell me whom you are expecting." Then they all tell him. Then he goes into the telegraph-office, and sends a message down the line to know how long the troops will be. Bell rings. Answer handed out on a slip of paper. "Delay on the line. Troops will not arrive for a quarter of an hour." General disappointment. "But we have this brave electric telegraph, my friends," says the marquis. "Give me your little messages, and I'll send them off." General rush round the marquis. Exclamations : "How's Henri?" "My love to Georges." "Has Guillaume forgotten Elise?" "Is my son wounded?" "Is my brother promoted?" etc., etc. Marquis composes tumult. Sends message — such a regiment, such a company. "Elise's love to Georges." Little bell rings, slip of paper handed out — "Georges in ten minutes will embrace his Elise. Sends her a thousand kisses." Marquis sends message — such a regiment, such a company — "Is my son wounded?" Little bell rings. Slip of paper handed out — "No. He has not yet upon him those marks of bravery in the glorious service of his country which his dear old father bears" (father being lame and invalided). Last of all, the widowed mother. Marquis sends message — such a regiment, such a company — "Is my only son safe?" Little bell rings. Slip of paper handed out — "He was first upon the heights of Alma." General cheer. Bell rings again, another slip of paper handed out — "He was made a sergeant at Inkermann." Another cheer. Bell rings again, another slip of paper handed out — "He was made color-sergeant at Sebastopol." Another cheer. Bell rings again, another slip of paper handed out — "He was the first man who leaped with the French banner on the Malakhoff tower." Tremendous cheer. Bell rings again, another slip of paper handed out — "But he was struck down there by a musket-ball, and — troops have proceeded. Will arrive in half a minute after this." Mother abandons all hope; general commiseration; troops rush in, down a platform; son only wounded, and embraces her.

As I have said, and as you will see, this is available for any purpose. But done with equal distinction and rapidity, it is a tremendous effect, and got by the simplest means in the world. There is nothing in the piece, but it was impossible not to be moved and excited by the telegraph part of it. . . .

CHARLES DICKENS.

WHITTIER WITH THE CHILDREN.

THE child-soul is born to live in every man. But it often tries its delicate wings too soon, alas! before the boy has emerged into what we call the world, and taken his place among the eager multitudes that throng the highways of life. The most beautiful thing thus being allowed to escape, the boy-man looks about him; his young eager eyes pierce the varying forms of individuality he sees; he is searching for the hero he would copy, the man whom he would take as his model. And all the while the hero is within his own breast, waiting for that divine and human summons to action, that can come from none other than his individual self impelled by God who sent him into the universe.

Thus it is that the marked individuality that men call character, comes by preserving as nearly as possible the soul fresh from the hand of God, its divine afflatus unspoiled by anything that would assert itself between these agencies. In other words, the soul, good conditions being around it, is let alone to do its own growing. And the garden of virtues thrives, not so much by the system of grafting, as by all those gentler and slower processes that leave much to Nature.

"My own dear Mr. Whittier," as he is called by the child of our household (who from baby days has by frequent visits been allowed to bask in the sunshine of the poet's presence), was so markedly a man who retained his child-soul to the very last, that it is a delightful task to go back over a long stretch of years to notice how he came by it in the first place; then to learn how he kept it, and allowed it to grow, till God took him with that soul of such native purity as to contain little that could be of alien growth in the Immortal Land.

To be all this, and go out of life with soul so fresh, Mr. Whittier must have had, silently or otherwise, much sympathy with child-nature. We know that he had; and it is to pick up the various links of the chain that bound together this man with the child-soul and the little ones whom he so closely resembled in purity and in simplicity of faith, that this imperfect sketch is written.

Lovingly we bring with tender and reverent hand this wreath of bays for his memory; each leaf is gathered from the carefully guarded reminiscence of relative and friend, garnered till the circle of years was completed and the poet passed on to the larger life that knows no ending.

The child, John Greenleaf Whittier, had a heritage from an ancestry singularly free from taint of any kind. Simple of creed, direct of purpose, virile of effort, it kept its long line unswervingly to good aims and healthful pursuits. So that when the baby opened his eyes on that winter day in 1807, in the old, homestead, under the shadow of Job's Hill, two miles or more from the center of Old Haverhill, he opened them to an horizon not hemmed in by the narrow bounds of his rural life.

What a happy home the sun smiled on there! Out upon those who would call its conditions hard! As well say that the New England winter air, crisp with recent snow, and smelling of frozen salt marshes and "passed orchard delights," as some one has termed it, is cold and cruel! Muscles tense as steel, wills made strong by battling with and beating adverse elements, and that sweet serenity that comes from self-conquest, were born of the New England winter and the New England home of the early part of the century, "barren" though you of the century's ebb, may be pleased to call it.

What an altogether delightful place is Mr. Whittier's old homestead, mellow now with its happy memories. Here is the room in which the baby boy, our poet, first saw the light; here is the quaint staircase, down which, wrapped in a blanket, he was projected by the infant hand of his sister, two years older, who held ideas of her own as to the manner of achieving descent to the room below; here are the "whitewashed wall and sagging beam," "the crane and pendent trammels," the "warm hearth blazing free," in whose reflected light

"The old, rude-furnished room,
Burst, flower-like, into rosy bloom."

Here are cupboards, redolent of juicy pie, doughnuts, and all the toothsome train of New England delicacies of that day. Crowded in upon each other, some with upper door to hide the treasures secure from the younger portion of the household, paneled, and with many a quaint device as to lock and button and hinge, they stand in the old kitchen, as who should say, "We guarded the family life; look upon us, for we are good to see."

And outside the small many-paned windows where the lilac bushes waved their sweet incense to usher in the long summer day, or the poplars, grim and straight, pointed without wavering to the sky; where the old well invited to the cool draught, more welcome than nectar to the parched lip, and the little brook ran down Job's Hill and danced across the house-place to sing its way over the road — here were packed myriad delights for the growing boy, as soon as he could toddle through the quaint doorway, and over the big flat stone that served as a step, to the limitless world of "out-of-doors."

When Mr. Whittier was seven years old, one day as he was driving the cows to pasture, a revelation, swift and unerring, came to him. He had let down the bars and the cows had just passed through, when a flash of thought struck him: "Why am I different from those cows? What have I got to do in life? What is life?"

And he never lost the influence of that hour and that revelation; it affected his whole life.

The love of animals was indeed a dominating quality of Mr. Whittier's mind. It contributed largely to the strength and sweetness of his verse, and from his companionship with the patient "beasts of the field" he received, as we shall show later, much that kept "his silences" from becoming misanthropic and too long-continued.

Nature opened up her secrets to him readily. Alert, indeed, she must have been to successfully hide her treasures from the sensitive boy whose keen eyes, even then, had the lambent gleam of the seer. His early work on the farm brought him into quick and sympathetic touch with every mood and tense of

"SWEET KENOZA FROM THE SHORE, AND WATCHING HILLS BEYOND."

Nature, until she was really his mother, to whom he would go for sweet counsel, for pleasure and for stimulus.

The books that influenced this child-soul were few in number, as the libraries in farmers' households were necessarily limited in those days. The Bible was his first choice, and over its pages he pored, gathering in rich material for those talismanic poems that were to draw all Christendom to him in loving sympathy. It is this love of the Holy Scriptures that made John Greenleaf Whittier pre-eminently the poet of the people; through his study of that book he came into the knowledge of human needs, and there he imbibed the spirit of the old prophets and reformers who thundered through the pages of the Old Testament, while the love breathed in every line of the New Testament was to bear fruit in him — another John whom the Saviour loved.

Look in at the old kitchen on one of little Greenleaf's childhood days. More than likely, if the morning hour be near to "sun-up," you will find him with mother, whom in after years he lovingly, in words of light, describes thus: "All that the sacred word mother means in its broadest, fullest significance our mother was to us — a friend, helper, counselor, companion, ever loving, gentle and unselfish." Perhaps he is busy over the homely household tasks, while she is spinning or weaving, as all the woollen cloth the family required must be made by her untiring fingers; it is more than probable that the boy has the big Bible where he can take a peep at some open page in the midst of his dish-washing, or his sweeping of the old kitchen floor, for little Greenleaf read his Bible as children nowadays hang over their toy-books and fairy tales — or perhaps it was the Farmer's Almanac or John Woolman's Journal, that talismanic narrative that stirred his little soul. By and by comes in one day "a paukie auld carle" of a Scotchman, one of the droppers-in at the Whittier house. He hungers, as usual, after the cheese and doughnuts, and thirsts after the mug of cider. When he has received them he sings in a generous fashion, "Highland Mary," "Auld Lang Syne" and "Bonnie Doon," till the old kitchen rings from floor to rafter. Later, when our poet was fourteen, his school teacher, Joshua Coffin, brought to the house a copy of Burns's poems. Greenleaf entreated him for the loan of the book a while; and then every spare moment was passed in conquering the dialect of Burns, and in rhyming and imagining all sorts of tales.

On First Day, as Sunday is called in Quaker homes, the old chaise would be brought to the door, and as many of the household as could, would stow away in its depths, and away they would drive over the hills to the quaint little Quaker meeting-house in Amesbury, eight miles away. It was sometimes Greenleaf's luck to be crowded out. But what cared he, to whom the groves and streams were his pulpit, and Nature the preacher? So he spent the lonely delicious days on Job's Hill, that silent watcher that overhung the lonely farmhouse. Up to this height the boy's eyes were often turned, as to a monitor who should point through "nature up to nature's God;" and with what a keen relish he spoke in after life, when the shadows of many graves fell across his pathway, of the early delights of "climbing Job's Hill which rose abruptly from the brook which rippled down at the foot of our garden." And then he tells over delightedly the different mountain peaks he could see from this same dear Job's Hill, "and Great Pond" (afterward named by him Kenoza — the lake of the pickerel) "stretched away from the foot of the hill." Or he wandered through fragrant, piny woods, and haunted

"Sweet Kenoza from the shore,
And watching hills beyond;"

or he followed patiently the course of the little brook that danced and gurgled its way along through many a tangled covert, till his young soul was ablaze with the glory of the Lord.

MARGARET SIDNEY.

INDEX TO AUTHORS.

ADAMS, JOHN (1735–1826).
 The Fourth of July 299
ADDISON, JOSEPH (1672–1719).
 Sir Roger de Coverley . . 44
ALCOTT, LOUISA M. (1833–1888).
 The Little Women's Romance . 92
ALDEN, MRS., G. R. (Pansy)
 The Temperance Preacher . . 57
ALLEN, JAMES LANE.
 "De Baptizin' in Elkhorn Creek" . 116
ARNOLD, MATTHEW (1822–1888).
 Sweetness and Light . . . 231
ARNOLD, THOMAS (1795–1842).
 At Rugby 11
ASCHAM, ROGER (1515–1568).
 An Apology for English . . 253
BACON, FRANCIS (1561–1626).
 On Studies 43
BANCROFT, GEORGE (1800–1890).
 Roger Williams 199
BEECHER, HENRY WARD (1813–1887).
 Deacon Marble's Trout . . . 16
BELLAMY, ELIZABETH WHITFIELD.
 Tilly Bones 347
BILLINGS, JOSH (See Shaw, Henry W.).
BLACKMORE, R. D.
 John and Lorna 64
 Lorna Doone 106
BOYESEN, HJALMAR HJORTH.
 Bergerson and Moe . . . 131
BRIGHT, JOHN (1811–1889).
 On England's Foreign Policy . 118
BRONTE, CHARLOTTE (1816–1855).
 To H. S. Williams . . . 216
BROOKS, ELBRIDGE S.
 The American Indian . . 311
BROWN, CHARLES BROCKDEN (1771–1810).
 An Encounter with a Panther . 294
BROWNE, SIR THOMAS (1605–1682).
 On Pride 41
BROWNE, CHARLES FARRAR (1834–1867).
 The Tower of London . . 344

BUNCE, OLIVER BELL (1828–1890).
 Is Gardening a Pleasure ? . 122
BUNYAN, JOHN (1628–1688).
 Christian in Doubting Castle 17
BURKE, EDMUND (1730–1797).
 Conciliation 81
BURNETT, FRANCES HODGSON.
 Joan. 303
BURNS, ROBERT (1759–1796).
 To Robert Ainslie . . . 215
CABLE, GEORGE WASHINGTON.
 The Last Train North . . 89
 Mr. Tarbox and Zozéphine . 291
CARLYLE, THOMAS (1795–1881).
 London 29
 Carlyle to his Mother . . 65
 To Thomas Murray . . . 103
 Fashionable Life at Kinkaird House . 232
CARLYLE, JANE WELSH (1800–1866).
 Mrs. Carlyle to her Husband . 230
CHANNING, WILLIAM ELLERY (1780–1842).
 Every Man Great . . . 207
CHATHAM, LORD (1708–1778).
 On Affairs in America . 13
CHAUCER, GEOFFREY (1328–1400).
 Upon Riches . . . 23
CHILD, LYDIA MARIA (1802–1880).
 Unselfishness 266
CHOATE, RUFUS (1799–1859).
 The Private Character of Webster . 300
CLAY, HENRY (1777–1852).
 On the War of 1812 . . 177
 An Appeal for the Union . 189
COBDEN, RICHARD (1804–1865).
 Protection 175
COLERIDGE, MRS. SARA (1803–1852).
 An English Sunset . . . 126
COOKE, JOHN ESTEN (1830–1886).
 The Rose of Glengary . . 124
COOPER, JAMES FENIMORE (1789–1851).
 An Encounter with the Iroquois . 257

INDEX TO AUTHORS.

Cox, Samuel Sullivan (1824-1889).
  Secession 126
Craik, Dinah Maria Mulock (1826-1887).
  "Stay" 277
Crane, Thomas Frederick.
  Aunt Maria and the Autophone . . 239
Curtis, George William.
  Mrs. Potiphar's "Cabinet Shop" . . 186
Cuyler, Theodore L.
  The Premier Gladstone . . . 298

Davis, M. E. M.
  News from the Front . . . 73
De Foe, Daniel (1661-1731).
  The Footprint on the Shore . . 273
De Quincey, Thomas (1785-1859).
  On Conversation 182
Dickens, Charles (1812-1870).
  Twenty-three 114
  Mr. Barkis 134
  The Death of Little Nell . . . 226
  Captain Cuttle's Island . . . 321
  Dramatic Realism 356
Dodge, Mary Abigail.
  My Garden 12
Dodge, Mary Mapes.
  Miss Maloney on the Chinese Question . 223
Douglas, Stephen Arnold (1813-1861).
  On the War 228
Dryden, John (1630-1700).
  Ben Jonson 42

Eliot, George (See Lewes, Mary Ann Evans).
Emerson, Ralph Waldo (1803-1881).
  Compensation 30
  Obedience to Law 161
Erskine, Lord (1750-1823).
  Limitations of Free Speech . . 15

Fiske, John.
  The Tyranny of Andros . . . 110
Fox, Charles James (1749-1806).
  War the Destroyer . . . 161
Franklin, Benjamin (1706-1790).
  The Whistle 77
Froude, James Anthony.
  Selfishness versus Nobility . . 153
Fuller, Thomas (1608-1661).
  The Good Wife 32

Garfield, James Abram (1831-1881).
  On American Institutions . . 227
Gavarre, Charles Etienne Arthur.
  The Legend of the Date-tree . . 317
Gibbon, Edward (1737-1794).
  The Invention of Gunpowder . . 146

Gladstone, William Ewart.
  Kin Beyond Sea 244
  Progress 287
Goldsmith, Oliver (1728-1774).
  Scotchmen 117
  To Mrs. Jane Lawder . . . 225
Hale, Edward Everett.
  A Lesson in Patriotism . . . 148
Hamilton, Gail (See Dodge, Mary Abigail).
Hardy, Thomas.
  A Question of Loving . . . 304
Harland, Henry.
  A Talent for Music . . . 178
Harris, Joel Chandler.
  The Wonderful Tar-Baby Story . . 97
  How Mr. Rabbit was too Sharp for Mr. Fox. 98
Harte, Bret.
  Melons 68
Hawthorne, Nathaniel (1804-1864).
  Little Pearl in the Forest . . 59
  The Town Pump 337
Haydon, Benjamin Robert (1786-1846).
  To Miss Mitford 269
Hayne, Robert Y. (1791-1840).
  On Mr. Foot's Resolution . . 35
Helps, Arthur (1808-1881).
  On the Art of Living with Others . 145
Henry, Patrick (1737-1799).
  Speech before the Virginia Convention . 87
Hildreth, Richard (1807-1865).
  Continental Congress . . . 138
Higginson, Thomas Wentworth.
  Spring in New England . . . 137
Holland, Josiah Gilbert (1819-1881).
  The True Track 278
Holmes, Oliver Wendell.
  Talk 48
  The Long Path 111
Hood, Thomas (1798-1845).
  Gradle 243
  To a Child 296
Howard, Blanche Willis.
  Philip and Leigh 266
Hughes, Thomas.
  Personal Influence . . . 289
Hume, David (1711-1776).
  On the Middle Station in Life . . 66
  To William Robertson . . . 276
Hurd, Frank R.
  For Freedom of Trade . . . 229
Irving, Washington (1783-1859).
  Wouter Van Twiller . . . 49
  The Alhambra by Moonlight . . 209
James, Henry.
  Spiritual Emancipation . . . 279

INDEX TO AUTHORS.

JEFFERSON, THOMAS (1743-1826).
   The Essential Principles of Government . 79
   The Rights of Man 270
JEWETT, SARAH ORNE.
   The White Rose Road . . . 220
JOHNSON, SAMUEL (1709-1784).
   Letter to Mrs. Thrale . . . 51
JOHNSON, RICHARD MALCOLM.
   Nipped in the Bud . . . 210
JONSON, BEN (1574-1637).
   On Bacon 44

KEATS, JOHN (1796-1821).
   John Keats to William Reynolds . 229
KING, THOMAS STARR (1824-1863).
   Sight and Insight 252
KINGSLEY, CHARLES (1819-1875).
   A Country Parish . . . 14
   The Miracle of Nature . . 184

LAMB, CHARLES (1775-1834).
   On the Death of an Old Friend . 22
   To Bernard Barton . . . 206
   A True Caledonian . . . 230
LANDOR, WALTER SAVAGE (1775-1864).
   Petition of Thugs . . . 233
LANIGAN, GEORGE T. (1845-1886).
   The Grasshopper and the Ant . 92
LEWES, MARY ANN EVANS (1819-1880).
   The Gift of Gold . . . 162
   Mr. Casaubon's Romance . . 218
   Adam and Dinah . . . 329
LINCOLN, ABRAHAM (1809-1865).
   The Gettysburg Address . . 75
LONGFELLOW, HENRY W. (1807-1882).
   Footprints of Angels . . 61
LOTHROP, HARRIET MULFORD.
   Joel at Work 742
   Old Concord 170
   Whittier with the Children . 357
LOWELL, JAMES RUSSELL (1819-1891).
   Democracy 152
LUSKA, SIDNEY (See Harland, Henry).
LYTTON, LORD (1805-1883).
   The Justice of Rienzi the Tribune . 254

MACAULAY, THOMAS BABBINGTON (1800-1859).
   On History 151
MACDONALD, GEORGE.
   Making a Friend . . . 203
MACLEOD, NORMAN (1812-1872).
   The Fishwife 125
MARVELL, ANDREW (1620-1678).
   Parody on the Speech of Charles II. . 72
McDOWELL, KATHARINE SHERWOOD BONNER (1849-1883).
   Hieronymus and Tiddlekins . 155

MELVILLE, HERMAN (1819-1891).
   A Scene in the Forecastle . . 318
MILTON, JOHN (1608-1674).
   The All-conquering Power of Truth . 10
   Reform 309
MONTAGU, LADY MARY WORTLEY (1690-1762).
   Italian Life 286
MORLEY, JOHN.
   Popular Culture . . . 141
MOTLEY, JOHN LOTHROP (1814-1877).
   The Siege of Leyden . . 139
MULOCK, MISS (See Craik).

PAINE, THOMAS (1737-1809).
   The Advent of Peace . . 215
PALFREY, JOHN GORHAM (1796-1881).
   The Culture of the Puritans . 338
"PANSY" (See Alden, Mrs. G. R.).
PARKER, THEODORE (1810-1860).
   Greatness and Ability . . 195
PEABODY, ANDREW P.
   Cuvier 121
PHILLIPS, WENDELL (1811-1884).
   Justice for the Slave . . 191
PITT, WILLIAM (1708-1778).
   On Refusal to Negotiate with Napoleon . 70
POE, EDGAR ALLAN (1809-1849).
   Torture 100
POPE, ALEXANDER (1688-1744).
   Homer's Inventive Power . . 217
PRESCOTT, WILLIAM HICKLING (1796-1859).
   The Battle of Tlascala . . 234
   Isabella of Spain . . . 315

RALEIGH, WALTER (1552-1618).
   Raleigh's Last words to his Wife . 192
RANDOLPH, JOHN (1773-1833).
   The Militia Bill . . . 181
READE, CHARLES (1814-1884).
   Lucy and the "Rajah" . . 111
RICHARDSON, SAMUEL (1689-1761).
   The Good Man \
   The Good Woman / . . 63
ROBERTSON, FREDERICK W. (1816-1853).
   The Sabbath . . . 302
ROGERS, SAMUEL (1763-1855).
   In Venice 350
ROMILLY, SIR SAMUEL (1757-1818).
   Palm Sunday . . . 133
RUSKIN, JOHN.
   Of King's Treasuries . . 168
SCOTT, SIR WALTER (1771-1832).
   The Dominie and Meg Merrilies . 51
   The Story of "Waverley" . . 56
   The Tournament . . . 332
SHAW, HENRY W. (1818-1885).
   The Ethics of Laughter . . 197

INDEX TO AUTHORS.

SHELTON, FREDERICK WILLIAM (1815-1881).
 A Question of Supremacy 142
SIDNEY, MARGARET (See Lothrop, Harriet Mulford).
SIDNEY, SIR PHILIP (1554-1586).
 In Praise of Poetry 271
SIMMS, WILLIAM GILMORE (1806-1870).
 A Sudden Hurricane 280
 Bess and the Snake 340
SMILES, SAMUEL.
 Indolence 331
 Home 343
SMITH, SYDNEY (1771-1845).
 To Lady Holland 206
SOUTH, ROBERT (1633-1716).
 Covetousness 131
SOUTHEY, ROBERT (1774-1843).
 To Grosvenor C. Bedford . . . 202
SPENSER, EDMUND (1553-1599).
 The Irish Bard 28
STANLEY, ARTHUR PENRHYN (1815-1881).
 To his Mother 104
STEELE, SIR RICHARD (1671-1729).
 The Strength of True Love . . . 47
 To his Wife 176
STEPHENS, ALEXANDER HAMILTON (1812-1883).
 The Destiny of the Republic . . . 73
STOCKTON, FRANK R.
 Annie and Lawrence . . . 197
STOWE, HARRIET BEECHER.
 Sam Mends the Clock . . . 24
SUMNER, CHARLES (1811-1874).
 On the Kansas-Nebraska Bill . . 83

SWETT, SOPHIE.
 Barberry Island . . . 326
SWIFT, JONATHAN (1667-1745).
 Country Hospitality 310
TAYLOR, BAYARD (1825-1878).
 The Midnight Sun 76
TAYLOR JEREMY (1602-1667).
 Death, the Conqueror 193
THACKERY, WILLIAM MAKEPEACE (1811-1863).
 Nil Nisi Bonum 105
 Tourists on the Continent . . . 184
 The Death of Colonel Newcome . . 200
THOREAU, HENRY DAVID (1817-1862).
 Solitude 33
 Spring Prospects 34
WARD, ARTEMUS (See Browne, Charles Farrar).
WARNER, CHARLES DUDLEY.
 Garden Ethics 58
WASHINGTON, GEORGE (1732-1799).
 The Pleasure of Private Life . . 21
 Inaugural Address 79
WEBSTER, DANIEL (1782-1852).
 Reply to Hayne 36
 The Constitution and the Union . 38
WEBSTER, NOAH (1758-1843).
 The Standard of Speech . . 353
WHITTIER, JOHN G.
 Virtue alone Beautiful . . . 120
WIRT, WILLIAM (1772-1834).
 Burr and Blennerhasset . . . 313

www.ingramcontent.com/pod-product-compliance
Lightning Source LLC
Chambersburg PA
CBHW032044220426
43664CB00008B/852